Janell
1018 South
Knoxville, Iowa
50138

PHYSICAL SCIENCE

PHYSICAL SCIENCE

F. BUECHE

University of Dayton

Worth Publishers, Inc.

Physical Science

Copyright © 1972 by Worth Publishers, Inc.

All rights reserved. No part of this publication may be reproduced, stored in a retrieval system, or transmitted, in any form or by any means, electronic, mechanical, photocopying, recording, or otherwise, without the prior written permission of the publishers.

Printed in the United States of America

Library of Congress Catalog Card No. 73-182927

ISBN: 0-87901-019-3

Design by Malcolm Grear Designers, Inc.

Fourth printing, April 1980

Worth Publishers, Inc.

444 Park Avenue South

New York, New York 10016

TO THE INSTRUCTOR

The teacher of a science course for non-science students is confronted with a formidable challenge and a magnificent opportunity. He has a captive audience whose attitudes towards science range from strong interest through indifference to mistrust and even fear. His first task is to persuade this audience that science represents a very human curiosity about the physical world.

The most effective way to accomplish this is to arouse a similar curiosity in each student. Fortunately for the teacher, each of us is born with an eagerness to explore and learn about our surroundings. The challenge, therefore, is to revive this original interest and to show the many and ingenious ways in which science explains the riddles of the physical world. If we meet this challenge successfully, we can help our students achieve the lifetime gift of an informed and objective interest in their surroundings.

In writing this text, I have tried to show that physical science is not simply a collection of facts to be memorized. Though facts are necessary, memorization of seemingly meaningless formulas is hardly what science is about. Instead, we want to show the student that science is a way of making meaningful and coherent the knowledge which, in large part, he already possesses. Moreover, by extending his horizons, by applying the understanding of nature which he already has, we can help the student to gain new insight into the world around him.

The laws of nature need not remain a mystery to those who are not mathematically inclined. No formula is needed to explain the physical basis of a home run or a sunset or any of the myriad other examples of natural laws we encounter in our daily lives. In my opinion, the student who can explain why car bumpers should have shock absorbers but cannot state Newton's first and second laws is to be commended for his understanding of physics, even though he may have a poor memory for statements of theory. Such students may never in their lives be in a position where they need to recite a definition or use a formula. Yet they will have to cope daily with practical problems that are governed by physical laws. They will be well served if we succeed in giving them a background sufficient for the qualitative interpretation of the physical world around them.

Although this text was designed primarily for a one-semester course, it has more material in it than can be dealt with in that time. The teacher should cover chapters 3 through 6 and parts of 7 and 8, since later chapters are dependent on them. However, he can easily delete one or more of the other chapters if necessary. The text can also be used in a two-semester course. In that case, some topics can be supplemented by paperbacks. Suggested titles are listed in the references at the end of each chapter and in the Teacher's Manual. The text will also serve well for a short (one-quarter, two-quarter, one-semester) physics course if the last four chapters and possibly the first two are omitted.

Since classes at this level vary widely, some sections of the book are optional. These are indented and marked by a vertical line on the left side of the column. Should you feel that your students will be dismayed by simple algebra, then I suggest you tell them at the outset that they will not be held responsible for these optional sections. If you do this, the numerical problems at the end of each chapter should not be assigned. There will still be ample end-of-chapter work in the Topical Check List and in the Questions.

If you decide to have your students work the problems, you will find solutions to the odd-numbered problems and answers for all the problems at the end of the book. The solutions of the even-numbered problems are in the Teacher's Manual. In addition, Appendix I gives a short review of mathematics.

In my own classes, the students are made aware of my feelings about the Check List and the Questions. They realize that the Check List material is a "must." But the crucial test of their understanding comes with the Questions. We spend a good deal of class time considering these and similar questions, because I find that they interest the students and provide them with opportunities to reinforce their understanding and confidence. Since my tests consist, for the most part, of essay questions like the end-of-chapter questions, the students soon begin asking themselves similar questions.

In my experience, students in a course such as this are enthusiastic about very simple, meaningful demonstrations. Many possibilities of this kind are included in the Teacher's Manual. There you will find several additional teaching aids as well.

Many people have contributed to this text, and I am most grateful to them. In particular, the comments of students and consultants on early drafts were of great value. I relied on the typing expertise of Miss Patricia Wright and Mr. Anibal Taboas, who successfully met the challenge presented by my handwriting. In addition, the following reviewers made a great many valuable suggestions, for which I am indebted:

JAY MARTIN ANDERSON, Bryn Mawr College
JAMES ARDINI, Diablo Valley College
WILLIAM BARNES, Mount San Antonio College
ALEXANDER CALANDRA, Washington University
JAMES P. COFFMAN, Edinboro State College
ERNEST COLEMAN, University of Minnesota
EUGENE COMMINS, University of California, Berkeley
RUSSELL COVERDALE, Purdue University
MARIA LUISA B. CRAWFORD, Bryn Mawr College
G. T. CREWS, Oregon State University
FRANCIS E. DART, University of Oregon
DELBERT W. DEVINS, Indiana University
VERNON J. EHLERS, Calvin College
NATHAN FEIFER, San Francisco State College
D. S. GORSLINE, University of Southern California
HENRY O. HOOPER, Wayne State University
DOROTHEA KATZ, New York City Community College
JAMES KUDRNA, California State College, Dominguez
D. R. LE GALLAIS, College of San Mateo
CHESTER LONGWELL, Stanford University

ROBERT MANLOVE, City College of San Francisco
CHARLES A. ROBERTS, California State College, Long Beach
R. J. RUCH, Kent State University
RANIER SACHS, University of California, Berkeley
RUTH L. SIME, Sacramento City College
HERSCHEL SNODGRASS, San Diego State College
WILLIAM STARBIRD, East Los Angeles College

A. A. STRASSENBURG, S.U.N.Y., Stony Brook
EDWARD WHEATFILL, Long Beach City College
G. M. WISSINK, Mankato State College
EDWARD ZGANJAR, Louisiana State University

Finally, I extend my heartfelt thanks to the staff of Worth Publishers for their constant help and suggestions during the preparation of the manuscript.

F. BUECHE
Dayton, Ohio
January 1972

TO THE STUDENT

In a sense, you are a scientist. If this sounds a little far-fetched, consider the usual definition of a scientist: "A scientist is one who actively seeks to discover and understand the laws of nature." You have been doing this since you were a baby. There are many laws of nature that you first learned through personal experience. It is true that countless others have made the same discoveries you have made. Nevertheless, you *have* discovered laws of nature and you continue to do so. That is what science is all about.

Unfortunately, mankind's knowledge has become so great that most of us have never experienced the excitement of a truly original discovery. It was not so in Isaac Newton's time. In that era, even the busy politician Benjamin Franklin could make valuable scientific discoveries while dabbling in science as a pastime. So many things were still unnoticed about nature that original discoveries were not difficult to come by.

But we should not let the complexity of today's research obscure the universality of science. To be sure, the details of modern research are often too complex for the non-scientist to understand. Fortunately, once the specialist has himself finally understood the details of his work, he usually can state the basic law he has discovered in a way that the layman can understand and appreciate. Indeed, the Nobel Prize-winning physical chemist P. J. W. Debye once said about physical theory, "If you cannot describe it in pictures which the common man can understand, then you yourself do not understand it."

What I am trying to say is that science is for everyone. We all enjoy the comforts it has brought into our lives. We all are subject to the laws of nature which scientists have discovered and described to us; we all can enjoy the beauty of its theories and its mechanisms. Perhaps it is the very universality of science that prompts some people to mistrust it. Admittedly, there are those who say they hate science (you may even be one of them). But is it really science they hate, or the dangerous and destructive offshoots of scientific discoveries? Is air pollution, for example, the fault of science or of human shortsightedness? Science is difficult and threatening only if you do not understand it. The laws of nature now known to us are beautifully simple in broad concept; they can be understood by all of us.

"What about all that mathematics, though?" you may ask. My answer is, "What mathematics?" True, as you flip through the pages of this book you will encounter a few equations and some arithmetic. However, the basic laws of nature described in this text can be understood reasonably well without resorting to mathematics. Since mathematics worries some of you, the material has been so organized that you can bypass most of the mathematical details. These optional paragraphs are marked by a vertical line printed to the left of them.

Don't misunderstand me though. You are certainly able to grasp the simple arithmetic in these optional sections. If your teacher concludes that these sections will trouble you, or that time is too limited, he will allow you to omit them. Barring some traumatic shock by exposure to a vicious fraction in the third grade, however, this should not be necessary. In any case, you should master the basic ideas whether or not they are stated in mathematical language. The reason scientists prefer the mathematical statement of a natural law is because most laws can be stated more concisely and precisely in mathematical terms.

What I am asking of you, then, is that you try to suspend your prejudices and preconceptions, especially if they are anti-science. If you give yourself a chance, I am sure you will find much to enjoy in the science you will learn in this text.

CONTENTS

Chapter 1
EARLY HISTORY OF THE UNIVERSE 1
The Primeval Fireball 1
The Formation of Galaxies 3
The Evolution of Galaxies 4
The Formation of Stars 8
Novae, Supernovae, and Quasars 10
Neutron Stars and Pulsars 12
Origin of the Solar System 13
Other Solar Systems 14

Chapter 2
ASTRONOMY AND THE SOLAR SYSTEM 19
The Distant Stars 19
Constellations and the Seasons 20
The Moon 24
Phases of the Moon; Eclipses 26
The Wandering Stars; Planets 30

Asteroids, Meteors, and Comets 37
Kepler's Laws and Quantitative Science 38

Chapter 3
FORCES AND MOTION 44
Science versus Superstition 44
Speed, Velocity, and Acceleration 45
Freely Falling Bodies 49
Newton's First and Third Laws of Motion 51
Newton's Second Law of Motion 53
Mass and Weight 54
The Metric System 56
Negative Acceleration—Deceleration 58
Linear Momentum and Its Conservation Law 59
More About Vectors 63

Chapter 4
WORK, ENERGY, AND ORBITAL MOTION 69
The Scientist's Definition of Work 69

Work and Kinetic Energy	72
Gravitational Potential Energy	73
Conservation of Energy	75
Terminal Speed	77
Power	78
Energy in Vibrating Systems	79
Newton's Law of Gravitation	82
Spaceships and Weightlessness	84

Chapter 5
HEAT ENERGY AND MOLECULAR MOTION	**91**
Rumford's Discovery	91
Temperature and Its Measurement	92
Gas Pressure and Absolute Temperature	94
Molecular Interpretation of Gas Pressure	96
The Kinetic Theory	99
The Earth's Atmosphere	101
Heat Energy in Liquids and Solids	103
Evaporation and Boiling	104
Crystalline Solids	106

Chapter 6
ELECTRICAL CHARGES AND CURRENTS	**113**
The Quantum of Charge	113
Coulomb's Force Law	115
Conductors and Insulators	116
Induced Charges	116
The Electric Field	118
Electric Current—The Ampere	120
Batteries and Electromotive Force Sources	121

Electric Potential Difference—Voltage	123
Ohm's Law	126
Electrical Power	128
House Lighting Circuits	129
Electrical Safety	130

Chapter 7
ELECTROMAGNETISM	**138**
Magnets	138
Sources of Magnetic Fields	140
Atomic Interpretation of Magnets	143
Forces on Currents in Magnetic Fields	143
Magnetic Forces on Moving Charges	146
Induced EMFs	149
The Electric Generator	151
The Transformer	152
Measurement of Atomic Masses	153

Chapter 8
WAVES	**160**
Waves on a String	160
Reflection of Waves on a String	162
Standing Waves	164
Resonance of Waves	165
Resonance and Vibration	166
Vibration of Complex Systems	168
Compressional Waves; Sound Waves	169
Resonance of Sound Waves	171
Interference of Sound Waves	174
The Doppler Effect	175
Noise Pollution	178

Chapter 9
ELECTROMAGNETIC WAVES — 184

Maxwell's Discovery	184
Radio Waves	185
The Magnetic Wave	187
Radio-Wave Detection	188
The Electromagnetic Spectrum	189
Reflection of Light	191
Refraction of Light	193
The Prism and Color	194
Interference from Slits	196
Diffraction by an Aperture	201

Chapter 10
THE TWENTY-FIVE GOLDEN YEARS — 209

Relative Motion	209
Einstein and Relativity	211
A Limiting Speed Exists	214
The Mass-Energy Interconversion	214
Time Dilation and the Twin Paradox	215
Planck's Discovery	216
The Photoelectric Effect	219
The Nuclear Atom	221
The Hydrogen Atom	224
Bohr's Model for the Hydrogen Atom	226
De Broglie Waves	228
Particle-Interference Experiments	230
The Heisenberg Uncertainty Principle	231

Chapter 11
THE ATOMIC NUCLEUS — 240

Measurement of Nuclear Masses	240
Isotopes and Nuclear Structure	242
Binding Energy of the Nucleus	244
The Mass Defect and Binding Energy	246
The Fusion Reaction	247
The Fission Reaction	248
Radioactivity	251
The Tunnel Effect	253
Nuclear Bombardment	254
Particles	258
Uses of Radioactivity	259

Chapter 12
ATOMS AND THE PERIODIC TABLE — 265

The Hydrogen Atom	265
Atomic Energy Levels	266
Emission and Absorption of Light	268
The Helium Atom and Electron Spin	270
Lithium and the Exclusion Principle	271
The Periodic Table of the Elements	274
Spectra of Complex Atoms	278
The Laser	279

Chapter 13
THE CHEMISTRY OF INORGANIC SUBSTANCES — 285

The Hydrogen Molecule	285
The Covalent Bond	287
The Ionic Bond	289
Bonding in Solids and Liquids	291
Types of Crystalline Solids	293
Ionic Solutions in Water	298
Acids	299

Bases and Hydroxides	300
Minerals in Water	301
A Summary of Chemical Names and Symbols	301

Chapter 14
THE MOLECULES OF LIVING THINGS: ORGANIC CHEMISTRY — 309

Hydrocarbon-Chain Molecules	309
Nomenclature	312
Alcohols	313
Organic Acids and Esters	314
Benzene and Related Compounds	314
Amines and Amides	316
Polymers	317
Proteins	318
Nucleic Acids	320
Mutations from Cell Damage	322
Photosynthesis and the Energy of Life	323

Chapter 15
THE EARTH'S ATMOSPHERE AND WEATHER — 327

Early History of the Earth	327
The Atmospheric Layers	328
The Ionosphere	331
Large-Scale Air Motion	332
The Earth-Air Boundary	334
Weather and the Troposphere	335
Localized Air Motions	338
Seasonal Changes in Weather	342
The Effect of the Oceans on Climate	343
Weather Prediction	345
Pollution of the Atmosphere	345

Chapter 16
THE EARTH — 353

Materials of the Earth's Surface	353
Seismic Waves and Earth Structure	360
Isostasy	362
The Continents	363
The Ocean Basins	368
Continental Drift	370
Erosion	373
Geologic Time Determination	377
Development of Life on Earth	377
Man	381

APPENDIXES	**389**
Appendix I: Mathematics	390
Appendix II: Temperature Scales	394
Appendix III: Laws and Formulas	395
Appendix IV: Table of the Elements	398
Appendix V: Solutions to the Odd-Numbered Problems	400
Appendix VI: Answers to the Problems	408
GLOSSARY	**411**
INDEX	**421**

PHYSICAL SCIENCE

Chapter 1

EARLY HISTORY OF THE UNIVERSE

You and I are a tiny part of a vast experiment: the birth, life, and death of the universe. It is a unique experiment in many ways; its scope is perhaps forever beyond comprehension; its progress lies almost totally outside the control of man; it began billions of years ago, and its future course is still not fully understood. Even though human beings did not devise or initiate this experiment, from the earliest times we have been curious about its course. Seeking to satisfy this curiosity, we become involved in science, the game whose goal is to learn how the universe and everything within it behaves.

We will never know with certainty how creation began. The best we can do is to examine the universe as we see it today and infer what must have happened in the distant past. But be forewarned; as in all science, we are making deductions from limited knowledge. As more knowledge becomes available in the future, we reserve the right to modify our deductions. In this, even more than in most areas of science, we must keep an open mind. With these words of caution, let us see what we can surmise about the origin of our universe.

The Primeval Fireball

Over the centuries it has become clear that the universe is expanding. Although our own solar system, consisting of the sun, the earth, and the planets, is no larger now than it was billions of years ago, the distance to the farthest stars in the universe is constantly increasing. We infer this from the color of the light sent to the earth by the atoms within these stars. The light from these receding stars has shifted in color; the blues and greens have been changed to yellows and reds. This shift of the colors toward the red end of the spectrum, the *red shift*, can be used to measure the speeds at which the stars recede from the earth. These recession speeds are found to be close to the fastest speed possible, the speed of light, for the most distant stars.

Since the universe appears to be expanding, it is tempting to extrapolate backwards in time and deduce that the universe was, at its beginning, an exploding ball. This is called the *big-bang* theory of creation. Other alternatives are possible, however, and the exploding-ball hypothesis was a subject for considerable dispute until several years ago. Even now, scientists are not unanimous in their support of it. However, increasing evidence obtained from measurements of the radiation reaching the earth from space makes alternative theories less likely.

From the experimental evidence available to us today, we conclude that about 10 billion years ago the universe was compacted into a ball having a diameter perhaps 10 times as large as that of our sun. You may well ask how the millions of stars, each comparable in size to our sun, together with all the rest of the universe, could be compressed into such a small volume. Surely atoms and molecules as we know them could not be packed this densely. However, atoms are composed of much smaller particles —electrons, protons, and neutrons—and these tiny particles can be pressed into a volume one millionth of one billionth the size of the atom. The atom itself is mostly empty space. Presumably, all atoms were collapsed in the extremely compacted material of the original universe.

The tightly compressed universe must have been incredibly hot; its exact temperature is unknown, but it was at least several billion degrees. At such high temperatures, atoms and molecules cannot exist, and even basic particles such as the electron, proton, and neutron live tenuous lives. Although we cannot yet achieve a controlled temperature this high in the laboratory, we know that the basic particles literally tear each other apart when they collide with the high speeds and energies characteristic of billion-degree temperatures. Thus it was in the original fireball, when countless fragments, moving at tremendous speeds, caused the fireball to explode. This explosion represented the beginning of time for our universe.

Much of the behavior of our universe since the primeval explosion can be explained in terms of gravitational force. This is the force that causes objects to fall toward the earth, the force that holds the planets in their orbits around the sun. The first clear understanding of this force and how it causes the effects we attribute to gravity was provided by Isaac Newton two centuries ago. He showed that all objects in the universe attract each other with a gravitational force, and the magnitude or strength of this force is proportional to the massiveness of the objects involved. For example, the attraction of the massive earth for a tennis ball is large enough to cause the ball to fall to the ground. But a stone placed on a tabletop close to the ball exerts so small an attractive force on the ball that it causes no perceptible motion. The attractive force exerted by the stone upon the ball is much smaller than the pull of the earth upon the ball because the stone is so much less massive than the earth. Clearly, the massive fireball universe exerted tremendous gravitational attractive forces upon the high-speed particles trying to escape from its surface.

In spite of the huge gravitational forces holding the particles of the primeval fireball together, the universe expanded into space. The situation is similar to the launching of a rocket from the earth. The controlled explosion of the rocket engines overcomes the earth's gravitational force and thrusts the rocket away from the earth. If it has enough energy, a high-speed rocket can pull loose permanently from the earth and escape into space. To escape from the earth's gravitational pull, the rocket must be shot from the earth with a speed of at least 25,000 mi/hr (miles per hour), or it will eventually be stopped and pulled back to the earth. We call this minimum speed which an object must have to free itself the *escape velocity*.

The escape velocity for particles trying to leave the primeval fireball was much faster than 25,000 mi/hr because the fireball's mass was so much greater than the earth's. When the fireball exploded at time zero, the particles in the ball shot out into space. Each escaping particle was like a rocket launched from the earth. If the particles had enough energy and speed, they exceeded the escape velocity and escaped the fireball forever. If their energies were too low, they would eventually slow down, reverse their direction of motion, and fall back once again to re-form the original fireball.

We do not yet know whether the particles of the universe will escape. The original fireball has expanded into a tremendously large region; but each particle of the universe is still attracting all others, trying to pull the universe back into a tight ball. To learn whether the particles of the universe have enough energy to escape from each other, we need to know the total mass of the universe and the energies of its particles. We can now estimate each of these quantities, but our estimates are still too imprecise to enable us to decide the question.

If the particles of the universe do not have enough energy to escape, then eventually the universe will cease to expand. Like rockets falling back to the earth, the particles that compose the universe will then begin to fall back to its center, speeding up and attaining tremendous energies as they fall. The exceedingly high-energy collisions that will occur as the universe comes flying back into its original center will generate such vast heat that all atoms and molecules will be torn apart. At the final instant, the primeval fireball will be formed again.

Perhaps this is the history of our universe. It may have pulsated over and over again in the distant past, existing for an instant as a fireball, exploding, expanding for billions of years, contracting an equal length of time, and then re-forming the fireball. If, indeed, this is the proper picture of our universe, we shall never be able to learn its history outside our own fireball era. All traces of previous history would be consumed in the incredible cauldron of each fireball.

The Formation of Galaxies

We picture the infant universe as an exploding ball of gas composed of elementary particles and other forms of high-energy radiation. As the dense gas expanded out from its origin, the temperature of the fireball dropped rapidly. After only a few hours, the temperature had dropped enough so that the particles of the gas no longer had enough energy to tear atomic nuclei apart. However, the drop in temperature was so rapid that hardly any large nuclei had time to form.

The formation of atomic nuclei, which are the center portions of the atoms, requires the coming together of protons and neutrons. In the case of the smallest atom, a hydrogen atom, the nucleus consists of a single proton, an elementary particle with a positive charge equal in magnitude to that of the negatively charged electron. (Electrons circle the nucleus like satellites and make up the outer "shell" of atoms.) The second smallest atom, helium, has a nucleus which is made up of two protons and two neutrons. (A neutron has about the same mass as the proton but has no charge. Both the neutron and proton have masses about 1,840 times larger than that of the electron.) Lithium, the next in size, has three protons and three neutrons in its nucleus. The largest atoms that exist in nature have many neutrons and protons in their nuclei. For example, the gold nucleus contains 79 protons and 118 neutrons packed tightly together.

To pack protons together, one must force them close to each other. But you probably already know that "like charges repel each other," and since all protons are positive, they repel each other. The fact that they do stick together and form a nucleus indicates that a much stronger attractive force overcomes the repulsion between like charges, and binds the protons and neutrons together in the atomic nuclei. This _nuclear attraction force_ has a short range; it is not felt unless the particles are extremely close to each other. Therefore, two protons must hit each other with very high energies in order to overcome the repulsion force between their like charges; only then can they come close enough so that the nuclear attraction force will hold them together.

Only for a short time did the protons in the expanding gaseous dust of the fireball have energies high enough to be able to fuse together. As the temperature of the expanding dust cloud dropped, the energies of the protons dropped too low for them to fuse together upon collision. The drop in temperature appears to have been too rapid to allow the formation of many nuclei larger than the nucleus of helium. In fact, the cloud at this stage still consisted mainly of single protons, neutrons, helium nuclei, and electrons, together with high-energy radiation such as x-rays. The nuclei of atoms larger than helium were almost totally absent from the universe as it cooled rapidly below the temperature necessary for their formation.

After 100,000 years, the temperature of the expanding cloud had dropped several billion degrees to about 5000°F (close to the temperature of the white-hot wire in an incandescent light bulb). At these lower temperatures, the protons could capture electrons to form hydrogen atoms. (It may prove helpful to picture the electron and proton in the hydrogen atom to be analogous to the earth and sun in the solar system. Like the earth circling the sun, the electron can be thought of as circling the proton. However, the earth is held in orbit about the sun by the sun's gravitational attraction force, while the negative electron is held in orbit about the positive single-proton nucleus by the attraction between unlike charges.)

As with any cloud this expanding dust cloud was not uniform. In some places the dust was more dense than in others. These regions of higher density exerted higher than average gravitational attraction forces upon the dust nearby. As a result, the dust began to gather more thickly in various regions of the expanding cloud. One such region contained the dust that later became our solar system; but our dust aggregate was much larger than just the solar system. It was the region of the universe we now call the Milky Way.

The Milky Way, then only a cloud of hydrogen and other tiny particles, now stretches over a distance of about 1,000,000,000,000,000,000 miles. (We often write this in shorthand as 1×10^{18} or simply 10^{18} mi to indicate we should write 1 with 18 zeroes after it.) To appreciate the size of the Milky Way, consider that nothing can travel faster than light and that a light pulse would require about 100,000 years to travel from one edge of the Milky Way to the other. Yet the Milky Way is only a tiny portion of the still-expanding universe. Since distances in the universe are so enormous, the mile is nearly useless as a unit of length. A more useful unit of length is the *light year*, the distance a light pulse would travel in one year (at 186,000 mi/sec, the speed of light). Thus, the distance across the Milky Way is about 100,000 or 10^5 light years.

During the time that gravitational forces were drawing the dust particles in the Milky Way together, dust aggregates were forming elsewhere in the universe. These regions, like the Milky Way, are called *galaxies*. Looking out into space with the most powerful telescopes available, we can see approximately 10^9 other galaxies (recall that 10^9 is 1,000,000,000), and there are certainly more, still farther out in space. These galaxies show red shifts, which tell us that the most distant ones are receding from us at speeds close to the speed of light. (We shall see later how the age of the universe—about 10 billion, or 10^{10} years—is estimated using the recession speeds of the galaxies.)

The expanding universe was perhaps only 100,000 years old when the galaxies became distinct. Since then, the galaxies have changed from the original dust clouds to large regions filled with stars, planets, asteroids, and all forms of particles, still including several percent of the mass as free atoms and molecules. Let us now examine the course of events within a typical galaxy during the billions of years of its development.

The Evolution of Galaxies

Several galaxies as viewed by telescope are shown in Figures 1.2 to 1.5. The bright points in the photo-

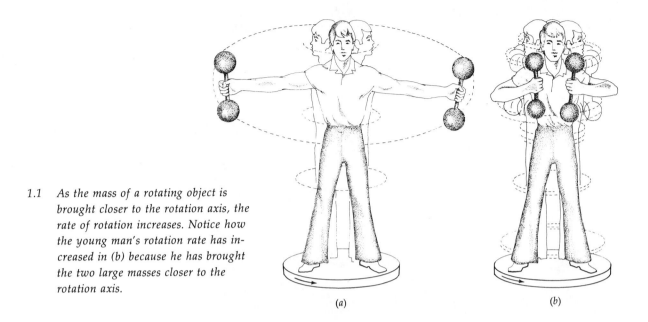

1.1 As the mass of a rotating object is brought closer to the rotation axis, the rate of rotation increases. Notice how the young man's rotation rate has increased in (b) because he has brought the two large masses closer to the rotation axis.

graphs show the positions of stars such as our own sun. In our galaxy, there are an estimated 100 billion stars, one of which is our sun. We think the Milky Way looks like the galaxy shown in Figure 1.6, with our sun at the position indicated. However, we can not be sure, because we are immersed in the galaxy and the dense central region hides portions of the galaxy from our view. From other angles, our galaxy would resemble those shown in Figures 1.2 and 1.3.

From the photographs, we can see that our galaxy is a disklike spiral. This shape is typical of many galaxies, and its origin is of interest. Many of the spiral shapes we are accustomed to seeing in fluids and dust clouds are the result of rotation. The water spiraling into the drain of a sink and the funnel of a tornado or waterspout are two typical examples. We believe that the spiral shapes of some galaxies are the result of their rotation. Our own galaxy makes one complete rotation in about 200 million years (2×10^8 years). Although this rotation rate seems negligibly small, the galaxy is so large that the sun, about two-thirds of the way to the edge, is circling the galactic center with a speed of about 9,000 mi/hr. The cause of this rotation is of interest, especially since the same phenomenon occurs as the planets move about the sun. We can learn something about this effect in the following way.

Let us see what happens to a rotating object when its shape is changed. Figure 1.1a shows a young man on a platform which can be rotated. The massive weights in his hands increase the mass of his own body. A push from the outside starts him rotating slowly with weights in his extended hands as shown in part (a). Notice that the massive weights are rather far from the axis or center of rotation. If he now pulls the weights in close to his body as shown in part (b), he begins immediately to rotate much faster. When he once again extends his arms, he slows down to his original rotation rate. The important _law of conservation of angular momentum_ illustrated by this experiment can be stated loosely in the following form:

> A freely rotating system will rotate more rapidly if the mass of the system is brought close to the axis of rotation.

This has great importance for the galactic cloud. We recall that as the galaxies flew outward from the center of the original fireball, each galaxy became a region of higher than average density. Gravita-

THE UNIVERSE   5

1.2 Typical spiral galaxies. (a) NGC 3031 (M81), a spiral galaxy in Ursa Major; (b) NGC 4565 seen edge on. Both photographs were taken with the 200-in. telescope. (Mount Wilson and Palomar Observatories.)

1.3 The Andromeda galaxy, Messier 31. (Mount Wilson and Palomar Observatories.)

1.4 Barred spiral galaxy NGC 2523. (Mount Wilson and Palomar Observatories.)

1.5 NGC 4486 (M87), giant elliptical galaxy in Virgo, photographed with the 200-in. telescope. (Mount Wilson and Palomar Observatories.)

1.6 The spiral galaxy shown here (NGC 1232) is similar in shape to our own galaxy, the Milky Way. To show this similarity, a portion of the Milky Way is superimposed on the galaxy; it appears as the bright stars on the upper spiral arms of the figure. Our sun (and solar system) is about half way out from the galactic center. Its position is indicated by the 0° designation, a conventional astronomer's representation. Notice how the spiral arms of the Milky Way fit nicely on the spirals of NGC 1232. (Photograph prepared by Prof. Wilhelm Becker, University of Basel, Switzerland.)

tional forces pulled hydrogen and other nearby particles toward the center of this high-density region, and as a result, the material composing the galaxy began to contract. In many cases, the material in this vast region of space may have had a slight rotation, since perfectly uniform outward flow of the gas in the universe is unlikely. Thus in its early stages the galaxy was a highly expanded, slightly rotating cloud that began to collapse.

As the galactic cloud was pulled into a smaller volume, the law of conservation of angular momentum tells us that the rotation rate of the cloud must have increased. In many cases, the original rotation rate was large enough so that the galaxy attained rotation rates comparable to that of our own galaxy. If you recall what happens to a child on the end of a line of children playing "crack the whip," it is easy to understand what happened to the outer portions of the collapsing cloud. Required to "snap" faster and faster around the inward-spiraling path, the outer portions of the cloud became extended and the cloud flattened into a disklike spiral. We believe that this explains the shape of the spiral galaxies shown in Figures 1.2 to 1.4.

Other galaxies undoubtably had very little rotational motion at the outset. They did not acquire large rotation speeds as the galaxy collapsed under gravitational forces, and so they were not thrown into spiral form. The elliptical galaxy shown in Figure 1.5 is an example of this, but even that galaxy is rotating. As a result, its outer portions, like those in the spiral galaxies, are subject to crack-the-whip forces. These forces became larger as the rotation rates of the galaxies increased during the gravitationally induced contraction of the galaxies. Eventually the rotation rate of each galaxy became large enough so that crack-the-whip effects balanced the gravitational pull toward the galactic center. At that time, each galaxy stabilized at its present size.

Thus, during the first billions of years after its formation, the expanding cloudlike universe changed into a more granular universe, in which each granule is a galaxy. The universe is still expanding, but now it consists of galaxies of stable size, shooting out into space from the primeval point of origin.

The Formation of Stars

Within each galaxy the density of the gas cloud varied. The regions of high density attracted the surrounding gas of neutrons, protons, electrons, and hydrogen atoms by gravitational forces. As the particles fell toward these centers, they acquired high speeds. Since the total mass of the gas involved was comparable to the mass of our own sun (about 3×10^5 times as massive as the earth), the gravitational forces were extremely large; the energies of the particles falling into the center increased and caused the temperature of the aggregating-particle gas to rise dramatically.

By the time each center had captured the particles surrounding it, the aggregate had become a flaming fireball as dense and hot as our own sun. (However, our sun did not form at that stage of the universe, as we shall see later.) Even though each galaxy then contained millions of flaming suns, the mass of these white-hot objects, which we call hydrogen stars, consisted almost entirely of protons, neutrons, and electrons. Elements such as iron and carbon were still missing from the universe.

As the gaseous material of these stars continued to contract under gravitational forces, their temperatures continued to rise, and eventually reached nearly 30 million degrees Fahrenheit. At this temperature the protons had enough energy to fuse together upon collision. As they fused, large amounts of energy were released. (This is the process used in the hydrogen bomb to release the explosive energy we all fear may destroy mankind.) In the interior of such a hydrogen star, the hydrogen nuclei and protons were fusing to form helium nuclei. The hydrogen star was a large-scale hydrogen bomb, prevented from exploding into space by gravitational attraction forces.

The vast quantities of energy given off in the nuclear fusion reaction caused the star's temperature to rise still further. However, since a gas expands as it gets hotter, the tendency to expand counterbalanced the contraction effect of gravitational forces. At that time the star became a stable entity. As heat radiated out into space from the star, more heat was formed in its core by protons fusing to form helium nuclei.* A hydrogen star can exist for billions of years as it slowly uses up the protons in its core. Many stars in the universe appear to be still in this stage of development.

We have said that the stars are made up of gas composed mainly of protons, neutrons, and electrons. The temperature is so high that all atoms have been destroyed, and consequently the particles within the gas can exist in a more dense form than a gas of hydrogen atoms could. The density of matter at the center of a star is hundreds of times greater than in the heaviest metals.

Eventually, as a result of the fusion reaction, the protons in the core of a star decrease in number to the point where the proton fusion reaction begins to die out. At that time, the star begins to contract once again under gravitational forces, and the temperature of the core begins to rise. The core now has a large amount of helium nuclei within it since they are the products of the hydrogen fusion reaction. (Recall that the helium nucleus is composed of two protons and two neutrons.) These helium nuclei are also capable of a fusion reaction. However, since each helium nucleus contains two positive charges because of the two protons, higher temperatures are needed to provide the energy required to force these larger like charges together. As the core of the star reaches about 200 million degrees Fahrenheit, a fusion reaction involving the helium in the core becomes operative. This reaction is called the _triple alpha process_ because three helium nuclei (also called alpha particles) fuse to form a carbon atom nucleus. Through this reaction, the helium nuclei in the core of the star provide the heat energy necessary to keep the star alive.

While this process within the core of the star is going on, changes are also occurring in the star's

---

* People sometimes say "hydrogen is burned to form helium" in the fusion reaction. Although it is true that the reaction resembles burning in the sense that heat is given off in the process, "burning," as chemists use the term, is a combining of atoms into molecules, not the fusing of nuclei.

exterior. Much of the hydrogen gas remaining in the star has diffused to the outer portion and has begun to cool, turning from white-hot to red-hot in the process. Moreover, the gas has diffused far into space from the star's core. From a distance, the star now looks like a giant red ball, and therefore is called a *red giant* star. Nearly all the star clusters seen in the sky contain at least one red giant.

During the red giant stage, the temperature at the star's center slowly increases as the fusion reaction continues. Eventually the temperature becomes so high that the carbon nuclei formed in the triple alpha process begin to undergo further reaction. By the step-by-step addition of alpha particles (i.e., helium nuclei), the nuclei of other elements can be formed. Presently we believe that most nuclei of atoms other than hydrogen and helium are and were formed in the interior of stars at this and later stages of development. The proportions of each type of nucleus will depend greatly on the stability of the nucleus and the number of avenues that lead to its production. There seems to be substantial agreement between the observed abundances of elements and what we would calculate from our current knowledge of nuclear reactions. In the universe as a whole, however, all but a small percentage of the material is still hydrogen and helium.

Eventually the nuclear fuel in the core of the red giant is exhausted. Although there are large amounts of hydrogen and helium in the outer portion of the star, the temperature there is far too low for fusion. As the nuclear fusion reaction begins to die out, the star begins to contract once again under the gravitational attraction force. During this period, the star frequently shoots mass out into space and often loses a considerable fraction of its mass over the centuries. The contracting star is heated once again as the gravitational force draws in its remaining mass, and is transformed into a much smaller, hotter star called a *white dwarf*.

Since the nuclear fuel within the white dwarf has been depleted during its early stages, nuclear fusion reactions cannot provide further heat to the star. It is believed that such a star cools slowly and disappears from sight unless other processes come into play. A few of these possibilities are discussed in the next section.

Novae, Supernovae, and Quasars

From time to time, astronomers see a star's brightness suddenly increase by a factor of thousands. Early astronomers, unable to detect the faint parent star, thought they were seeing a star being born and called such an event a *nova*, meaning "new." Although a nova comes to full brilliance in a day or even less, it requires years to return to its normal brightness. A typical nova is shown in Figure 1.7.

Exactly how a nova is born is still disputed, but the following possibility seems likely. Theory predicts that a white dwarf is not stable unless its mass is less than about 1.2 times the mass of our sun. Therefore, if any mechanism adds mass to such a star, the star should explode to eject the mass. It appears that all novae occur in so-called *binary stars*, two stars very close together. In approaching the white dwarf stage, a star often ejects mass into the space around it. If one of the binary stars adds mass to its companion star, the companion may exceed its stable mass and violently eject its outer layers in order to regain stability. We know that the great brightness of a nova occurs when the star suddenly throws off its outer layer and we know that a given star may do this several times. Whether or not the binary star leakage theory correctly explains the beginning of a nova is still open to question.

A much more violent stellar explosion, called a *supernova*, occurs less frequently. Unlike the nova, whose brightness increases by a factor of only thousands, a supernova occurs when a star's brightness suddenly rises by a factor of tens of millions. These highly visible events occur only once every few hundred years in a typical galaxy. Often, they

1.7 In 1934 the star shown by the arrow at left suddenly became thousands of times brighter. It was the nova Herculis shown at the right. (Yerkes Observatory.)

outshine the whole galaxy until, after a few months, they fade away. In our own galaxy, 14 supernovae have been observed in the past 2,000 years and others, hidden by the core of our galaxy, must certainly have existed. Famous supernovae were observed in the years 1054, 1572, and 1604, and traces of these supernovae are still visible.

For example, the Crab nebula shown in Figure 1.8 is the remnant of the supernova of 1054. It can be used to test our ideas concerning the red shift since the speed of expansion of the nebula (a cloud of interstellar dust) can be computed from the red shift of the light coming from it. If we assume that the nebula has been expanding at this same speed over the centuries, we compute that it should have exploded in 1054. In this case, at least, our interpretation of the red shift appears correct.

In addition to sending out light, supernovae and their remnants are strong sources of radio waves and x-rays. Radar techniques developed during World War II made it possible to construct huge radio telescopes which can scan the sky for radio- and radar-wave emitting sources. With their use it has been possible to locate many other strong sources of radio energy in the universe. Although the data

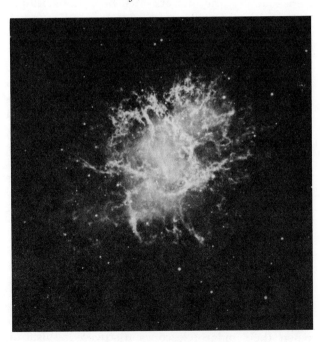

1.8 The Crab Nebula is the remnant of the supernova explosion observed in our galaxy in 1054 A.D. (Mount Wilson and Palomar Observatories.)

THE UNIVERSE 11

provided by this additional tool have greatly increased our understanding of supernovae, we are still not certain about their cause.

Radio astronomers have also discovered tiny but extremely powerful sources of radio waves. These puzzling radio-wave sources are called quasi-stellar radio sources, or _quasars_. Soon after their discovery, they were found to be associated with faint galaxies, galaxies that show exceedingly large red shifts. If we believe that the large red shifts are the result of recession speeds close to the speed of light, we are faced with a serious problem. Anything we observe in the heavens that achieved its large recession speed in the primeval fireball will have continued with nearly that speed for the 10 billion years in which the universe has existed. Consequently, the quasars must be billions of light years away from the earth. But to send as much energy to the earth as we observe, they must be sending out radiation much more strongly than any known galaxy.

After the discovery of quasars, light telescopes were used to search for similar objects. Many systems were found which resemble quasars in red shift and luminosity but which do not emit strong radio waves. Although astronomers are still unable to explain how such a distant source can send out the tremendous energies required to give rise to the radiation reaching the earth, no satisfactory way has yet been found to reconcile the observed red shifts with a source closer to the earth.

Neutron Stars and Pulsars

In 1968, another puzzle confronted astronomers when their radar telescopes discovered pulsating radio-wave sources. Many such sources, called _pulsars,_ are now known; each sends out pulses of radio waves at regular intervals, mostly in the neighborhood of one pulse every few seconds. A well-studied pulsar found in the Crab nebula has a period of 0.033 sec and the period is slowly increasing.

The discovery of pulsars started a rash of speculation about their origin. Theories ranged from pulsating white dwarfs to attempts at communication by intelligent beings in other solar systems. Unfortunately for those of us who seek real excitement, this latter possibility appears to be wrong. However, there are still enough exotic explanations under consideration to keep interest high. One theory that is gaining support from the experimental data being acquired involves the concept of a _neutron star_.

You will recall that a neutron is much like a proton except it has no net charge. If the interior of a star were composed of neutrons rather than a mixture of neutrons, protons, and electrons, in the white dwarf stage the star would be much more compressed than the usual white dwarf. Calculations indicate that a neutron star having twice the mass of our sun would be only about 100 mi in diameter. (The earth's diameter is about 8,000 mi, whereas our sun's diameter is about 860,000 mi.) The temperature at the surface of such a dense star would be around 5 billion degrees Fahrenheit, and its interior would be even hotter. In spite of their very large mass and high temperature, these stars would probably be almost invisible to even our best telescopes. Their spheres would be so small that they would appear as small pinpricks of light in the sky. Moreover, much of their radiation would be in the form of X-rays rather than light. Neutron stars such as this would be difficult to detect; thus far, only a very few possible neutron stars have been tentatively identified.

If they do exist, it is conceivable that a binary neutron star system (i.e., two neutron stars in close proximity) could give rise to the pulsar phenomenon. If two such stars were rotating as a unit around a common axis, they would send out pulses of radiation at a frequency comparable to the rotation frequency of the star system. Or, by some mechanism still not understood, a single rotating neutron star may be the source of these pulses. It remains uncertain whether either of these theories is correct.

Origin of the Solar System

Even in discussing our own solar system we cannot leave the realm of speculation and speak with certainty. Although much new knowledge has been acquired from exploration of the moon and planets, great gaps exist in our understanding of the solar system. Therefore, we shall present the current thinking about the origin of the solar system with the understanding that you will modify these ideas as the news media report pertinent discoveries.

The solar system consists of our sun and nine planets, together with lesser bodies, captured in the gravitational field of the sun. We believe that it was formed by the gravitational aggregation of stellar dust, together with hydrogen and helium from the primeval fireball. Since the planets as well as the sun contain some elements having large nuclei which we think were manufactured in stellar interiors, it appears that this portion of the material in the solar system must be the dust ejected into space by older stars. We believe the solar system evolved from this dust between 5 and 10 billion years ago.

As in the formation of stars and galaxies, local regions of high density in the gas and dust cloud acted as centers to which the particles were pulled by gravitational forces. We infer from the rotation of the present solar system that the collapsing cloud was rotating slightly. The law of conservation of angular momentum tells us that the speed of rotation increased as particles drew closer to the center of the cloud. Like our spiral galaxy, our embryonic solar system was whipped into a flat spiral disk of gas and dust.

Eventually, small parts of the spiral broke loose and continued to circle the center region without falling into it. However, 99.86 percent of the total mass fell into the center, acquiring extremely high energy as it fell. When this mass collided at the center point, a hot star was formed; it is our sun.

Our sun already contained a small percentage of heavier elements, but was otherwise identical to the hydrogen stars discussed earlier. At present the sun's heat is largely furnished by the proton fusion reaction, and it appears to have been in this stage of development for at least 5 billion years.* It is estimated that another 5 billion years will go by before the sun undergoes any major change.

While the sun was forming, each wisp of gas and dust torn loose from the whirling dust spiral collapsed into its own center, while still preserving its rotation in a near circle about the sun. Since the wisps themselves were in rotation, the comparatively tiny spheres into which they collapsed continued rotating on their own axes. This is the origin of the earth's rotation about its own axis, and that of the other planets as well.

Unlike the sun, the material of the wisps that formed the planets did not heat enough to ignite the proton fusion reaction. Since the masses of the wisps were small, the gravitational forces pulling the dust cloud together were too small to furnish the required energies and temperatures. In fact, the gravitational forces of most of the planets were too small to pull the very light hydrogen and helium atoms to the planet. The earth, for example, has few free hydrogen or helium atoms. On the other hand, the largest planet, Jupiter, was able to capture large amounts of hydrogen and helium. If Jupiter had been much more massive, the hydrogen nuclei within it might have begun to fuse and it too might have become a sun. We have noted that many stars exist as pairs; our sun nearly had a twin. (If this had happened, our solar system would behave as it does now. The mass and speed of each body, not the number of suns, determined its course.)

---

* The age of the solar system has been estimated using a method involving the examination of radioactive decay products (see Chapters 11 and 16).

How the moon formed is still uncertain. Rocks brought back from the moon on early expeditions indicate that it has about the same history as the earth, contrary to some previous theories. This fact narrows the number of acceptable theories about its origin, but none has been conclusively proved correct.

The earth will exist as it does now for at least a few billion years more, unless some unforeseen catastrophe occurs. Eventually, though, the sun will enter the red giant stage, and will expand out toward the earth, perhaps even reaching it. The oceans will vaporize, and all life on earth will disappear. In the end, the earth will be only a bleak cinder as the sun contracts into a white dwarf.

Other Solar Systems

We see that the universe contains billions of stars similar to our sun. If we were looking at our own sun from the vicinity of the nearest star, we would be unable to see the planets even with the best telescopes available. Even though they shine in the reflected light from the sun, they are just too tiny and nonluminous to be seen from such large distances. It follows from this that we cannot ascertain the existence of other solar systems by direct visual sighting. Moreover, the nearest star, Alpha Centauri, is 4.3 light years away from us, and so it is unlikely that we will be able to visit even this solar system nearest to us in the foreseeable future.

In spite of these difficulties, it seems probable that other solar systems like ours exist. Since processes similar to the birth of our sun and planets occurred in billions of places in the universe, we can reasonably expect that our own solar system has been duplicated somewhere else.

Whether or not there are intelligent beings (such as we credit ourselves with being) on planets in these solar systems is a more complicated question. In the first place, we do not know enough about how life was created. We know that the protein molecules necessary to life can be formed by repeated lightning discharges in atmospheres like those we believe existed on the earth at early times. However, this is a far cry from the creation of a living cell, and we do not yet know just how complicated that process is. It seems so intricate that the existence of even the simplest forms of life elsewhere in the universe is uncertain. But we should not let our pride blind our vision. We tend to think that we are the most advanced beings in the universe. But can we truthfully look at ourselves and think that some beings elsewhere might not be superior?

BIBLIOGRAPHY

1. ABELL, G.: *Exploration of the Universe,* 2d ed., Holt, New York, 1969. A readable text on astronomy for those who have little background in science and mathematics. It is quite detailed and contains many beautiful pictures. Easy to browse through.

2. BIRNEY, D. S.: *Modern Astronomy,* Allyn and Bacon, Boston, 1969. Much the same as reference 1, but considerably shorter. Can be read easily by the nonscientist. Essentially no mathematics.

3. HOYLE, F.: *Galaxies, Nuclei and Quasars,* Harper & Row, New York, 1965. Written for the layman by noted astronomer. A little out of date in this era of rapid increase in knowledge.

4. OVENDEN, M. W.: *Life in the Universe* (paperback), Anchor Science, Doubleday, Garden City, N.Y., 1962. Written for the layman. You might find it better to wait with this until you have read Chapter 14 of this text if you have had no chemistry, but it will be understandable to you even now.

5. JASTROW, R.: *Red Giants and White Dwarfs* (paperback), Signet Books, New York, 1967. Contains a highly readable account of the material implied by the title as well as a discussion of the planets.

SUMMARY

The big-bang theory indicates that the universe was compressed into an extremely hot and dense fireball about 10 billion years ago. This primeval fireball expanded under the explosive forces generated within it because of its high temperature. After 100,000 years, the temperature of the fireball had dropped from its original billions of degrees to about 5000°F. Only then was the temperature low enough so that atoms could form.

The still-expanding cloud of hot gases and dust began to condense in certain regions. The material in these regions, impelled by gravitational attractive forces, fell together into what became the sites of embryonic stars. As the swiftly falling material aggregated into massive bodies, the impacts of the colliding matter raised the temperature of each aggregate to tens of millions of degrees. At this temperature, the proton fusion reaction was ignited within the core of the body and it became a star much like our own sun. At least a billion billion such stars formed in the expanding universe.

Looking out into the universe, we see huge clusters of stars called galaxies. Our own galaxy, the Milky Way, has perhaps 100 billion stars, one of which is our sun. The Milky Way galaxy has a diameter of about 10^{18} mi. In terms of the unit of length equal to the distance light travels in one year, called a light year, the galaxy diameter is of the order of 10^5 light years.

According to the law of conservation of angular momentum, the rotation speed of an object increases as the object contracts. Because of this, the spiraling dust cloud, falling into the object that was to become a star, began to rotate very rapidly about the site of the star as it fell toward it. In the case of our sun, this rotation was so rapid that portions of the cloud pulled apart, leaving cloud fragments circling the sun. Each fragment was pulled together into a solid object because of the gravitational attraction forces between its various portions, and these fragments became the planets.

The proton fusion reaction within the stars is the major source of heat within many of them, including our sun. As time progresses, the proton supply is depleted and the star undergoes transformation. In successive stages, each billions of years long, the star expands into a red giant and then contracts to form a white dwarf. Under certain conditions it becomes an extremely hot, dense object called a neutron star. During this progression, some stars undergo violent explosions, giving rise to a nova or supernova.

Our universe is still expanding. It may continue to do so forever. However, its expansion is slowed by gravitational forces. If these forces are strong enough, the universe will eventually stop expanding. The gravitational forces will then pull the universe back to re-form the original fireball. In that event, the universe will pulsate; it will become a fireball, expand, and contract to recreate the primeval fireball over and over again.

TOPICAL CHECKLIST

1. What is meant by the primeval fireball and what evidence is there for it?
2. What is the red shift and what is its importance? Define light year.
3. Outline the big bang theory.
4. What factors influence whether or not ours is a pulsating universe, and why?
5. Why does the proton fusion reaction occur only at high temperatures?
6. What does the symbolism 2×10^7 stand for? What about 10^4? 5×10^3? 1×10^3?
7. What is a galaxy and how did they come about?
8. What causes some galaxies to take on a spiral shape?
9. State the law of conservation of angular momentum as used in this chapter, and give an example to illustrate it.

10. About where is our solar system located in the Milky Way galaxy?
11. How does a hydrogen star form and evolve into a red giant?
12. Where and when were the heavy nuclei elements probably formed?
13. What is a white dwarf? Trace its history.
14. What is a nova and how does it differ from a supernova?
15. What is a quasar? A pulsar?
16. How did the Crab nebula originate?
17. What is a neutron star?
18. Outline the history of the formation of the sun and planets.
19. Give a probable reason for the fact that the planets rotate in the same direction and in nearly the same plane about the sun.
20. Discuss the possibility of the existence of other solar systems. Of other inhabited planets.

QUESTIONS

1. What evidence do we have that the universe is expanding?
2. Suppose a tiny bug is sitting on a broken rubber band, held unstretched but straight. When the rubber band is stretched, he notices that the pieces of dust on the band recede from him on both the left and right sides. Does this mean he is sitting at the center of the rubber band? Reasoning from this, does the fact that the universe is receding from us in all directions mean we are at its center? (Assume that, like us, the bug is unable to see far enough to observe the edge of his universe, the ends of the rubber band.)
3. The earth circles around the sun once each year and is held in its orbit by the gravitational attraction force of the sun upon it. If the sun's attraction force should suddenly increase so as to cause the earth to circle much closer to the sun, what does the law of conservation of angular momentum tell us would happen to the time required for one orbit around the sun?
4. Give several examples of situations in which a rapidly moving object causes heat to be generated when the object is suddenly stopped. How is this pertinent to the material discussed in this chapter?
5. As we shall see in Chapter 10, the average temperature of the universe has dropped until it is now very cold, −455°F. What does this imply for a planet far from any sun? For astronauts during future journeys into distant space?
6. One method for finding the distance to the nearest stars is the *method of parallax*, illustrated in Figure 1.9. The earth is shown in two positions on its yearly orbit around the sun. In order to see a distant star at times six months apart, a telescope must be aimed at different angles to the plane of the orbit as shown. This difference in angles tells us the distance to the

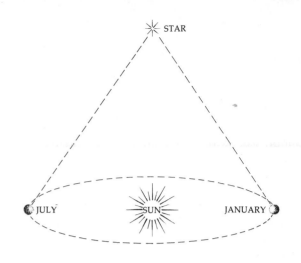

1.9 *The parallax method uses the fact that a telescope must be aimed in a slightly different direction to see a certain star at different times of the year.*

star. How? Try to draw to scale such a diagram for the nearest star, Alpha Centauri, which is about four light years away. (The earth-to-sun distance is about 0.00000016 light years away.) Why is this method impractical for very distant stars?

7 Suppose the nearest star, four light years away, exploded at noon today. When would we on earth become aware of the fact?

8 It is speculated that there are "black holes" in the universe, which are extremely massive, dense stars. As we shall see in Chapter 10, Einstein showed that a beam of light experiences gravitational forces as though it were a beam of particles. In terms of Einstein's discovery about light, explain why we might not be able to see these very massive stars even though they are white hot. Why is the name "black hole" appropriate?

Chapter 2

ASTRONOMY AND THE SOLAR SYSTEM

When we look into the cloudless sky we see the sun, moon, stars, and sometimes other distant objects tracing out paths across the heavens. Man has long studied these paths and has been able to predict major astronomical events since early in recorded history. Striking evidence for this is still visible at Stonehenge, where early Britons built a giant stone structure to help them predict eclipses and similar events, even though they had not yet developed a written language. In this chapter we shall discuss some of their observations and interpret them in terms of our knowledge of the universe. We shall then consider the structure and character of our solar system as revealed by our present-day space probes.

2.1 This rock structure at Stonehenge was used by ancient Britons in their successful efforts to predict eclipses and similar astronomical events. (Courtesy of the British Tourist Authority.)

The Distant Stars

We have seen that the earth is immersed in a vast cloud of stars which fills the universe. Even though many of these stars are receding from the earth, they are so far away that their motion is nearly imperceptible over the years. Their relative positions and separations in space have changed only slightly over the past few centuries. Yet, as we view these stars from the earth, we see them move large distances across the sky in a few short hours. This apparent motion is due to the rotation of the earth.

If astronomy were simply the recording of stellar motions in the heavens, the luckiest astronomer would be situated at either the North or South Pole of the earth. As Figure 2.2 shows, a polar astronomer would be standing at the earth's axis of rotation, and so his observations would be relatively simple to interpret. As our figure shows, Polaris, the North or Pole Star, lies on the (extended) axis upon which the earth rotates. Thus a man standing at our North Pole would find Polaris directly overhead, since both he and it are on the same axis. However, the star that he views in front of him at A will be behind him 12 hr (hours) later when the earth's rotation has turned

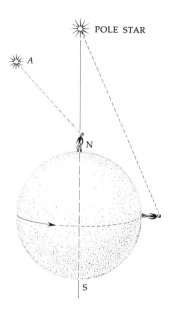

2.2 *From the north pole, the pole star appears to remain stationary while an off-axis star seems to follow a circle about the pole star.*

him around. Hence all stars but Polaris will appear to him to describe circular paths about the Pole Star. Figure 2.3 is a photograph showing how the stars streak across the heavens when viewed from the North Pole.

A man looking away from the earth at the South Pole would also notice that the stars seem to circle as in Figure 2.3. However, since he would be looking into the universe in the opposite direction, he would not see the Pole Star. None of the stars visible to him would be seen by a man at the North Pole, and vice versa, since half of the universe is hidden from each by the surface of the earth.

A person viewing the stars from near the equator sees a quite different path for the stars in the sky. Referring again to Figure 2.2, and recognizing that the Pole Star is much farther away than shown, we see that he will observe the Pole Star to be close to the horizon and directly northward. You should be able to show from the diagram that all observers in the earth's Northern Hemisphere will see the Pole Star in the Northern regions of the sky, but at different angles above the horizon. The observers at the North Pole and equator represent the two limiting cases, one seeing Polaris directly overhead, the other seeing it at the horizon.

The observer at the equator, unlike the polar astronomer, has the opportunity to look out in all directions into the universe as the earth rotates 360° each 24 hr. A star directly overhead at 9 P.M. will have moved to the Western horizon by 3 A.M., one quarter of a day later. For him, stars appear at the Eastern horizon, streak in straight lines across the sky, and finally disappear over the Western horizon. Star streaks as seen by an equatorial observer are shown in Figure 2.4.

As you might expect from the stellar paths observed in these limiting cases, we who are elsewhere on the earth see the stars following intermediate types of paths. They will appear over the Eastern horizon and disappear in the West, but their paths will be curved rather than straight. You might be interested in reasoning out the motion of stars as seen by a person in the United States. In doing so, a more realistic version of Figure 2.2 will help you.

Constellations and the Seasons

In the previous discussion we ignored the fact that sunlight makes it impossible to see stars during the day. As a result, an observer at the earth's equator sees only those stars that are visible to him at night. For example, Figure 2.5 shows that half of the earth is in darkness. Only those people on the dark side of the *terminator*, the dividing line between darkness and brightness, can observe the stars easily. As a result, they can see the stars only in a restricted region of the sky.

However, since the earth completes a nearly circular path around the sun once a year, the region of the sky visible at night varies as the year goes on. These various regions of the sky, and the *constellations* or groupings of stars in them, were given names centuries ago. Gemini is overhead at midnight in

2.3 Polar star trails. Can you estimate the exposure time for the photograph? (Lick Observatory photograph.)

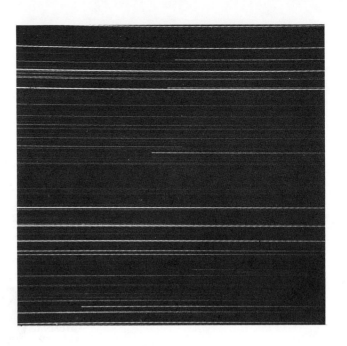

2.4 Equatorial star trails. (Lick Observatory photograph.)

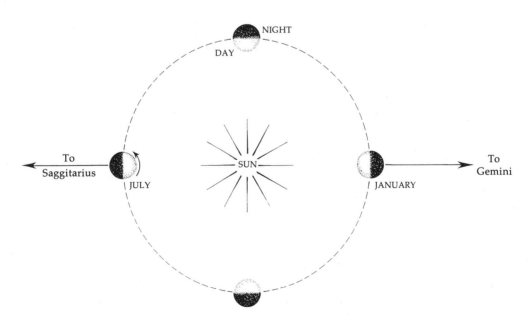

2.5 A person on the earth sees different parts of the universe in the night sky at various times of the year.

THE SOLAR SYSTEM 21

2.6 Because of the inclination of the earth's axis to the ecliptic plane, the full rays of the sun fall chiefly on the northern hemisphere in July and on the southern hemisphere in January.

January while Sagittarius is overhead in July. Unfortunately, the Gemini constellation no longer coincides with that region of the January sky which the astronomers defined as Gemini over 2,000 years ago. As we shall see later, this drifting apart is the result of the fact that our calendar was designed to keep in step with the seasons. (Those who are interested in astrology should note that horoscope readers usually fail to take this factor into account.) Star patterns, then, drift westward in the night sky because of the earth's yearly movement around the sun. A constellation that is overhead at midnight in July will be close to the Western horizon in September.

The earth's movement around the sun also brings about the change of seasons. Since the orbit of the earth is nearly a circle about the sun, with the earth-to-sun distance varying from 94,500,000 mi in July to 91,500,000 mi in January, very little of the change in earth's temperatures is the result of its motion around the sun. Figure 2.6 shows the additional factor responsible for the seasons: the earth's rotation axis is not perpendicular to the plane of its orbit, the *ecliptic plane.*

As the earth makes its annual circuit of the sun, the North and South polar regions undergo vast seasonal changes. As Figure 2.6 shows, when the earth is at the winter solstice position on its orbit (about December 22), the North Pole is in perpetual darkness while the South Pole has continuous daylight. In spite of the daily rotation of the earth on its axis, the North Pole is hidden from the sun and the South Pole is continuously exposed to sunlight (and so, in December, the South polar region is truly "the land of the midnight sun"). The figure also shows that in December the entire Southern Hemisphere is turned more toward the sun than the Northern Hemisphere is. As a result, at that time the Northern Hemisphere begins to experience winter while summer has reached the Southern Hemisphere; North America has its shortest day of the year while Argentina has its longest.

By the time the earth has reached the opposite side of its orbit, about June 21, the situation is reversed. The North Pole is then the land of the midnight sun, and summer has come to the Northern Hemisphere. (People sometimes wonder why the United States does not experience its hottest season around June 21 since the days are longest then and the sun's rays are striking most strongly. The time lag between this and the very hot period of late July is the result of the cumulative effect of many long, sunshine-filled days building up heat on this portion of the earth even after June 21.) At the times of the spring and autumnal equinoxes, March 21 and September 21, all portions of the earth have 12-hr days and 12-hr nights. What do you think is happening at the North and South Poles at these times?

Before leaving this section, we should point out that a fundamental law of nature is illustrated in Figure 2.6. Notice that the rotation axis of the earth remains in a fixed orientation relative to the distant stars even while the earth moves around the sun. This is not an isolated phenomenon; all spinning objects retain the orientation of their spin axes relative to the stars unless a twisting force is applied to the object. A spinning gyroscope demonstrates this same behavior. The following statement, usually considered a part of the law of conservation of angular momentum, summarizes this observed behavior:

> *In the absence of externally applied twisting forces, the axis of a spinning object will retain its orientation relative to the distant stars forever.*

This law raises deep scientific questions. It implies that the spinning object "communicates" in some way with the rest of the universe, or it would not oppose changes in its orientation. If so, does this mean that a change involving any body in the universe would affect the spinning object? If the external universe began to rotate, would this spinning object rotate also to keep its same relative orientation? These questions, the subject of much conjecture, will come up again in our study of Einstein's theory of relativity.

As the earth moves around the sun, a small twisting force actually does try to reorient the earth's axis of rotation. This results from the fact that the earth bulges somewhat at the equator and the sun's gravitational attraction tries to pull the bulge closer. As you can see from Figure 2.6 when the earth is at the winter solstice position, the attraction force of the sun on the equatorial bulge tries to twist the earth so that its rotation axis will become perpendicular to the ecliptic plane. Although small, this force has caused noticeable changes over thousands of years. One such change is the apparent displacement of the constellations during the past 2,000 years, an effect mentioned earlier.

The Moon

Even before man's first landings on the moon, telescopic examination revealed the mass of the moon (about 1 percent of the earth's), and its diameter (2,160 mi, about one-fourth the earth's diameter). We also knew that its distance from the earth varies from about 222,000 mi at its closest point (called the *perigee*) to 252,000 mi at its farthest point (the *apogee*). These latter distances can be determined to an accuracy of a few feet by measuring how long it takes for a pulse of light from a laser to travel the round trip to the moon.

Although the orbit of the moon around the earth is elliptic rather than circular, the extreme radii given above are so nearly the same that we can approximate the orbit by a circle, as shown in Figure 2.7. The figure also shows the position of a particular point on the moon's surface as the moon circles the earth. This same point always faces the earth, because the moon takes 27.3 days to circle the earth and exactly the same time to rotate once on its axis. But the time from one sunrise to the next on the moon is 29.5 earth days. This is a little longer than it takes for the moon to rotate once on its axis because it is moving with the earth around the sun at the same time.

We know that the mass of the moon is small, about 1 percent as large as the earth's. Since the gravitational attraction of the moon (or earth) for objects on its surface is proportional to the mass of the moon (or earth), you might expect an object on the moon to weigh about 1 percent as much as it does on the earth. But the radii of the earth and moon are not the

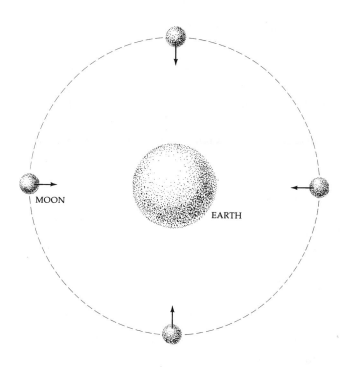

2.7 *The moon rotates once on its own axis each time it makes one orbit around the earth. Consequently, the same side always faces the earth. The period for both rotations is 27.3 days. Drawn to scale, the orbit diameter would be about six times larger than shown.*

same, and the gravitational force decreases rapidly as the distance from the center of a spherical mass increases. The shorter distance from the center of the moon counteracts the smaller effect of its mass, and causes the weight of an object on the moon to be about one-sixth as large as its weight on earth. A man who can just lift a 10-ft piece of pipe on the earth could lift 60 ft (six times as much) of the same pipe on the moon. Since the moon pulls him down only one-sixth as hard as the earth, he should be able to jump six times as high on the moon as on the earth. In fact, the moon's gravitational force is so weak that it cannot even retain air molecules. As a result, the moon has no air or other atmosphere.

Because it lacks a protective layer of atmosphere, and has days and nights approximately 15 earth-days long, the moon surface experiences extreme fluctuations in temperature. Temperatures on the moon rise slightly above the temperature of boiling water when the sun is in full view, which is why moon explorers carry refrigeration units on their backs. During the lunar night, the temperature drops to about −200°F, since there is no atmosphere or cloud blanket to hold heat to the moon. Our astronauts have discovered much about the physical makeup of the moon. They have brought back enough moon soil and rocks to make us quite certain that the earth and moon were formed at about the same time. The age of the oldest rocks on both the moon and earth is about 4 billion years. Moreover, the chemical makeup of the moon rocks is similar to that of rocks found on the earth. Since the earth is about twice as dense as the moon, the central cores of the two bodies must differ. The larger gravitational forces on the earth must cause higher internal pressures and a more dense core for the earth. We know that much of the earth's core is molten; whether or not this is true for the moon is still open to question.

Perhaps the most striking differences between the moon and the earth result from the lack of atmosphere on the moon. Because the moon's surface is in a high vacuum, water and wind erosion are totally absent. Aside from the effects of impinging meteors and interstellar dust, the moon's surface has not been disturbed for billions of years. Early in its history the moon probably had volcanic activity, but no active volcanoes are seen there now. Mountains as high as the highest on the earth have been observed, however. The large craters you can see in Figure 2.8 are thought to be pockmarks caused by large meteorites

2.8 A view (a) of the far side of the moon's surface as photographed by an unmanned Lunar Orbiter. (Courtesy of NASA.) The floor of the lunar crater Copernicus (b) photographed by Orbiter II from a height of 28 mi. (Courtesy of NASA.)

(a)

(b)

2.9 *The Barringer meteorite crater photographed from the air. It is located in Winslow, Arizona, and is 4,200 ft across by 600 ft deep. (Courtesy of Yerkes Observatory.)*

impacting the moon. Similar marks must also have existed on earth, but they soon eroded away. One such crater, presumably formed in the more recent past, is found on the earth near Winslow, Arizona (Figure 2.9); it is nearly 1 mi across. There is always a faint possibility that a meteorite as large as the one that caused the Arizona crater could strike a populated portion of the earth.

Since the moon is basically a ball of rock and rubble in a waterless vacuum, it is not surprising that no trace of life has been found there. Its surface is like a giant museum, showing us what the earth would have been like without the disturbing effects of wind, rain, and living things.

Phases of the Moon; Eclipses

One of the features we notice most about the moon is its changing shape as each month progresses, the so-called *phases of the moon*. If we are to believe the old legends, various phases can bring on such diverse things as romance, werewolves, lunacy, or a good harvest. Although it is highly unlikely that the phases of the moon affect most of the things attributed to them, the cause of the phases is of interest.

As with all nonstellar objects, the moon can only be seen because of light reflected from it. Light from the sun strikes the moon and is reflected to the earth. In Figure 2.10, we see that only half of the moon is lighted at any time. Depending upon where the moon is situated relative to the earth in the course of a month, we will see various portions of this lighted half, as shown at the bottom of the figure. Examine at least a few of these so you can see why they appear as they do.

We can see another important feature from Figure 2.10. If the moon, earth, and sun were all in the plane of the page, then when the moon was in position 1, it would be in the shadow cast by the earth. The moon would be dark; this is an *eclipse* of the moon. Alternatively, when the moon was in position 5, it would cast a shadow upon the earth. People in the shadow region would be unable to see the sun; this is an eclipse of the sun. If the sun, the moon, and the earth were all in the same plane, we would have one eclipse of each type once each month as the moon orbited the earth.

Actually, the sun is not in the same plane as the moon's orbit, as Figure 2.11 shows. As time goes on,

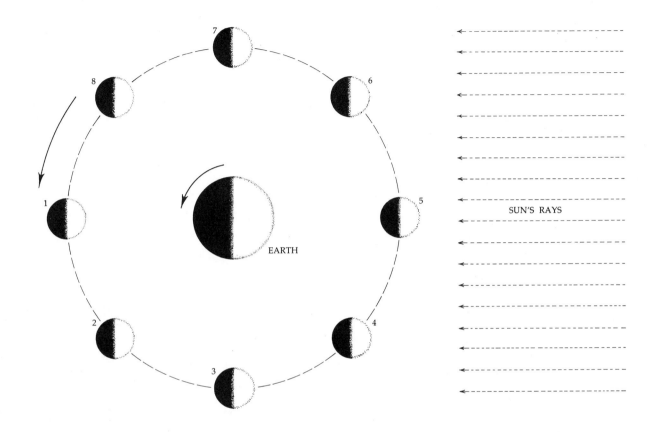

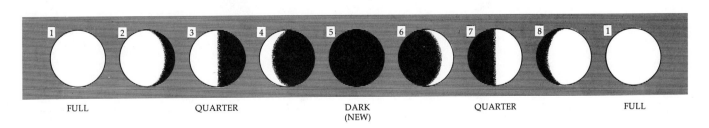

2.10 Phases of the moon. Only part of its lighted surface can be seen from the earth at various times during the moon's monthly orbit.

THE SOLAR SYSTEM

2.11 The line from the earth to the sun is not in the plane of the moon's orbit. If it were, an eclipse of the sun would occur each month. (Drawing not to scale.)

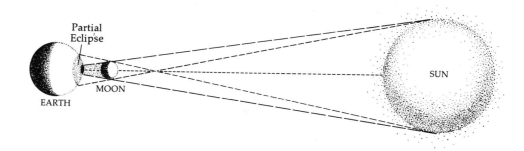

2.12 An eclipse of the sun. (Not to scale.)

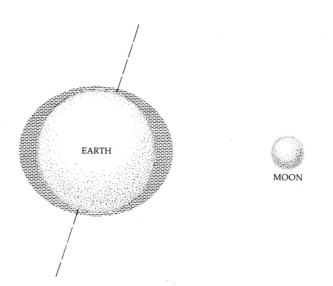

2.13 Tides rise simultaneously on opposite sides of the earth because of the gravitational forces of the moon.

the orientation of the moon's orbit with respect to the sun changes, and the orbital path crosses the line between earth and sun. However, the chance that the moon will cross the point in its orbit that strikes the earth-sun line is small. When it does, an eclipse of the sun occurs, as we see from Figure 2.12.

During such an eclipse, a small portion of the earth is hidden from the sun. The moon's shadow sweeps across the earth's surface as the earth and moon follow their courses. The path of total eclipse is never more than 167 mi wide and takes no more than 7.5 min to pass over a given point on the earth. Of course, the sun is partly obscured from the sight of people close to the region of total eclipse, as you can see from the figure. The band across the earth in which a partial eclipse can be seen may be as wide as 2,000 mi.

In addition to causing eclipses, the moon is responsible for the ocean tides. As you probably know, the level of the ocean rises and falls twice each day, often rising several feet above the low-tide mark. At the Bay of Fundy in southeast Canada, the tidal range sometimes exceeds 50 ft. Figure 2.13 shows the tidal pattern (the earth's land masses have been left out to simplify this illustration). Notice that the gravitational attraction of the moon draws the water from the sides of the earth into the region closest to the moon. The water also flows to the side of the earth facing away from the moon. Hence, high tide occurs at opposite sides of the earth simultaneously.* It is easy to understand the high tide closest to the moon, since the moon's gravitational forces are pulling in that direction. We must take a closer look at the situation to understand why the tide is also high on the side opposite the moon.

The tidal phenomenon depends on two facts: (1) the gravitational force of the moon varies from place to place on the earth, since the gravitational force decreases with distance, and (2) the earth is not a rigid body.

If the earth were perfectly rigid and not rotating on its axis, all points within it would move in the same path that the earth's center follows about the sun. Even in such a simplified case, however, the moon would influence this path slightly. When the moon came between the sun and earth, its gravitational attraction for the earth would pull the whole earth a little closer to the sun than normal. When it was on the side of the earth opposite the sun, it would pull the whole earth slightly farther away from the sun. But since the earth is not perfectly rigid (its seas, for example, are fluid) not all parts of it respond in the same way to the moon's pull. The earth's center follows a path around the sun determined by the combined forces exerted on all parts of the earth by the moon and the sun. This is not true of the outer parts of the earth.

Consider what effect the moon has on the earth as it tries to pull it out of its path around the sun. Because the earth is not perfectly rigid, the side of the earth closest to the moon is pulled more strongly toward the moon than the opposite side, and that portion of the earth is pulled into a position closer to the moon. This closest portion of the earth is distorted somewhat, as shown in Figure 2.13, and the oceans, being more fluid than the continents, show the distortion most vividly. Simultaneously, the side of the earth opposite the moon is pulled toward the moon, but with a force less than the average because of its greater distance away; its path is less deflected by the moon. Hence the side opposite the moon bulges *away* from the moon as shown. As a result, the more fluid portions of the earth bulge out both on the side nearest and on the side farthest from the moon. During the daily rotation of the earth, these bulges in the seas move around the earth, causing rising tides in the regions where they exist.

---

* The exact height and time of the tides depends upon the flow of the water around the land barriers. At the Bay of Fundy, the tidal flow is funneled into a small channel and this gives rise to the very large tide observed.

Table 2.1   Data on the Planets

Planet	Sun Distance (miles)	(ratio)	Diameter (miles)	(ratio)	Mass (m/m_e)	Density (water = 1)	Rotation Period (days)	Revolution Period (years)	Temperature (°F)	Gravity (g/g_e)
Mercury	0.36×10^8	0.39	3,170	0.40	0.055	5.6	59	0.24	600	0.40
Venus	0.68×10^8	0.73	7,750	0.98	0.81	5.2	243	0.62	~900	0.90
Earth	0.93×10^8	1.00	7,910	1.00	1.0	5.5	1.00	1.00	72	1.00
Mars	1.42×10^8	1.53	4,270	0.54	0.11	4.0	1.02	1.88	27	0.40
Jupiter	4.8×10^8	5.2	89,000	11.3	314	1.3	0.41	11.9	−220	2.40
Saturn	8.9×10^8	9.6	74,000	9.4	94	0.7	0.43	29.5	−230	1.20
Uranus	17.8×10^8	19.2	33,000	4.2	14	1.4	0.47	84	−350	1.00
Neptune	28×10^8	30	31,000	3.9	17	1.3	0.60	165	−350	1.00
Pluto	37×10^8	40	~4,000	~0.5	<0.2	?	6.6	248	?	?

The Wandering Stars: Planets

We have seen that the motions of the stars and the moon through the heavens are rather simply explained if we do not worry about minor effects such as the slow movement of the constellations over the centuries. By the time of Newton (1642–1727), astronomy already had become precise enough so that these complications had been noticed. In addition, astronomers had discovered that a few stars seemed to wander through the heavens in a much more puzzling way. We now know that these are not stars at all; they are the planets shining in the reflected light of the sun.

Unfortunately, however, scientists, including astronomers, are influenced in their judgments by preconceived notions and prejudices. It is not surprising that man, being the egotist he is, had difficulty in relinquishing the idea that the earth was the center of the universe. For many centuries astronomers had interpreted the motions of the stars in terms of the universe revolving around the earth. If you try to interpret the motions of the stars shown in Figures 2.3 and 2.4 in terms of such a concept, you will find it is quite easy to do. But the motions of the planets are almost impossible to describe in terms of an earth-centered universe.

Early scientists were so reluctant to relinquish the concept of an earth-centered universe that they devised exceedingly complicated schemes for the motions of the planets in their supposed orbits around the earth. Even though a few supported the theory that the planets revolve around the sun, most people rejected it. Over a period of centuries, the overwhelming evidence for a sun-centered solar system forced its acceptance. Unfortunately we cannot take the space here to give the intriguing history of this lengthy struggle; you may wish to read about it in the references at the end of this chapter.

We now accept the fact that the wandering stars noticed by the ancients are really the planets* in orbit around the sun, shining like stars in the night sky as light from the sun is reflected from them. Nine planets, including the earth, revolve around the sun in elliptical orbits. Most of the orbits are nearly circular and the planes of the orbits are nearly coincident. The orbit of the planet Pluto is the least

* The word *planet* is a transliteration of the Greek word for *wandering*.

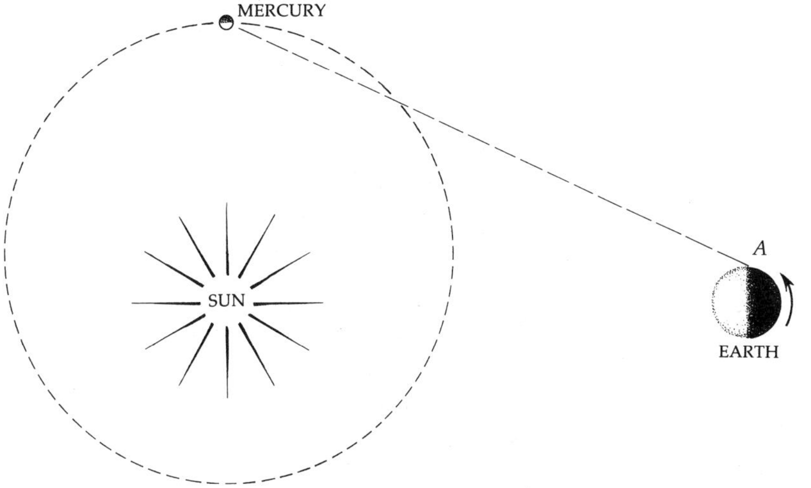

2.14 *Even in the optimum case, shown here, Mercury is visible only to the nighttime portion of the earth at A. It will be seen low in the east a short time before sunrise. Where will Mercury be when it is seen at sunset?*

circular, and its plane is tilted by the largest amount; the angle between its orbital plane and that of the earth is 17°. All the planets revolve in the same direction about the sun, which is not surprising if you recall our ideas about the formation of the solar system. Table 2.1 summarizes the features of the planets; we will discuss each planet in turn.

Mercury

Mercury is the closest planet to the sun, less than half as far away as the earth. Because of this closeness, the average temperature of the lighted side of Mercury is high, about 600°F. At such high temperatures, hot enough to melt lead, no form of life could be expected to exist. In addition, the gravitational force on Mercury is only 0.4 as large as that on earth, and so all water and gaseous substances have long since boiled from its surface. Like the moon, Mercury has no atmosphere, and therefore it also cools rapidly during its nearly month-long nights. It is estimated that its temperature dips to −450°F at night.

Although Mercury is very brightly lit, since it is so close to the sun, it is difficult to see. As Figure 2.14 shows, Mercury's orbit is so close to the sun that, whenever it is visible on earth, the sun can also be seen, or else it is close to sunrise or sunset. The light from the sun is so strong that our naked eyes have difficulty detecting the fainter light from the planet. However, powerful telescopes enable us to see Mercury during daylight hours.

Venus

Like Mercury, Venus orbits the sun with a radius smaller than the earth's orbit. Because Venus follows a larger orbit than Mercury, it can be seen more easily. It is particularly brilliant as the *Evening Star* seen in the Western part of the sky. Even so, Venus is visible only in the early evening or a short time before sunrise, depending upon where it is in its orbit around the sun.

Venus is similar to the earth in size and density, but there the similarity ends. It rotates slowly on its axis, its "day" being over half a year long. Venus has a dense, thick atmosphere, and is continuously shrouded in a cloud cover. Until the United States Mariner and Soviet Venera space probes penetrated the Venusian atmosphere in the late 1960s, we had no clear picture of the planet below or of the chemical composition of its atmosphere. The space probes have shown the atmosphere to be mostly (90 to 95

THE SOLAR SYSTEM

percent) carbon dioxide. Although the data are not completely reliable, Venus's atmosphere appears to have a fraction of a percent of oxygen, about 1 percent water vapor, and anywhere from 0 to 7 percent nitrogen.

The most startling revelation of the space probes was that the surface temperatures of Venus are high on both the light and dark sides of the planet; they are probably in the range near 900°F. Although the cloud layer has a much lower temperature on its outer surface (about −10°F), it is approximately 15 mi thick. Evidently, this thick cloud blanket traps the rays of the sun and keeps the heat within from radiating outward. As a result, the surface of Venus is almost certainly an unlivably hot desert, and it seems impossible that life could exist there. But our knowledge of this nearest neighbor to the earth is still scanty and you will be learning more about it as future space probes provide additional data.

Mars

Life is more likely to exist on Mars than on any other planet (besides earth) in our solar system. Our largest telescopes have enabled us to see the surface of Mars, and space probes have photographed it extensively. Even though it takes several months to reach Mars in today's spacecraft, man will almost certainly step on this planet before the end of the century.

Mars is farther from the sun than the earth is, and hence is considerably cooler than earth; its average temperature is close to −45°F, and temperatures vary from about 80°F at the warmest points to −150°F at the coldest. Because of Mars's thin atmosphere (only about 1 percent of that of the earth), temperatures fluctuate widely, perhaps dropping as much as 100°F when the sun sets. In keeping with its thin atmosphere, cloud cover is minimal.

The chemical composition of the atmosphere is at least 80 percent carbon dioxide, and if any oxygen at all is present, it is only a trace. Because plants on earth consume carbon dioxide and release oxygen, scientists believe that earth-type plants do not exist on Mars. Moreover, only a trace of water is found in the atmosphere, and so rivers and lakes are absent. There is no evidence to indicate that water ever existed in quantity on Mars.

Photographs of Mars taken by space-probe cameras such as those in Figure 2.15 show that the surface looks like that of the moon in many respects. Notice the white polar regions on the otherwise reddish sphere. We think these caps are carbon dioxide snow, a material familiar to us as dry ice. At the extremely low temperatures of the polar regions, carbon dioxide freezes out of the atmosphere just as water vapor forms snow on earth. We believe that these dry ice snow drifts are whipped to heights between 20 and 50 ft by the feeble winds in the thin Martian atmosphere.

As you can see in Figure 2.15, Mars has craters similar to the ones on the moon. However, Mars also has regions where erosion or some other mechanism has smoothed the surface. We have learned that "canals" and other imaginative structures, which some early astronomers claimed to have seen, do not exist. In addition, life as we know it on earth is absent from Mars, and life of any kind seems unlikely.

Two tiny moons orbit the planet; their diameters are too small to measure directly, but they must be approximately 5 to 20 mi. As befits their association with Mars, the god of war, they are called Phobos (fear) and Deimos (panic). It is interesting to note that in 1726, Jonathan Swift wrote in *Gulliver's Travels* that the Laputan astronomers had discovered ". . . (two) satellites, which revolve about Mars. . . ." The actual discovery of the Martian moons was made 150 years later, in 1877.

Jupiter

As discussed in Chapter 1, Jupiter nearly became a second sun in the solar system. You will recall that the hydrogen fusion reaction was ignited in the sun's core as a result of the high temperatures gen-

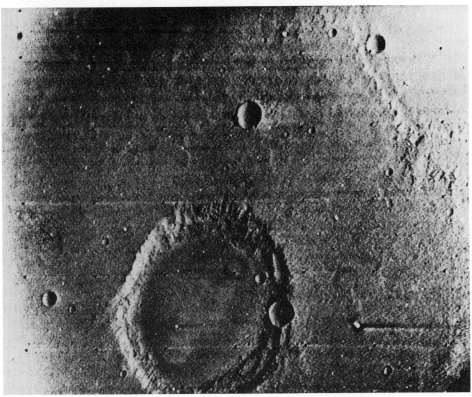

2.15 *Photographs of Mars taken by the space probe Mariner 6 in 1969. For comparison purposes the area shown in (b) is approximately 63 by 48 mi. (Courtesy of the Jet Propulsion Laboratory.)*

erated by the material of the sun collapsing under huge gravitational forces. Jupiter, the largest of all planets, is 318 times as massive as the earth and about 1/1,000 as massive as the sun. Scientists have speculated that this mass is nearly large enough to provide the gravitational forces necessary to make the fusion reaction go. Jupiter's gravitational forces are so large that its structure is far different from the other planets we have studied.

Unlike the earth, Jupiter is largely composed of hydrogen and helium; estimates have been made of 80 percent hydrogen and 10 percent helium. Because of its large gravitational force (about 2.4 times that on earth) and low temperature (estimated to be −200°F), Jupiter has retained these light molecules. Preliminary data indicate that Jupiter has elements in about the same proportions as the sun, which further confirms our assumption that the solar system was formed from a single giant cloud. If all the hydrogen and helium were to escape from Jupiter, a solid object with a chemical composition somewhat like the earth's would remain.

We do not know for sure at this time, but we think that Jupiter has an atmosphere a few hundred miles thick, composed mainly of hydrogen and helium, in which about 1 percent methane and ammonia is also present. The atmosphere is cloud-filled, but it is difficult to ascertain the composition of the cloud layer. At these low temperatures, crystallized ammonia may be the major cause of reflected light from the layer. One or two hundred miles below the cloud layer, the gravitational pressure becomes large enough to liquify the hydrogen, and so the "surface" of Jupiter is probably an "ocean" of liquid hydrogen with small amounts of other elements dissolved in it. This liquid layer is a few hundred miles thick. At still greater depths the pressures are high enough to solidify the material. Unfortunately, we do not yet know enough about the behavior of hydrogen and helium at high pressures to predict with certainty the nature of Jupiter's core.

The presence of methane and ammonia on Jupiter raises the possibility of life there. One theory for the evolution of life on earth proposes that lightning discharges in the earth's atmosphere caused the methane and ammonia to react and form protein molecules. If this is true, a similar situation may exist on Jupiter. Telescopic examination has provided no evidence for such reaction products; we must wait for space probes to Jupiter to settle the point.

Because Jupiter is so large, it is bright in the night sky. Circling it are at least 12 moons, four of them large enough for Galileo to see them with his crude telescope. They are approximately the size of the earth's moon. The remaining eight moons are much smaller, their diameters ranging from 15 to 150 mi. The smaller satellites are difficult to see even with the best telescopes, but the four larger ones can be seen with a good pair of binoculars.

Jupiter's surface has been the subject of speculation for centuries. The Great Red Spot in Figure 2.16 changes slightly in size, color, and shape and has been as large as 30,000 miles across. In addition, alternate dark and light cloud bands circle the planet parallel to its equator. These features, as well as many others, remain the subject of much speculation. As future space probes enter Jupiter's atmosphere, we expect to find their true cause.

Saturn

Unique among the planets, Saturn is distinguished by its three rings, as seen in Figure 2.17. The planet is second only to Jupiter in size and it is 95 times as massive as the earth; but its density is exceptionally low, being only 70 percent that of water. Like Jupiter, Saturn's gravitational forces, and its low temperature (about −300°F) resulting from its distance from the sun, hold hydrogen and helium in its atmosphere. At −300°F, the ammonia has crystallized and settled out of its atmosphere. Otherwise, its atmosphere is similar to that of Jupiter.

Astronomers believe that the interior of Saturn is similar to that of Jupiter. Under a thick layer of

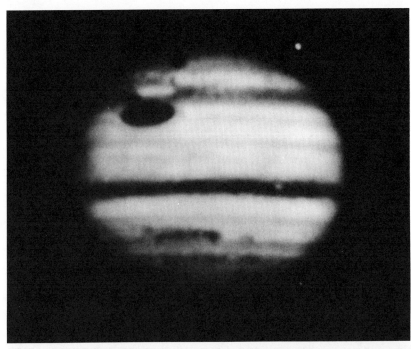

2.16 Jupiter as seen from the earth through a 200-in. telescope. The Great Red Spot is at the top left. (Mount Wilson and Palomar Observatories.)

2.17 Saturn and its rings as viewed through the 200-in. telescope; the innermost ring is visible only where it passes in front of the planet. (Mount Wilson and Palomar Observatories.)

gaseous hydrogen and helium is a vast ocean of liquid hydrogen, which covers the entire planet. Hundreds of miles lower, the hydrogen has been solidified by the planet's large gravitational pressures. Even though the density of the planet as a whole is low, the highly compressed center core is thought to be about 16 times as dense as water.

The rings of Saturn were first seen by Galileo, but his telescopes were too crude to make out their exact shape. Their true form was first described in 1655. The rings, which are of the order of 10,000 mi wide, are composed of tiny particles no larger than gravel. We do not know the exact chemical composition of the rings; but we do know that the rings, despite their width, are probably less than a few feet thick. In view of how the planets probably formed, we should not be surprised to find that rings were torn loose from the condensing cloud during the planet's birth. The planet spins rapidly (see Table 2.1), and the rings spin around the planet in the same direction.

You might well wonder why these rings did not condense further to form moons. A theorem proved mathematically in 1850 by E. Roche tells why: If a moon has the same makeup as the planet, and if it is held together only by gravitational forces, the moon cannot exist closer than 2.44 times the radius of the planet. This figure is called _Roche's limit_. Any moon closer to the planet than this would experience such large gravitational forces from the planet that the moon would be torn apart. Thus Saturn's rings represent embryonic moons which have been torn apart and spread on a ring around the planet.

At least 10 moons circle Saturn, and they all lie outside the Roche limit, as we would expect. The last of these moons was discovered in late 1966. It is close to the outside rim of the largest ring, and Roche's theory shows that it causes the gap between the two outer rings. (However, the moon is _not_ made of the condensed material of the ring; the theory is not that simple.) Titan, the largest moon of Saturn, is larger than our own moon. Because of its great distance from the sun, its temperature is low enough so that it retains an atmosphere. However, it seems unlikely that life could exist there.

Uranus

Uranus is so far away from the earth that it is no brighter than many of the stars in the sky. As a result, it was not known as a planet until 1781. It appears greenish in the telescope, presumably because its atmosphere contains considerable quantities of methane gas which reflect light in this region of the spectrum. Uranus is 14 times more massive than the earth. Its large gravitational forces and low temperature (about $-350°F$), enable it to retain a considerable amount of hydrogen in its atmosphere. We believe it is much like Jupiter and Saturn in structure. The planet has at least five moons. Its axis of rotation is nearly in the plane of its orbit and it is peculiar in this respect (see Question 7).

Neptune

After the discovery of Uranus, it was found that its orbit was not quite as predicted. In order for Uranus to travel the path it does, a nearby planet must be exerting attractive forces upon it. In 1845 this planet was sighted by telescope and was given the name Neptune. As you can see from Table 2.1, it is similar to Uranus; we believe they have the same general structure. Unlike Uranus, however, its axis orientation is normal.

Pluto

Pluto's existence, like that of Neptune, was first predicted by the discovery of inconsistencies in the orbit of Uranus. It was found in 1930. We know little about Pluto because it is so far from us and the sun. Pluto has the most elliptic orbit of all the planets and it is inclined at the largest angle to the plane of the earth's orbit. Presently we believe that Pluto is about twice as dense as the earth and about half as large in diameter. Its temperatures must be very low, probably near $-400°F$.

Asteroids, Meteors, and Comets

In addition to the planets, smaller objects, called *asteroids* are trapped in the gravitational fields of the solar system. The largest of these asteroids (or planetoids), Ceres, was discovered in 1801; its diameter is about 500 mi. About a dozen asteroids have diameters larger than 100 mi; perhaps a few hundred have diameters as large as 20 mi. About 1,600 of the estimated 44,000 asteroids have been discovered and named.

The composition of a typical asteroid is thought to be like that of earth's crustal rocks. It has been suggested that an asteroid would serve as a valuable site for a space laboratory. Since the mass of an asteroid is small, gravitational forces would not affect landing and takeoff seriously but would be large enough so that a person would not float away. A man might be able to jump a thousand yards high and fall back slowly to the asteroid in a few minutes. The asteroid would be large and stable enough to serve as a spacious laboratory site. Astronomers have charted the paths of many asteroids; they usually move in highly elliptic orbits about the sun. Some of them come comparatively close to the earth. For example, Eros passed within 14 million miles of the earth in 1931; periodically the asteroids Icarus and Geographos come within 4 and 6 million miles, respectively, of the earth.

Also scattered through the solar system are streams of small particles called *meteors;* most of them are much smaller than a bean, and tens of millions of them strike the earth's atmosphere each day. As they enter the earth's atmosphere, they become white-hot, and their glowing vapor trails are the "shooting stars" often seen in the dark, moonless sky. Most of them are completely vaporized before reaching the ground, but a few survive and strike the surface of the earth. The remnants, cold, rocklike aggregates, are called *meteorites*. Some meteorites have been very large. A 36-ton meteorite was found in Greenland and later moved to the American Museum of Natural History in New York City. Another landed in Arizona less than 5,000 years ago and left the crater shown in Figure 2.9.

In this century two large meteorites have landed in Siberia. The impact of the first, in 1908, was recorded by earthquake monitors as far away as central Europe; it killed nearly 1,500 reindeer and knocked unconscious a man standing 50 mi away. Its mass is estimated to have been 10^5 tons. The second meteorite struck near Vladivostok in 1947. Although it shattered before striking the ground, over 5 tons of meteorite fragments were recovered from the 2 square mile area it devasted. If either of these had hit a populated area, a tragedy of major proportions would have occurred.

Meteors are believed to be material left behind in the wake of *comets*. Typically, a comet has three parts. The nucleus (or head) is an aggregate of hard particles ranging from dust specks to rocks a yard in diameter. Some of the particles may be imbedded in an ice made of methane, water, and ammonia. The diameter of the nucleus usually ranges from 10 to 100 mi. Surrounding the nucleus is a cloud of gas and dust called the coma. The coma may be as much as a few hundred thousand miles across. Finally, a tail of gas and dust trails for millions of miles along the comet's path.

Some comets retrace elliptical paths through the solar system; others enter the solar system, pass through, and leave, never to be seen again. A comet becomes most visible when it swoops close to the sun and shoots back into the distance. As it nears the sun, the sun's heat and high-energy radiation vaporize some of the comet's particles, and they, together with the attendant gas, begin to glow. In addition, the comet reflects sunlight. A series of photographs showing the passage of a typical comet and its trails is shown in Figure 2.18.

Each year, as the earth swings through space in its orbit about the sun, it moves through the "tracks," composed of meteors, left behind by past comets. As a result, we can predict rather well which parts of the year will show strong meteor activity. The remains of Halley's comet are seen near May 4 and October 20

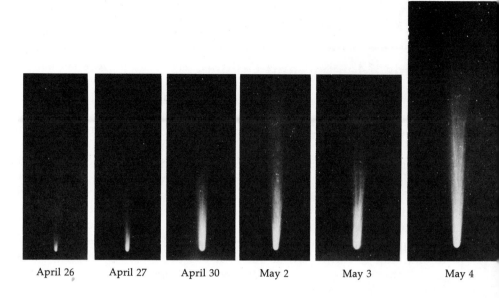

| April 26 | April 27 | April 30 | May 2 | May 3 | May 4 |

each year. The largest meteor shower, the Perseid shower, appears for about three days around August 11.

We can learn about the history of comets and meteors from meteorites found on earth. Using a technique called radioactive dating, (described in Chapter 11), we find that meteorites appear to be about 4.5 billion years old, the same age we give to the earth and the sun. From this we surmise that these materials were formed in the same condensation phenomenon that started the solar system.

Kepler's Laws and Quantitative Science

Centuries before the invention of the telescope, astronomers had successfully charted and predicted the paths of planets and stars through the heavens. Often their methods were based on their observations alone, without any fundamental understanding of what they observed. Some schemes, such as Ptolemy's (about 140 A.D.), which described an earth-centered solar system, were founded upon false premises. Finally, when Copernicus proposed a sun-centered solar system, (about 1500 A.D.), astronomers began to understand the rules that underlay their observations.

But even then the reasons for the rules were not clear. Why should the planets move in elliptic orbits? Why should their masses and speeds and distances from the sun be as they are? No one had answers to these questions.

One famous and fundamental set of rules was formulated (but not explained) soon after Copernicus. Johannes Kepler (1571–1630) studied the planetary-motion data acquired by his mentor, Tycho Brahe, and summarized his observations in three statements known as Kepler's laws:

The elliptical orbit law:
All planets move in elliptical orbits about the sun. The sun is at the focus point of the ellipse.

The equal area law:
The line joining the sun and the planet sweeps out equal areas in equal times (see Figure 2.19).

The harmonic law:
The square of the time taken to make one trip around the orbit (i.e., the square of the period of motion) is directly proportional to the cube of the average orbit radius.

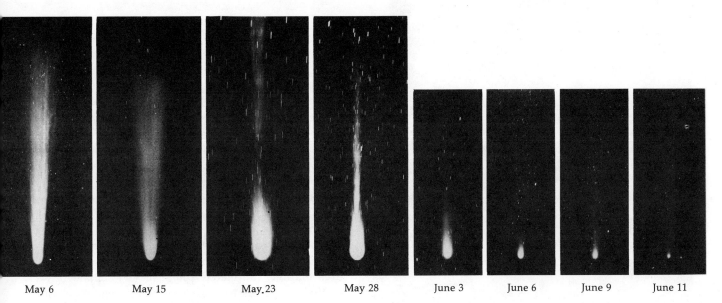

May 6   May 15   May 23   May 28   June 3   June 6   June 9   June 11

2.18 *Photographs showing the History of Halley's Comet from April 26 to June 11, 1910. The same comet will come into earth's view again in 1986. (Mount Wilson and Palomar Observatories.)*

Figure 2.19 shows an elliptical orbit with the sun at its focus. The second law says that if areas $A_1 = A_2 = A_3$, the times taken for the planet to move from A to A', from B to B', and from C to C' are identical. In other words, the closer the planet comes to the sun, the faster it moves. The third law states that the closer a planet is to the sun, the less time it takes to circle the sun. These observations, accurate in their predictions, suggest the law of conservation of angular momentum, but Kepler could give no fundamental reason for their validity.

Science often follows this same pattern: discovery, careful observation, and then systematization of observation by rules such as Kepler's laws. But if the pattern stopped there, we would soon have millions of unconnected, unrelated rules to learn and apply. Today's technological society would be impossible if scientists had gone no further than Kepler's laws and other systematized observations.

But to go further, man had to resort to abstractions, to mathematics, to more comprehensive laws. Galileo and Newton were among the first to recognize this. Galileo showed that the motion of objects along straight-line paths depends on four quantities: speed of the motion, time for which the motion persists, distance covered by the object, and accel-

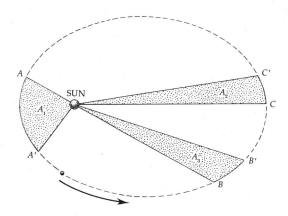

2.19 *Kepler's law of equal areas tells us that if areas $A_1 = A_2 = A_3$, then the times taken for the planet to pass sections $\overline{AA'}$, $\overline{BB'}$, and $\overline{CC'}$ are equal.*

THE SOLAR SYSTEM

THE BETTMANN ARCHIVE

JOHANNES KEPLER (1571–1630)

Kepler was born in Germany 28 years after Copernicus first challenged the idea of an earth-centered universe. As a young man, he became convinced that the planets circled the sun, as Copernicus said. Though he made his living as a court astrologer, Kepler also worked as an assistant to the renowned astronomer, Tycho Brahe. Driven by a mystic conviction that the universe reflected the perfect order of the mind of God, Kepler worked for years to prove his belief in mathematical terms. But Brahe's observations were so accurate that Kepler, despite his mathematical genius, could not reconcile the planetary orbits with the perfect circles that were considered the signs of God's order. In a daring leap of courage and imagination, Kepler followed the scientific data beyond the limits set by tradition and belief. His courage was rewarded when he discovered the natural laws that bear his name, but science owes him a debt of gratitude as well for pressing on to facts rather than falling back on superstitions. This is the essence of the scientific method; it cost Kepler much criticism to follow it, and this may have contributed to his poor health and his troubled family life. It also made him one of the great explorers of his age.

eration, a measure of how the motion speeds up. Newton showed how forces influence the motion of objects; he learned exactly how the gravitational force behaved; and he used these two pieces of knowledge to explain Kepler's laws.

Modern science strives to learn the meaning of all observational laws such as Kepler's. Over the years it has become possible to explain nearly all the behavior we notice in the physical world in terms of a comparatively small number of basic physical laws. As a result, we do not learn millions of seemingly unrelated physical facts. Instead we relate many facts to these few basic laws. As a consequence, when we read that a new scientific discovery has been made, we can fit it into our understanding of the physical world and it becomes a coherent part of the whole, just as a jigsaw puzzle piece fits nicely into the picture once its proper place is found.

In Chapter 3, we shall begin to study these fundamental laws, shifting our emphasis from a description of events to understanding why they happen. As you will see when you have learned the meaning and interrelations of the laws of nature, these laws fit together to form a beautiful picture of our physical world.

BIBLIOGRAPHY

1 ABELL, GEORGE: *Exploration of the Universe,* 2d ed., Holt, New York, 1969. (See description in Chapter 1.)

2 BIRNEY, D. SCOTT: *Modern Astronomy,* Allyn and Bacon, Boston, 1969. (See description in Chapter 1.)

3 HOLTON, G., and ROLLER, D.: *Foundations of Modern Physical Science,* Chapters 6–12, Addison-Wesley, Cambridge, Mass., 1959. A physical science text containing a great deal of historical material presented in a mostly nonmathematical way.

4 MARSDEN, B.: "The Great Comet of 1965," *Sky and Telescope,* vol. 30, no. 6, p. 332. An interesting and readable account of the appearance and behavior of a recent comet.

5 OHRING, G.: *Weather on the Planets* (paperback), Anchor Science, Doubleday, Garden City, N.Y., 1966. This is a readable treatment of the nature of the climate and atmosphere of the planets. It is at about the same level as the material in this chapter, but is more detailed.

6 The *National Geographic Magazine* (August, 1970) has a series of beautiful color pictures dealing with the solar system.

SUMMARY

For practical purposes, the distant stars may be considered fixed in the heavens. As the earth revolves on its axis, Polaris, the Pole Star, remains directly overhead for an observer at the North Pole while the other stars appear to follow circular paths about it. For an observer at the equator, the stars appear in the East and disappear over the horizon in the West as the earth revolves.

Since we see the stars only at night, the portion of the universe we observe depends upon the position of the earth relative to the sun. As the earth travels around the sun once each year, we are able to observe all portions of the sky. Various portions of the sky contain groups of stars called constellations. During the course of a year, the constellations seen at night change as different portions of the universe are exposed to view during the hours of darkness. Over the centuries, the positions of the constellations in the night sky have changed because our calendar is based upon the seasons.

Because the rotation axis of the earth is not perpendicular to the earth's orbit plane (the ecliptic plane), a given point on the earth is exposed differently to the sun's rays at various times of the year. As a result, when the Northern Hemisphere is experiencing summer, it is winter in the Southern Hemisphere and vice versa. Since the earth's axis remains at a fixed orientation relative to the stars as the earth circles the sun, the exposure to the sun's rays varies during the year, accounting for the seasons.

The moon moves in a nearly circular path around the earth; one circle is completed each 27.3 days. Since the moon revolves on its axis at exactly this same rate, the same side of the moon is always facing the earth. Because of its small mass, the weight of an object on the moon is only about one-sixth its weight on earth. As a result of this small gravitational attraction, all gas molecules have escaped, and the moon has no atmosphere. Therefore, it suffers the effects of full exposure to the sun's rays by day, and cools to about $-200°F$ at night. As far as we know, the moon has about the same composition as the earth and is about the same age. When the moon is in line between the earth and sun, it casts a shadow on the earth. That portion of the earth in shadow experiences a total eclipse of the sun. Similarly, an eclipse of the moon occurs when it is in the shadow of the earth. The phases of the moon occur because the side of the moon opposite the sun is in shadow and we see only the portion of the moon that is lighted by the sun's rays. The moon is also responsible for the tides observed on earth.

Like the moon, the planets shine in the sun's rays as they rotate around the sun. The nine planets, in order of increasing distance from the sun, are Mercury, Venus, Earth, Mars, Jupiter, Saturn, Uranus, Neptune, and Pluto. Since Mercury and Venus are so close to the sun, they have very high temperatures, and life as we know it could not exist there. Mars has a temperature variation more like that of the earth, but its atmosphere is only about 1 percent as dense as the earth's. Most of that is apparently carbon dioxide. No more than traces of water seem to exist there.

Jupiter and Saturn are so massive that their huge gravitational forces retain both hydrogen and helium gases. The outer portions of both consist of a deep, cold sea of liquid hydrogen and helium. Saturn is unique because of the rings that encircle it. The remaining planets are cold, and difficult to see because of their large distance from the sun.

Comets, asteroids, and other smaller objects also circle the sun or enter the solar system and escape once again into space. Debris from comet trails frequently enters the earth's atmosphere and give rise to luminous meteors streaking to the earth.

Kepler's three laws first quantified the motion of the planets in a coherent way. Newton later gave them a physical basis.

TOPICAL CHECKLIST

1. What sort of pattern do the stars trace out when observed from the poles? From the equator? Why?
2. Why does the Pole Star move only slightly in the sky?
3. How do the constellations behave in the sky from hour to hour? From month to month?
4. Why is the summer hot and the winter cold?
5. Explain why the seasons are reversed in the two hemispheres.
6. What is meant by each of the following: ecliptic plane; winter and summer solstice; spring and fall equinox?
7. What is the cause of the phases of the moon? Eclipse of the moon? Eclipse of the sun?
8. Describe the features of the moon.
9. List the planets in order of increasing distance from the sun. In order of decreasing mass.
10. Describe the general features of each of the planets.
11. What is Roche's theorem and how does it apply to Saturn's rings?
12. What are asteroids? Meteors? Meteorites?
13. What are comets and how are they related to meteors?
14. What are Kepler's three laws?

QUESTIONS

1. A particular star configuration is seen directly overhead in June at the equator. Where will the same person see it in August?
2. While standing on the earth looking eastward at midnight, you raise your left arm upward and point out a star from your side at an angle halfway between vertical and horizontal. Describe the motion of this star through the night sky if you are (*a*) at the equator, (*b*) at the South Pole, (*c*) in the United States.
3. Describe how Mercury will appear from the earth as it passes between the sun and earth. Do Mercury and Venus have phases like the moon? Does Mars?
4. Describe the following as seen from the moon: the earth; sunset on the moon; the night sky; the daytime sky.
5. Suppose you opened a bottle of Pepsi on the moon. What would happen?
6. If the earth's axis was perpendicular to the ecliptic plane (i.e., if the plane of the equator was in the ecliptic plane), what would the seasons be like on earth?
7. The axis of the planet Uranus is nearly in the ecliptic plane. Describe what the climate on earth would be like if its axis were similarly oriented.

8 Suppose the planet Mars had a mass 1,000 times larger than it is but was the same distance from the sun. Describe the properties you would expect for such a planet.

9 The craters on the moon and on Mars show that meteorites have often struck these bodies. Compare what an observer in either of these places would see when a meteorite strikes with what an observer on earth would see in a similar situation.

10 Suppose the planet earth suddenly changed its orbit to one like that of Mercury. What effect would this have on the earth? What would happen if its orbit became similar to that of Jupiter?

11 Why is Venus never seen very high above the horizon?

12 Planets are never seen very far north or very far south in the sky; they are always close to a line extending from the eastern to the western horizon. Why is this?

13 What determines whether Venus occurs as the Evening or Morning Star? Using a diagram, show the two different cases.

14 Halley's comet travels about the sun in a very elongated ellipse, with the sun at a focal point near one end. What can you say about the speed of the comet as it moves along the orbit?

15 People who have touched meteorites soon after they hit the ground say that the meteorites are cold. Yet they were flaming brightly as they streaked through the atmosphere. How can this be? How is this related to the mode of reentry of a spaceship from the moon?

Chapter 3

FORCES AND MOTION

Nearly everything we do and see involves forces and the motions they cause. Whether it is an event as majestic as the sunrise or as prosaic as pouring milk from a bottle, the physical laws of motion can account for it. In this chapter we will become acquainted with these laws, which were first discovered by Galileo and Newton. We shall see that they describe, with wonderful brevity and simplicity, a large fraction of the physical phenomena we encounter in our daily lives.

Science versus Superstition

Many people believe that if you cut a rattlesnake's head off, its tail will continue to rattle until sundown. This is true if you kill it close enough to sundown. However, it is the deterioration of the snake's nervous system, not the advent of sundown, that makes the rattling stop. In this example, people have correctly observed a phenomenon (the end of the headless snake's rattling) but ascribed the wrong cause (sunset) to it. Unfortunately, few basic events occur in nature for which we can easily find the cause.

Almost always, an event is preceded and accompanied by several possible causes; experience teaches us which causes are related to which events. Our skill in determining cause-effect relationships is a measure of our progress in the transition from pseudoscience and superstition to science. This transition has come slowly in the history of mankind, and it is far from complete.

Science as we know it began to develop around 1600. Before that time many physical phenomena had been observed but few had been precisely described or understood. For example, the earliest men knew that an object falls to the earth when dropped. But not until about 1600 did anyone measure precisely *how* an object falls.

The Greek philosopher Aristotle (384–322 B.C.) had stated what he considered to be the law of falling bodies. He claimed that lighter objects always fall more slowly than heavier objects, and supported his claim with an ingenious philosophic proof which went unchallenged for the next 1,900 years. After all, it does seem like a reasonable law; feathers do fall more slowly than a rock. A leaf falls less swiftly than a nut from the same tree. But these examples do not prove the general law Aristotle thought he had proved. It is important that we understand the fallacy involved here, since it is the major difference between science and pseudoscience. It is the same fallacy that has led well-meaning people to embrace all sorts of quack cures and curious superstitions.

When an experiment is performed to test a theory about nature, the experiment must be a controlled experiment; that is, we must compare two (or more) situations which are identical except for the one factor we intend to test. In the case of the effect of weight on the way objects fall, we must compare two objects that are identical except for their weights. Clearly, no real test is possible when we compare the rate of fall of a feather and a stone. These objects are different in so many ways that, if they do not fall at the same rate, we could not possibly say from this experiment alone that the weight difference was the cause. In fact, we now know that the friction of the feather with the air causes it to fall more slowly than a stone. When both are allowed to fall in the absence of air, they fall at the same rate.

Galileo Galilei (1564–1642) first fully appreciated the flaw in the feather-stone experiment. His experiments led him to the law that actually governs the behavior of freely falling objects. But before we discuss the physical law that he discovered, we need to learn the meaning of the terms speed, velocity, and acceleration.

Speed, Velocity, and Acceleration

Many words we use from day to day have slightly different meanings in science. Often this difference is simply a matter of preciseness in definition. Such is the case with *speed*, *velocity*, and *acceleration*. We all have a good idea what these words mean in a qualitative sense, but the scientist makes them more precise by giving them a quantitative meaning.

For example, in science we define speed in the following way:

THE BETTMANN ARCHIVE

GALILEO GALILEI (1564–1642)

Galileo's reputation as the father of experimental science is well deserved, for his method of exploration was as revolutionary as his discoveries. By combining inductive reasoning with precise observations and mathematical proofs, he established the basic framework of the scientific method. Galileo developed his method while working out his laws of falling bodies as a young mathematics professor in Pisa, Italy, after which he moved to Padua and became a noted astronomer. In order to make precise observations, he developed more powerful and accurate models of the newly invented telescope. His astronomical discoveries made him an outspoken believer in the sun-centered solar system, and this offended the authorities, who were committed to the earth-centered view. Neither influential friends nor his fame could protect him from the Inquisition; he was forbidden to speak further of such heresies. After nearly 15 years of silence and work, he defiantly published his observations and conclusions, which still contradicted the Church's doctrines. The result was trial and imprisonment, until Galileo, by now an old man broken in health and spirit, denied his discoveries.

FORCES AND MOTION 45

If an object moves a distance s in a time t, its average speed $\bar{v}$ is the distance traveled divided by the time taken. In symbols,*

$$\bar{v} = \frac{s}{t} \tag{3.1}$$

You have often used this defining relation to answer questions such as "How fast must a car go if it has to go 120 mi in 2 hr?" In this case, the distance traveled s is 120 mi, and the time t taken is 2 hr, so that

$$\bar{v} = \frac{s}{t}$$

gives

$$\bar{v} = \frac{120 \text{ mi}}{2 \text{ hr}} = 60 \frac{\text{mi}}{\text{hr}}$$

We read this as sixty miles per hour, or 60 mi/hr. Notice how the units, miles and hours, are carried right along with the numbers.

As another example, how far will a car going at 70 mi/hr go in a time of $\frac{1}{2}$ hr? In this case we wish to know s where $\bar{v} = 70$ mi/hr and $t = \frac{1}{2}$ hr. To find it, we notice that if we multiply both sides of the equation

$$\bar{v} = \frac{s}{t}$$

by t, we get

$$s = \bar{v}t$$

---

* Average quantities are frequently represented by placing a bar above the symbol, hence $\bar{v}$ (said "vee bar") for average speed.

and so

$$s = \left(70 \frac{\text{mi}}{\text{hr}}\right)\left(\frac{1}{2} \text{hr}\right)$$

or

$$s = 35 \text{ mi}$$

Again, we carry the units along with the numbers and cancel the "hr" in the numerator with the "hr" in the denominator.

In still another example, suppose the edge of a glacier creeps across the earth at the rate of 2 in./day. How far will it move in 5 years? We wish to find the distance it will go, s, in a time $t = 5.00$ years; we know the average speed $\bar{v} = 2.00$ in./day. We shall substitute into the formula:

$$\text{Distance} = (\text{average speed})(\text{time})$$
$$s = \bar{v}t$$

However, we are using two different units of time; $\bar{v}$ is given in inches per day and t is in years. If we substitute these values, we find

$$s = \left(2.00 \frac{\text{in.}}{\text{day}}\right)(5.00 \text{ years}) = 10 \frac{\text{in.-years}}{\text{day}}$$

But no one measures distances in inches-years per day, and so even though this answer is formally correct, it is not useful. The difficulty arises because we have used both years and days for units of time in the same problem. To avoid answers in uncommon units, we do not mix units. In the present case, since 1 year = 365 days, we replace 5 years by 1,825 days. Then

$$s = \left(2.00 \frac{\text{in.}}{\text{day}}\right)(1{,}825 \text{ days}) = 3{,}650 \text{ in.}$$

This is a more useful result since everyone knows what a distance given in inches means.

Frequently we are not interested in the average speed $\bar{v}$ but in the *instantaneous speed v*, the speed an object has at a particular instant. For example, a patrolman is not impressed at all when you tell him you have been driving for 2 hr and have gone 40 mi so your average speed was 20 mi/hr. He is more interested in his observation that your instantaneous speed was 35 mi/hr as you went through a 20 mi/hr school zone. The distinction between instantaneous and average speed is transparent. Which one is being used is usually obvious from the context.

The layman uses the words *speed* and *velocity* interchangeably. A scientist distinguishes between them. He says that *instantaneous velocity* is equal in magnitude (i.e., numerical value) to instantaneous speed, but velocity has a direction associated with it as well. For example, if we say a car is moving at 50 mi/hr, we are stating the car's speed. However, if we say the car is moving *eastward* at 50 mi/hr, we are stating the car's velocity, since we have specified both the magnitude and direction of the car's motion. Quantities such as velocity that carry connotations of directions are called *vectors* while those that do not, such as speed, are called *scalars*.

This may seem like a minor technicality to you at this point, but it is often really important. For example, notice in Figure 3.1 that even though the earth moves around the sun with nearly constant speed, its direction of motion through the universe is constantly changing. Its velocity today is opposite in direction to what it was six months ago. Newton showed that a force exerted on an object necessarily changes the direction of motion (and therefore the velocity) even if no change in speed occurs. A familiar example is a tennis ball which bounces off a wall at the same speed with which it hit the wall. The speed of the ball does not change, but the impact reverses its velocity. *Velocity, a vector, has direction, while speed, a scalar, does not.*

When a car is speeding up, we say it is *accelerating*. In science we make the further distinction that acceleration is a *vector*, i.e., it has direction. When

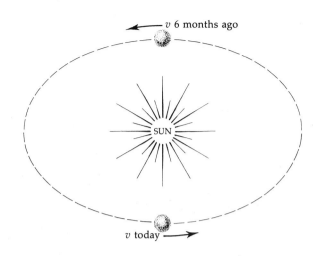

3.1 *The speed of the earth today is nearly the same as it was six months ago. Its velocity is different, however, since it is now moving through the universe in the opposite direction.*

we say that a car going eastward is increasing its speed by 2 mi/hr in each second, we are stating its acceleration. An acceleration of 2 mi/hr each second means that the car's speed is increasing by 2 mi/hr every second. If the car is going 30 mi/hr at a certain instant, 1 sec later it will be going 32 mi/hr. Precisely, we define acceleration as follows:

The acceleration of an object is equal to the amount its velocity changes in a unit of time, usually a second. The direction of the acceleration is the same as the direction of the velocity change.

If we concern ourselves with motion in a single direction, the distinction between velocity and speed loses its importance. Then the defining equation for acceleration

$$\text{Acceleration} = \frac{\text{change in velocity}}{\text{time taken for the change}}$$

becomes

$$a = \frac{v - v_0}{t} \qquad (3.2)$$

FORCES AND MOTION

In writing this, v_0 (read "vee sub zero") is the initial speed, the speed the object had at the first instant, while v is the speed the object had t sec later, the final speed.* Frequently people write $v - v_0$, the change in speed, as Δv (read "delta vee"), which is a shorthand way of writing "change in v."

As an example, suppose an impatient driver at a traffic light accelerates from rest to a speed of 40 mi/hr in a time of 10 sec after the light changes. To compute his acceleration, we note that $v_0 = 0$ (since he was initially at rest), that $v = 40$ mi/hr, and that the time taken $t = 10$ sec. Forgetting for the moment that we are mixing time units, we substitute into Equation (3.2) to find the acceleration

$$a = \frac{(40 - 0) \text{ mi/hr}}{10 \text{ sec}} = 4 \text{ mi/hr/sec}$$

which we read as four miles per hour per second. This statement says that the car speeded up at the rate of 4 mi/hr every second. The following data show how the car's speed increased with time after the traffic light changed:

Time (sec) →	0	1	2	3	4	5	etc.
Speed (mi/hr) →	0	4	8	12	16	20	

Clearly, the car speeded up (or accelerated) 4 mi/hr every second; therefore $a = 4$ mi/hr/sec. These tabulated data are shown graphically in Figure 3.2. You should examine the graph to make sure that you understand how it shows that the acceleration is 4 mi/hr/sec.

If we had followed our rule and changed to consistent units, we would not have used both hours and seconds in the same problem. When changing speed units, it is convenient to know that a car going 60 mi/hr is going 88 ft/sec. By proportion,

---

* We shall frequently use this equation for a in a slightly different form. If we multiply both sides of the equation by t and transpose $-v_0$ to the other side, we find $v = v_0 + at$.

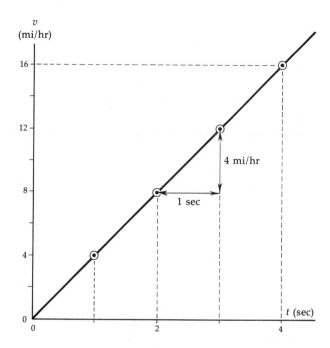

3.2 What is the acceleration for the motion depicted in this graph?

$$\frac{40 \text{ mi/hr}}{60 \text{ mi/hr}} = \frac{v}{88 \text{ ft/sec}}$$

we cancel mi/hr and cross-multiply to get

$$40(88 \text{ ft/sec}) = 60v$$

which gives

$$v \cong 59 \text{ ft/sec}$$

Using this figure in Equation (3.2) we obtain

$$a = \frac{(59 - 0) \text{ ft/sec}}{10 \text{ sec}} = 5.9 \text{ ft/sec/sec}$$

This means, of course, that the car speeded up 5.9 ft/sec each second. We often write it as 5.9 ft/sec². Although the previous answer, 4 mi/hr/sec, looks different from 5.9 ft/sec², both give the same information about the acceleration of the car though it is expressed in different units. A yardstick provides a similar example; we can say it is 1 yd, 3 ft, or 36 in. long.

We can obtain another useful equation from Equation (3.1) ($s = \bar{v}t$). If an object has a constant acceleration, it is speeding up smoothly and the average speed $\bar{v}$ is simply the average of the initial speed v_0 and the final speed v; that is,

$$\bar{v} = \frac{v_0 + v}{2}$$

When substituted into $s = \bar{v}t$, this gives

$$s = (\tfrac{1}{2}v_0 + \tfrac{1}{2}v)t$$

or

$$s = \tfrac{1}{2}v_0 t + \tfrac{1}{2}vt$$

But we know from Equation (3.2) that

$$v = v_0 + at$$

and so we find

$$s = \tfrac{1}{2}v_0 t + \tfrac{1}{2}(v_0 + at)t$$

or

$$s = \tfrac{1}{2}v_0 t + \tfrac{1}{2}v_0 t + \tfrac{1}{2}at^2$$

or

$$s = v_0 t + \tfrac{1}{2}at^2 \qquad (3.3)$$

To see the utility of this equation, let us compute how far a car will go in the first 10 sec after a traffic light changes if the car is accelerating at 6 ft/sec². Since the car started from rest, its initial speed v_0 is zero. Then, since $a = 6$ ft/sec² and $t = 10$ sec, we have from Equation (3.3)

$$s = 0 + \frac{1}{2}\left(6\frac{\text{ft}}{\text{sec}^2}\right)(10 \text{ sec})^2$$

or

$$s = 300 \text{ ft}$$

This is the distance beyond the traffic light the car will have gone by the end of 10 sec.

Freely Falling Bodies

One of the simplest yet most important examples of accelerated motion is the free fall of an object under the action of gravity. Figure 3.3 shows this situation. From this picture, taken with a strobe-light flashing at a fixed time interval, it is possible to compute the acceleration of the falling ball. We will find that it speeds up as it falls by the same equal amount in each of the equal time intervals.

As an example, consider the following data for a freely falling object:

t (sec) →	0	0.1	0.2	0.3	0.4	0.5	0.6
v (ft/sec) →	0	3.22	6.44	9.66	12.88	16.10	19.32

It is clear that in each 0.10-sec time interval, the falling object speeds up by 3.22 ft/sec. Therefore its acceleration is [from Equation (3.2)]

$$a = \frac{\Delta v}{t} = \frac{3.22 \text{ ft/sec}}{0.10 \text{ sec}}$$

$$= 32.2 \text{ ft/sec}^2$$

where Δv is change in velocity.

Galileo first showed that all freely falling bodies speed up with this same acceleration, 32 ft/sec². He called this the *acceleration due to gravity*, and it is customarily represented by the symbol g. We can state Galileo's discovery in the following way:

> In the absence of all forces except gravity, every object accelerates toward the earth with the acceleration $g = 32$ ft/sec².

A falling feather does not fall with this acceleration because another appreciable force, the friction of the air, keeps it from falling as fast as it should. Since the feather is so light, even a small disturbing force influences its motion. Careful measurements show that g, the acceleration due to gravity, changes slightly from place to place on the earth, as the ex-

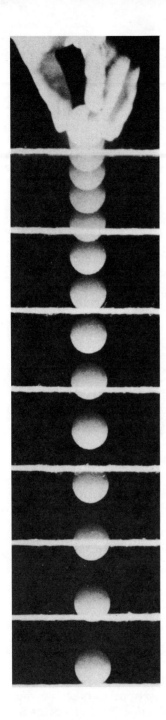

3.3 *The strobe light exposes the ball at equal time intervals. (Courtesy of Fundamental Photographs.)*

Table 3.1   Acceleration Due to Gravity, g

Place	Elevation (feet above sea level)	g (ft/sec²)
Beaufort, N.C.	3	32.143
New Orleans	6	32.130
Galveston	9	32.128
Seattle	190	32.176
San Francisco	370	32.151
St. Louis	500	32.152
Cleveland	680	32.160
Denver	5,280	32.139
Pike's Peak	14,000	32.118

perimental data in Table 3.1 show. We shall see later that g should decrease as we go farther from the center of the earth's core, provided all other factors remain unchanged. On the moon, the acceleration of a freely falling object is much less than on earth; g is only about 5 ft/sec² there. This result follows because an object weighs less on the moon than it does on earth; we will see why in Chapter 4.

Sometimes we want to know how far an object will fall in a certain length of time. For example, suppose we drop a stone from a bridge and determine that it takes 3 sec to hit the water below. From this experiment, we can tell how high the bridge is above the water. We note that $a = 32$ ft/sec² since the stone is falling freely. When we released it, its speed was $v_0 = 0$ and we know it took a time of 3 sec to fall the required distance s. Therefore, using Equation (3.3),

$$s = v_0 t + \tfrac{1}{2} a t^2$$

we find

$$s = 0 + \frac{1}{2}\left(32 \frac{\text{ft}}{\text{sec}^2}\right)(3 \text{ sec})^2$$

or

$$s = 144 \text{ ft}$$

As we see from this example, an object falls a large distance even in such a short time as 3 sec. Galileo had difficulty measuring the rate of fall of objects, since the clocks of his time could not measure such short intervals accurately. According to legend, he performed one of his experiments by dropping spheres from the top of the leaning tower of Pisa. Actually, in most of his experiments he rolled balls down an inclined plane. By increasing the angle of the incline, he was able to infer, from the measured accelerations at various incline angles, what the acceleration of a freely falling body would be. It is a tribute to his scientific ability that he was able to discover the properties of freely falling bodies in this indirect way.

Newton's First and Third Laws of Motion

So far we have discussed types of motion and some of their characteristics. Now we will consider the causes of motion. Even in Galileo's time it was known that a force must be exerted on an object if the object is to be speeded up or slowed down. But Isaac Newton (1642–1727) was the first to describe the interrelation of force and motion in quantitative statements. His discoveries are summed up in his three laws of motion.

Newton's First Law

The first of Newton's laws of motion is subdivided into two parts and may be stated as follows:

(1) A body at rest will remain at rest and (2) a body in motion will remain in motion in the same direction unless an unbalanced force is applied to the object to cause it to change its state of rest or motion.

The first part of this law seems reasonable and agrees with our everyday experience. We know that bodies at rest will not start to move until some

CULVER PICTURES

ISAAC NEWTON (1642–1727)

Before the age of thirty, Newton had invented the mathematical methods of calculus, demonstrated that white light contained all the colors of the rainbow, and discovered the law of gravitation. His was a lonely and solitary life. His mother died giving birth to him, and he was raised by his father, the owner of a small manor near the town of Grantham, England. In 1661 he was admitted to Cambridge, where he worked for the next eight years, except for a year at home to escape the plague. During those years, he made his major discoveries, though none of them were published at that time. Nonetheless, his genius was recognized and in 1669 he was appointed Lucasian Professor of Mathematics at Cambridge, a position he retained until 1695. His major scientific work was completed prior to 1692, when he suffered a nervous breakdown. After his recovery, he determined to lead a more public life and soon became Master of the Mint in London. He was elected president of the Royal Society in 1703 and held that position until his death.

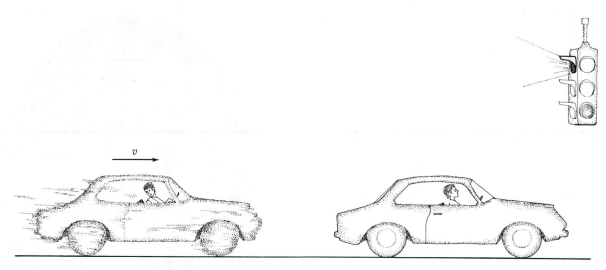

3.4 *Reasoning from Newton's first law, what will happen when the cars collide? Which driver should have a seat belt? A head rest?*

external force causes them to move. The second part of this law is more subtle. Restated, the law says that a moving body will continue moving in a straight line forever unless some external force is applied to it. Common experience seems to contradict this; we know that nothing continues to move forever without change. A ball rolled across the ground soon stops; a metal object sliding across a smooth table eventually slows and stops. You know many other examples.

However, none of the examples cited is a valid test of Newton's law. A force is acting on each of these bodies, trying to stop its horizontal motion. It is the force of friction. We know that the more precautions we take to eliminate this force, the more slowly the moving object comes to rest. Newton mentally generalized this observation to the case where *no friction exists* and concluded that, if this case were possible, *the moving object would never stop.* Although an example of perfect, unchanging motion has never been attained, all our experience leads us to believe that Newton's conjecture is valid. To test your understanding of the first law, you may wish to analyze the situation shown in Figure 3.4. It shows the effects of *inertia,* the tendency of a body at rest to remain at rest and of a body in motion to remain in motion.

We will postpone Newton's second law until last; his third law is as follows:

Newton's Third Law

For every force exerted on one object (the action *force), an equal but opposite force (the* reaction *force) is always exerted on some other object.*

This law is often called the action-reaction law, since it says, in effect, *the action force is equal and opposite to the reaction force.*

Notice that two objects are always involved; the *action* force pushes or pulls on one of them while the reaction force is exerted on the other. Newton amplified the law in his *Principia*:

> *Whatever draws or presses another is as much drawn or pressed by that other. If you press a stone with your finger, the finger is also pressed by the stone. If a horse draws a stone tied to a rope, the horse (if I may so say) will be equally drawn back towards the stone: for the stressed rope, by the same endeavour to relax or loose*

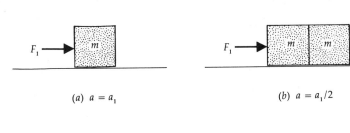
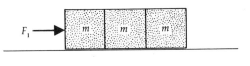
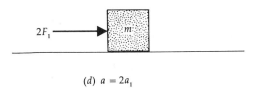
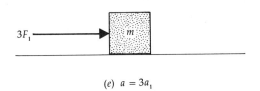

3.5 The experiments shown indicate $a \sim F/m$. Friction between the surface and block must be negligibly small to obtain the results shown.

itself, will draw the horse as much towards the stone, as it does the stone towards the horse, and will obstruct the progress of one as much as it advances that of the other.

To understand this law more clearly, let us consider a few more examples. When a car hits a tree, it exerts a force on the tree. But at the same time, the tree exerts an equal and opposite reaction force on the car. When you slam your hand down upon a table to emphasize a point, not only does your hand exert a large force on the table, but the table exerts an equal and opposite force on your hand (and that is what causes your hand to sting if you hit the table hard enough). *In every situation in which a force is exerted on one object, an equal and opposite reaction force is exerted on some other object.* We will use this law whenever we wish to compare two forces that can be shown to be an action-reaction pair.

Newton's Second Law of Motion

We know that a large force is needed to set a massive object in motion, whereas a smaller force is required to start or stop a less massive object. In a more technical sense, we attribute a large inertia to a massive body. By this we mean that a massive object is difficult to set in motion and equally difficult to stop if it is already moving. Newton's second law quantified these qualitative observations.

Newton showed that the acceleration given to an object by a force is inversely proportional to the mass of the object. For example, in Figure 3.5a we see a block of mass m being accelerated by a force F_1. Let us call its acceleration a_1. Now if the mass is doubled, as in (b), the same force F_1 gives the block an acceleration only half as large, $a_1/2$. Similarly, as in (c), three times the mass receives only one-third the acceleration. We therefore conclude that, for constant applied force, the acceleration a is proportional to $1/m$.

If the experiment is modified so that the force is increased as in parts (d) and (e), we find that the acceleration of the block increases in proportion to the applied force. In symbols, this is expressed as $a \sim F$. Both these results, which can be summarized in the relation

agree with our qualitative impressions that the acceleration should increase with increasing force and decrease with increasing mass. We can write this proportion as an equation provided we use a proportionality constant c. Then we have

$$\frac{F}{m} = ca \quad \text{or} \quad F = cma$$

To find the proportionality constant's numerical value, we need only measure how large a force F is required to give a certain mass m an acceleration a. If we substitute these measured values for F, m, and a, we can obtain a numerical value for c and it will forever after have this value. But here we have a difficulty. We know how to measure a in feet per second per second and the force F in pounds. However, we have not yet decided on a unit in which to measure the mass of the object, m. This is fortunate because we can do as we please about it. Let us measure m in such a unit that the proportionality constant c is simply one, $c = 1$. Then we can write

$$F = ma \qquad (3.4)$$

It is understood that we still have to see what units m should be measured in.

Equation (3.4) is a mathematical statement of Newton's second law. It summarizes the statements we have made previously, namely:

1 When a given mass is accelerated, the acceleration is proportional to the applied force.

2 The force required to provide a given acceleration is proportional to the mass of the object being accelerated.

We also notice that the acceleration is in the same direction as the unbalanced force which causes the acceleration. In order to proceed further, however, we must settle the question concerning the units of mass m.

Mass and Weight

The *mass* of an object is a measure of its inertia. If an object has a large mass, it is difficult to set in motion and it is also difficult to stop. Even though we tend to think of massive objects as being heavy, *mass and weight are entirely different concepts*. We define weight as follows:

> The *weight* of an object is the force of gravitation pulling on it.

Clearly, *weight is a force.*

When you weigh an object, you are really measuring the force with which the earth attracts the object. For example, in Figure 3.6 we see an object supported in a person's hand. If the hand were not supporting the object, the object would accelerate toward the earth with the acceleration due to gravity, g. The earth's gravitational attraction force acting on the object would give it this free-fall acceleration. The gravitational attraction force, pulling the object toward the center of the earth, is the weight W of the object. In the figure, the weight of the object is balanced by the supporting upward push (or force) exerted on it by the person's hand. Since the vertical forces on the object cancel each other, the object remains motionless. Similarly, when you stand on a scale, the scale platform pushes up on you with a force large enough to balance the earth's force of gravity pulling down on you. The scale indicates the force it must exert to support the object on its platform, and thus shows the weight of the object.

Now that we understand that the weight of an object is the force of gravity pulling on it, let us see how an object's weight is related to its mass. Consider a freely falling body; as we know, it will experience a downward acceleration equal to g, the acceleration due to gravity. This acceleration is caused by the pull of gravity, the weight of the object, W. If we apply Newton's second law to this situation,

$$F = ma$$

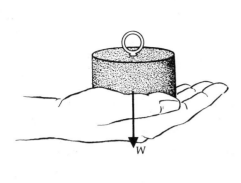

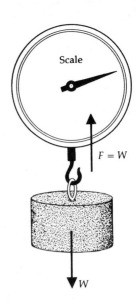

3.6 The gravitational force W is called the weight of the object.

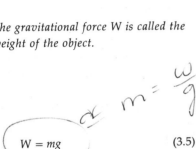

becomes

$$W = mg \qquad (3.5)$$

This equation tells us *the relation between weight and mass.* Whenever the mass of an object is required, we need only measure the weight of the object and its free-fall acceleration resulting from its weight. Then the mass is simply W/g.

Let us review the meaning of mass so there will be no misunderstanding. Mass is that intrinsic property of an object which causes it to have inertia. If the object is very massive, it has large inertia and, consequently, it is hard to stop when moving and hard to set in motion when at rest. The following examples will help consolidate our understanding of Newton's laws.

EXAMPLE 3.1

What do Newton's three laws tell us about a book sitting at rest on a table?

Method: Since the book does not move, Newton's first law tells us that no unbalanced force acts on it. Even though the earth pulls down on it with the force of gravity, its weight W, the table must push upward to support the book. This upward force on the book must be equal in magnitude to W if the forces are to cancel as they do. The second law reinforces the first law in a case like this, since with $a = 0$, $F = ma$ tells us that F must be zero. But we already knew this from the first law, because the upward push of the table balances the downward pull of gravity.

The action-reaction law involves two bodies, as you recall. In this case, the action force of the book pushing on the tabletop is equal to the reaction force of the table pushing on the book.*

EXAMPLE 3.2

How large a force is required to give a 2,400-lb auto an acceleration of 2 ft/sec²?

Method: To find the force F needed to accelerate the car, we make use of $F = ma$. Since the weight

---
* We can reverse the action-reaction labels. Which force we choose to call the action force makes no difference.

of the car is 2,400 lb, and since $g = 32$ ft/sec², we find that

$$F = ma$$

becomes

$$F = \frac{2,400 \text{ lb}}{32 \text{ ft/sec}^2}(2 \text{ ft/sec}^2)$$

We have used Equation (3.5) to replace m by W/g. After canceling the units "ft/sec²" from numerator and denominator and doing the arithmetic, we find

$$F = 150 \text{ lb}$$

EXAMPLE 3.3

Assuming that the car of the previous example started from rest, how fast will it be going and where will it be 7 sec after starting?

Method: Since its acceleration was 2 ft/sec² and its initial speed was zero, we find from $v = v_0 + at$, which is Equation (3.2) rewritten, that

$$v = 0 + (2 \text{ ft/sec}^2)(7 \text{ sec}) = 14 \text{ ft/sec}$$

To find how far it went in the 7 sec, we can use $s = v_0 t + \tfrac{1}{2}at^2$ [Equation (3.3)] to give

$$s = 0 + \tfrac{1}{2}(2 \text{ ft/sec}^2)(7 \text{ sec})^2 = 49 \text{ ft}$$

Or we might equally well have noted that its average speed $\bar{v} = (v_0 + v)/2$, so that

$$\bar{v} = \tfrac{1}{2}(0 + 14 \text{ ft/sec}) = 7 \text{ ft/sec}$$

Then $s = \bar{v}t$ would give us

$$s = (7 \text{ ft/sec})(7 \text{ sec}) = 49 \text{ ft}$$

The Metric System

Until now we have measured lengths and forces in the units commonly used in the United States. This unit system, which uses the foot and pound, is called the *British system*. Most other parts of the world use the units of the *metric system*, which even Britain is adopting. The metric system is easier to use, since it is based upon multiples of 10 rather than the odd conversion factors we find between feet, inches, etc., in the British system. The United States will probably convert to the metric system in the next few years, in order to keep in touch with the rest of the world's measurements. We already use the metric system in the United States when we talk about the common electrical units, the volt and the ampere, since they are based on the metric units.

The basic unit of length in the metric system is the *meter*. The meter is about a yard long. More precisely,

$$1 \text{ meter (m)} = 39.4 \text{ in.} = 1.1 \text{ yd}$$

Instead of using miles, yards, feet, inches, etc., the metric length units are based on multiples of ten. These units are

1 kilometer (km) = 1,000 m = 0.64 mi
1 centimeter (cm) = 0.01 m = 0.39 in.
1 millimeter (mm) = 0.001 m
1 micrometer (or micron, μ) = 0.000001 m

In the metric system, certain prefixes always indicate definite multiples. As we see in the length units, "kilo" indicates a factor of 1,000, and so kilometer means 1,000 m. The common metric prefixes are

giga- billion (1,000,000,000)
mega- million (1,000,000)
kilo- thousand (1,000)

deci-	one-tenth (0.10)
centi-	one-hundreth (0.01)
milli-	one-thousandth (0.001)
micro-	one-millionth (0.000,001)
nano-	one-billionth (0.000,000,001)
pico-	one-thousandth of a billionth (0.000,000,000,001)

We shall see that these prefixes are applied to many different types of units.

The metric system (to be more precise, the mks or meter-kilogram-second system) emphasizes the unit of mass rather than the unit of force. A special block of metal kept near Paris is defined to have a mass of exactly 1 kg. All other masses are determined by comparison with it. It is convenient to know that this standard 1-kg mass weighs about 2.2 lb on earth. As a point of reference for you, this textbook has a mass of about 1.2 kg. In most countries, meat, potatoes, sugar, etc., are all sold by the kilogram (or gram, where 1 kg = 1,000 g) rather than by the pound.

Remember that the kilogram is a unit of mass, not a force unit. Therefore, when using Newton's second law, m is measured in kilograms. As a result, the proper units are

$$F = m \cdot a$$
$$\text{Unit} \rightarrow ? \quad \text{kg} \quad \text{m/sec}^2$$

The force unit, represented here by a question mark, is appropriately called the *newton*.

To understand the meaning of the force unit, let us look at the case of a freely falling object. In such a case, $F = ma$ becomes $W = mg$. We already know that $g = 32$ ft/sec² on the earth. In metric units the acceleration due to gravity, g is 9.8 m/sec² on earth. The W of a 1-kg *mass* is given directly by $W = mg$:

$$W = (1 \text{ kg})(9.8 \text{ m/sec}^2)$$
$$= 9.8 \text{ newtons}$$

That is, on the earth, a 1-kg mass weighs 9.8 newtons. Since a 1-kg mass weighs 2.2 lb, we see that 2.2 lb = 9.8 newtons, or

$$1 \text{ newton} = 0.225 \text{ lb}$$

Let us review the salient points of the metric system. Its basic unit of length is the meter, which is a little more than a yard. Other common length units are the kilometer (about 0.6 mi) and the centimeter (about 0.4 in.). Masses are measured in kilograms. The metric unit of force is the newton (about 0.2 lb) and, on the earth, a 1-kg mass weighs 9.8 newtons. The acceleration due to gravity, g, is 9.8 m/sec² on the earth. A few examples will illustrate the use of the metric system.

EXAMPLE 3.4

If you wish to buy about 5 lb of sugar in France, how many kilograms should you request?

Method: If we recall that a 1-kg mass weighs about 2.2 lb, it is simple to see that 2 kg weighs about 4.4 lb. You should therefore order 2 kg of sugar.

EXAMPLE 3.5

The speed limit on a main highway leading into Ankara, Turkey, is 40 km/hr. How fast, in miles per hour, is it legal to drive on this highway?

Method: Recalling that 1 km is about 0.6 mi, we know that a distance of 40 km is equivalent to about 25 mi. Hence the speed limit is about 25 mi/hr.

EXAMPLE 3.6

How large a force is needed to give a 1,500-kg car an acceleration of 3 m/sec²?

Method: To find the force, we need only substitute in $F = ma$. But, in doing so, we must recall that the *mass* of the car is 1,500 kg. (Frequently people say something *weighs* a certain number of grams

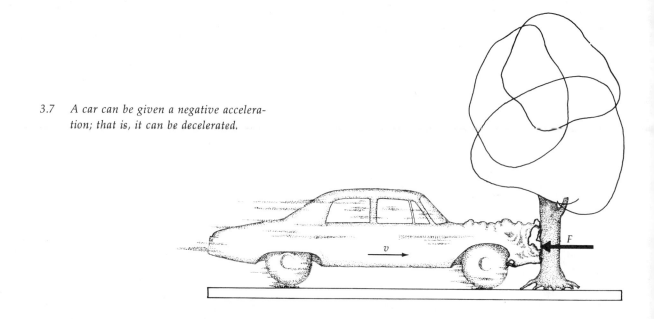

3.7 *A car can be given a negative acceleration; that is, it can be decelerated.*

or kilograms. In spite of their choice of words, they are stating the object's mass, not its weight.) Then

$$F = (1{,}500 \text{ kg})(3 \text{ m/sec}^2)$$
$$= 4{,}500 \text{ newtons}$$

Notice that the newton unit is an abbreviation for the kg m/sec² units. Since 1 newton = 0.225 lb, this force is roughly 1,000 lb.

Negative Acceleration—Deceleration

In our discussion of the relation between force and acceleration, we have assumed that the force is used to increase the speed of the object. But forces frequently decrease the speed of an object. For example, in Figure 3.7 a 2,400-lb car is stopped as it strikes a tree. Suppose the car had a velocity of 50 ft/sec just before it hit and came to rest in a time of 0.10 sec. From Equation (3.2), the acceleration of the car during the process is

$$a = \frac{v - v_0}{t}$$
$$= \frac{0 - 50 \text{ ft/sec}}{0.10 \text{ sec}}$$
$$= -500 \text{ ft/sec}^2$$

Notice that the acceleration is negative because the quantity $v - v_0$ is negative. If the object is slowing down, it has a negative acceleration which we call its *deceleration*.

Furthermore, if we use $F = ma$ to find how large a force the tree exerted upon the car, we find

$$F = \frac{2{,}400 \text{ lb}}{32 \text{ ft/sec}^2}(-500 \text{ ft/sec}^2)$$
$$= -37{,}500 \text{ lb}$$

The negative sign tells us the force is a stopping force, and it is drawn as such in Figure 3.7. Notice how large the force is; it is not surprising that a car is demolished in such an accident.

3.8 *The upward-moving ball is decelerated by the gravitational force W. Its deceleration is equal in magnitude to g.*

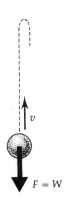

Figure 3.8 illustrates another example of deceleration, an object being thrown upward from the earth. Here the speed of the object is being slowed by the force of gravity, its weight, pulling down on it. Notice that this slowing down (or deceleration) of the object is caused by the force of gravity, mg, which is trying to stop its ascent. This force will give the object a deceleration which we can compute from $F = ma$. We have

$$F = ma$$

which becomes

$$mg = ma$$

and, after canceling m from both sides of the equation, we have $a = g$. In other words, the deceleration of a rising body is equal in magnitude to the acceleration of a falling body. Since we have agreed to call decelerations negative accelerations, we therefore can state that the acceleration of a freely falling object is g, while a freely rising object has an acceleration $-g$. The following example shows how we make use of this fact.

EXAMPLE 3.7

If the object in Figure 3.8 is shot upward with an initial velocity of 30 m/sec, how long will it go up before stopping? How high will it go?

Method: The force of gravity will cause the object to decelerate, and so its acceleration is $-g$, or -9.8 m/sec² since we wish to use metric units. But $v_0 = 30$ m/sec, and we are interested in the point where the object slows to a stop so the final velocity $v = 0$. To find how long it rises, we can use Equation (3.2), $v = v_0 + at$, to find t. Thus

$$t = \frac{v - v_0}{a} = \frac{0 - (30 \text{ m/sec})}{-9.8 \text{ m/sec}^2} = 3.1 \text{ sec}$$

Then, to find how high it went, we can use the relation $s = \bar{v}t$. However, $\bar{v}$ is the average speed, given by

$$\bar{v} = \tfrac{1}{2}(v + v_0) = 15 \text{ m/sec}$$

Then

$$s = (15 \text{ m/sec})(3.1 \text{ sec}) = 46 \text{ m}$$

The object rises to a height of 46 m.

Linear Momentum and Its Conservation Law

When Newton first presented his second law, he did not state it as we did. Instead of relating force and acceleration, he related force to a quantity we now call *linear momentum*. To reconcile his approach with ours, we need only write $F = ma$ in a different way. If we replace the acceleration a by $(v - v_0)/t$, its value from Equation (3.2), we find

$$F = m\frac{v - v_0}{t}$$

in place of $F = ma$.

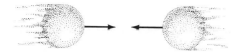

3.9 Since the forces are equal and opposite, the change in momentum of one object is equal and opposite to the momentum change of the other; the total change in momentum is zero.

After a little arithmetic, this can be written as

$$mv - mv_0 = Ft \qquad (3.6)$$

The quantity mv is called the <u>linear momentum</u> of the mass m moving with velocity v. Like the velocity itself, momentum is considered to have direction (the direction of motion) and is a vector. We can also state Equation (3.6) in words:

The change in momentum of an object, caused by a force F acting on it for a time t, is simply Ft.

All of us have experienced this relationship in a variety of ways. We frequently speak of an object as having a large amount of momentum. For example, a slowly moving car has large momentum because m is large; a pitched baseball has large momentum because of its large v. Of course, a rapidly moving car has extraordinarily large momentum because both m and v are large. In all these cases, a very large force is needed to change the momentum quickly. A car can be stopped quickly by the large force a tree can exert on it. Smaller friction forces can also stop it while it is coasting along the road, but in this case the force must act for a longer time. A much larger force is needed to change the momentum of an object quickly than to change the momentum of an object slowly. You might relate this fact to the following situations:

1   It is much nicer to dive into a filled swimming pool than into an empty one.

2   When jumping from a table to the floor, you should allow your legs to "give" when landing.

3   The most modern car bumpers are designed to "give" upon impact.

Let us turn to another property of momentum. Suppose we have a collection of objects that exert forces on each other when they collide. We will stipulate that no unbalanced forces are exerted on these objects from *outside*. We call such a group of objects an <u>isolated system</u>. It might consist of two balls rolling toward each other on a billiard table, or it might be 10^{18} molecules inside a closed box containing air. The law we are about to arrive at will be valid in either case.

Consider what happens when any two of these objects collide as in Figure 3.9. During the collision, they will exert forces on each other. From Newton's law of action and reaction, we know that the two forces will be equal and opposite. Since these forces are equal, they will change the momentum of each object by exactly the same amount. But the changes in momentum are opposite in direction, since the forces that cause them are opposite. As a result, even though one object has *lost* momentum in one direction, the other has *gained* an equal momentum in that direction. Therefore, the combined momentum of the two objects in that direction remains unchanged. No matter how complicated the system, as long as it experiences no unbalanced forces from outside, every force on one object will correspond to an equal but opposite force on another object. As a result, the momentum of the whole system in any given direction does not change. This is a fundamental law of nature, the <u>law of conservation of linear momentum</u>. It may be stated as follows:

> In the absence of unbalanced external forces, the linear momentum of a system will remain constant.

The following examples show how this law is applied.

EXAMPLE 3.8

The gun of mass M in Figure 3.10 has within it a bullet of mass m and is at rest on a frictionless table. The gun discharges, sending the bullet out with speed v. What will happen to the gun?

Method: Before the gun fires, everything is at rest, so the momentum of the system, gun and bullet, is zero. Since no unbalanced external forces act on the gun-bullet system, the momentum must still be zero after the bullet is fired. But the bullet has momentum mv toward the left, and so to cancel this and preserve zero momentum, the gun must have an equal momentum toward the right. Therefore

$$\text{Gun's momentum} = -\text{bullet's momentum}$$
$$MV = -mv$$

and so the gun's velocity must be

3.10 What happens to the gun as the bullet fires?

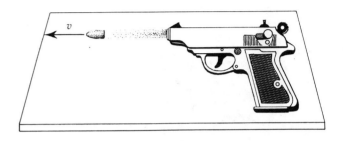

$$V = -\frac{m}{M}v$$

The negative sign tells us that the gun's velocity is opposite in direction to the bullet's. This backward motion of a gun is called its *recoil*, or "kick." Can you see any similarity between this situation and the way a jet or rocket engine operates? You can test your understanding by answering the question posed in Figure 3.11.

EXAMPLE 3.9

As shown in Figure 3.12, a 90,000-lb truck moving at 20 mi/hr strikes a 3,000-lb car standing still at an intersection. If the car and truck remain locked together after the collision, what can we say about their subsequent velocity?

Method: In most physical problems, we must make approximations to compute anything. We know that the present situation will be complex; the friction as the vehicles slide across the road will be large. However, it will not slow the vehicles appreciably until they have slid some distance, so we can ignore this friction force if we agree to talk about the situation just after the

3.11 Newton proposed the steam-driven wagon shown below. In what way is it like the rocket? [Part (a) courtesy of The Bettman Archive; part (b) courtesy of NASA.]

(a)

(b)

3.12 What will happen to the two drivers in an accident such as this?

collision. Thus we can say that the law of conservation of linear momentum applies at that moment. Therefore we can write

Momentum before collision = momentum after collision

In symbols,

$$MV + 0 = (M + m)v$$

where M is the truck's mass and m is the mass of the car. The car's momentum is zero before collision. We designate the initial and final velocities by V and v respectively.

Solving for v, the final velocity, we obtain

$$v = \frac{M}{M + m} V$$

But the ratios of the masses will be the same as the ratios of the weights, and so

$$\frac{M}{M + m} = \frac{90{,}000}{93{,}000}$$

from which

$$v = 0.97V$$

Since the speed of the truck V was 20 mi/hr, we find $v = 19.4$ mi/hr. Notice that the huge truck was hardly slowed at all in the impact. From this, what can you say about the injuries the two drivers might sustain in the accident?

More about Vectors

As you have seen throughout this chapter, it is convenient to represent quantities that have direction by arrows. We have done this for velocities, accelerations, forces, and momentum. All these are

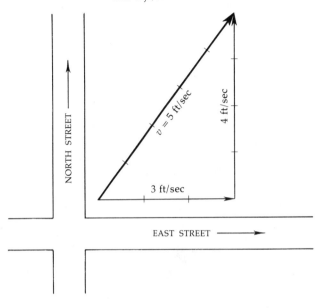

3.13 *If you run with a vector velocity as shown, you are running with component velocities of 3 ft/sec eastward and 4 ft/sec northward.*

examples of vector quantities; they have direction as well as magnitude. We can gain even more utility from vectors if we go one step further and speak about their *components*.

Suppose the streets in a certain location run east–west and north–south as shown in Figure 3.13. If you run across a vacant lot in the direction indicated with velocity v, you are really doing two things at once: you are running both east and north. We say that your velocity, a vector, has both an eastward and a northward component. These are shown as *component vectors* (arrows) in the figure. Using a ruler, if you measure the relative lengths of the velocity vector (which we will take to represent a velocity of 5 ft/sec) and its components, you will see that the northward component is 4 ft/sec and the eastward component is 3 ft/sec. You can appreciate the utility of the component representation by noticing that it tells you at once (from $s = \bar{v}t$) that you will move 30 ft east in a time of 10 sec if you continue with this velocity. To check your understanding of this, how far north would an airplane go in half an hour if it were flying in this same direction with a velocity of 500 mi/hr?

FORCES AND MOTION

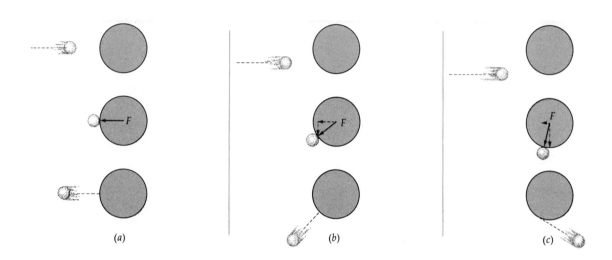

3.14 *By considering the components of the force vector exerted upon the ball, we can understand why it bounces off the pole as shown.*

Forces are also vector quantities of great importance. Let us examine the ways in which vector components can aid us in our understanding of forces. Consider the three situations shown in Figure 3.14, where a ball strikes a rigid pole in three different ways. In (*a*) we see a head-on collision. The force exerted on the ball by the pole is straight backward. Hence, the impact simply reverses the direction of motion, as was the case in Figure 3.9. However, in (*b*) and (*c*), a glancing blow is struck; the ball strikes the pole at an angle, and so the reaction force of the pole on the ball does two things in each case. The leftward force component tries to stop and reverse the motion of the ball, while the downward component deflects the ball downward. Notice that in part (*c*), the leftward force component is so small that the ball's horizontal motion is simply slowed a little rather than being reversed. Thus we see that the motions of the ball can be interpreted by an examination of the horizontal and vertical force components exerted on it by the pole. Situations such as the ones above can often be understood best by considering the components of a force rather than trying to think in terms of the force as a whole.

BIBLIOGRAPHY

1. MAGIE, W. F.: *Source Book in Physics,* Harvard University Press, Cambridge, Mass., 1963. This book contains excerpts from the nonmathematical writings of Galileo, Newton, and others. You will be able to understand much of these writings which, at their time, represented the frontier of science.

2. COHEN, I.B.: *The Birth of a New Physics* (paperback), Anchor Science, Doubleday, Garden City, N.Y., 1960. A readable history of the discoveries of Galileo and his contemporaries. Written by a recognized historian of science.

3. HOLTON, G., and ROLLER, H. D.: *Foundations of Modern Physical Science,* Addison-Wesley, Cambridge, Mass., 1958. Contains much relevant history. Extends the mathematical treatment of the topics covered in this chapter.

4 GALILEO: *Dialogues Concerning Two New Sciences* (paperback), McGraw-Hill, New York, 1963. Galileo's original presentation of his discoveries written for the layman of his time.

5 NEWTON: *Principia* (paperback), University of California Press, Berkeley, 1962. Newton's famous book in which he presents his discovery of the laws of motion.

SUMMARY

Science is the study of the laws of nature. These laws are discovered and tested with the aid of theories and experiments. To achieve a valid test of a theory, the experiment designed for the test must be controlled. This is done, if possible, by carrying out two experiments that are identical except for the feature being tested. The difference in results of the two experiments can then be attributed to the feature that was changed. Galileo was the first to show by such controlled experiments that a change in the weight of an object does not change the rate at which the object falls.

The average speed of an object is simply the distance traveled divided by the time taken. It is always measured in a distance unit divided by a time unit, miles per hour or feet per second, for example. In equation form, $\bar{v} = s/t$.

The instantaneous speed is, as its name suggests, the value of $\bar{v}$ when s and t are so short that you are sure this is the speed of the object at the instant of concern.

Instantaneous velocity is like instantaneous speed except that velocity has a direction associated with it. Numerically, instantaneous speed and velocity are identical. However, the velocity of an object changes when the direction of motion changes even though the speed may not change.

Quantities which have direction are called vectors, while those which do not are scalars. Velocity is a vector while speed is a scalar.

When an object is speeding up, it is accelerating. When it slows down, it is decelerating, or its acceleration is negative. Numerically, the acceleration is equal to the change in velocity divided by the time taken for the change. Its units are a velocity unit divided by a time unit, such as feet per second per second. This is alternatively written ft/sec². As an equation, $a = (v - v_0)/t$, where v is the final velocity and v_0 is the initial velocity for the time interval t.

On the earth a freely falling object accelerates at 32 ft/sec each second. This acceleration is called the acceleration due to gravity, g. Consequently, an object falling freely under the action of gravity alone speeds up 32 ft/sec each second. This is equivalent to 9.8 m/sec².

Newton's first law says that each object has inertia. It states that, in the absence of unbalanced forces; (1) an object at rest will remain at rest and (2) an object in motion will remain in motion in the same straight-line path with unchanging velocity.

Newton's third law states that for each action force on one body, an equal and opposite reaction force must exist on some other body.

The second of Newton's laws of motion relates the acceleration a of an object to the unbalanced force F that causes it. This relation is expressed as $F = ma$. The quantity m is the mass of the object and is a measure of its inertia. Since a force equal to the weight of an object W gives the object an acceleration g, $F = ma$ tells us that $W = mg$, which provides a relation between mass and weight. Although W, the weight or pull of gravity on an object, changes as the object is moved from the earth to the moon, the object's mass m is the same at all places. The mass, a measure of inertia, is a property of the object alone, while W, the object's weight, is determined by the object together with the other objects in its vicinity. In the British system, F is measured in pounds and a in feet per second per second. The metric system expresses F in newtons, m in kilograms (kg), and a in meters per second per second (m/sec²).

Another property of a moving object is its linear momentum, expressed by mv. If an unbalanced force F pushes for time t on an object, the object's momentum changes by an amount Ft. Momentum is a vector, and its direction is the same as the direction of the velocity. The law of conservation of linear momentum tells us that in the absence of unbalanced forces from outside an isolated system, the momentum of such a system remains unchanged. This law is usefully applied to phenomena such as recoil and collisions.

TOPICAL CHECKLIST

1. What is a controlled experiment and why does lack of a control often lead to pseudoscience?
2. What is the definition of speed? Of velocity? How are they different?
3. How does a vector quantity differ from a scalar?
4. State the following equations in words: $s = \bar{v}t$; $\bar{v} = \frac{1}{2}(v + v_0)$.
5. What is meant by acceleration and what is its defining equation?
6. The symbol g is used to designate what quantity? What did Galileo discover about it?
7. State Newton's three laws of motion.
8. What is meant by inertia and how does it enter into Newton's laws?
9. What is the action-reaction law?
10. How is mass related to weight? Why do they differ? What is the definition of each?
11. What are the metric units of length? List the metric prefixes and their meaning. What is the metric unit of mass?
12. Of what is the unit, the newton, a measure? How is it related to the corresponding British unit?
13. What is the meaning of a negative acceleration?
14. Define linear momentum.
15. What is the law of conservation of linear momentum and what are some examples of its use?

QUESTIONS

1. List a few pseudoscientific superstitions and try to guess how they may have originated.
2. Some people insist that blacks are, by virtue of their heredity, better athletes than whites. Others insist that whites are intrinsically more intelligent than blacks. Discuss the difficulties associated with a scientific test of either of these theories.
3. Many people attributed unusual weather patterns during the late 1950s and early 1960s to the effects of nuclear bomb tests. Do you believe this can be proved?
4. Some have claimed that women who use birth-control pills are more likely to die of blood clots than those who don't. Discuss the difficulties in obtaining a conclusive test of this point. How might a reporter reading experimental data concerning this matter inadvertently cause great harm?
5. An almost legendary party trick is to pull the tablecloth out from under the dishes on a table. Which of Newton's laws is fundamental to the successful completion of it?
6. It is possible to pound a nail into a small block of wood while the block is resting on one's head provided a heavy block of metal is used as a "cushion" between the wooden block and head. Explain why the metal block serves as cushion.
7. In ancient times when fathers still spanked their children, a father was likely to tell his child "This hurts me as much as it does you." Is there any scientific basis for such a statement?

8. Give five examples of the "remain at rest" part of Newton's first law. How many examples of the "remain in motion" part can you give?

9. Three outside forces act on a boy as he slides across a freshly waxed floor. What are they and in what direction does each act on the boy?

10. Describe the force exerted by the floor against your shoes as you start to walk across it. Is it possible to start to walk on a perfectly frictionless floor? To keep walking?

11. According to Newton's third law, for every force with which a person pushes on an object, the object will push back on him with an equal force. How is it possible for him to accelerate the object in view of these equal and opposite forces?

12. (a) An automobile moving at a high speed strikes a solid wall head-on. Discuss what sort of injuries the person in the seat next to the driver would sustain and explain how they are consequences of Newton's laws. Why are seat belts important in such a case?
(b) A car at rest is hit from behind by another car. Explain what sort of injuries the driver of the car originally at rest will sustain and how they are consequences of Newton's laws. Why are head rests important in cases such as this?
(c) Explain why the driver of a large truck is in far less danger than a driver of a small Volkswagen in accidents such as outlined in (a) and (b).
(d) Explain why the damage to the auto in part (a) will be less if the car smashes through the wall and goes perhaps 3 ft beyond the point of impact before stopping.

13. Suppose a 1-oz marble and a 1-oz sphere of jelly are dropped from the same height onto a bald man's head. Which would hurt most?

14. If a kilogram mass were dropped from a height of a meter onto a man's hand when the hand was flat on a table, it would injure the hand badly. However, the man would suffer no injury at all if he caught the falling weight with an unsupported hand. Why the difference?

15. Why does a gun "kick"? How does a rocket ship propel itself in outer space where there is no air?

16. Describe what will happen to a man standing on slippery ice if he tries to catch a fast-moving baseball.

17. Which has more inertia, a pail of water or a pail of feathers? What about their comparative masses? Their weights? Which will fall faster when dropped?

18. Suppose you have a parakeet hidden in a completely enclosed cage that is suspended from a sensitive scale which measures its weight. If you are hard of hearing, can you tell when the parakeet flies from place to place?

PROBLEMS

1. The air distance between two cities is 1,800 mi. At what average speed must an airliner fly in order to take 4 hr for the trip?

2. How fast must one run to finish a 100-yd dash in 9 sec?

3. A ball starts from rest and rolls down an incline; its speed is 40 in./sec after rolling for 3 sec. (a) What is its acceleration? (b) How far did it roll in the 3 sec? (c) How fast was it going 2 sec after its release?

4. A car moving at 80 ft/sec skids to rest in a time of 6 sec. (a) What was its average speed during the stopping process? (b) How far did it go before stopping? (c) How large was its deceleration?

5. A flower pot falls from a window of a tall building. (a) How fast is it falling 2.5 sec after dropping? (b) How far does it fall in 2.5 sec?

6. In order to have a speed of 50 ft/sec just before hitting the ground, from what height must an object be dropped? How long must it fall to achieve this speed?

7. A ball is thrown straight up in the air and reaches a height of 20 m. How fast was it thrown?

8. A 3,200-lb car moving at 40 ft/sec is to be stopped in 5 sec. Find its deceleration and the required stopping force.

9. A 10-gm (0.010-kg) bullet is accelerated from rest to a speed of 400 m/sec in a time of 0.002 sec. How large a force was used to accelerate the bullet?

10. A 64-lb boy running with speed 10 ft/sec slides across the ice and stops in a distance of 30 ft. (*a*) What was his average sliding speed? (*b*) How long did it take him to stop? (*c*) What was his deceleration? (*d*) How large was the friction force between him and the ice?

11. A 10-gm bullet moving with speed 400 m/sec hits a tree and imbeds 2.0 cm into the wood before stopping. (*a*) Find its average speed during the stopping process, (*b*) the time taken to stop, (*c*) the bullet's deceleration, and (*d*) the force exerted on the tree by the bullet.

12. Make an educated guess as to how large a force a tree would exert on you if you ran into it at a speed of 20 ft/sec. Assume you hit it with your chest and were stopped.

13. Two balls of equal mass are moving with equal but opposite velocities of 10 ft/sec when they collide head-on. A piece of gum on one makes the two stick together. How fast will they be moving off together after collision?

14. In the previous problem, if one ball bounces straight back with a speed of 4 ft/sec, what is the other doing after the collision? (The gum is absent in this case.)

Chapter 4

WORK, ENERGY, AND ORBITAL MOTION

With the discovery of Newton's laws, many seemingly unrelated phenomena of nature became unified and comprehensible. Basically, this is the purpose of science, to discover the unifying principles and laws with which we can describe the universe most concisely. We saw three such principles in the previous chapter: the simple equation $F = ma$, the concept of linear momentum, and its conservation law. We will introduce another unifying concept in this chapter, the quantity we call energy, which also obeys a conservation law.

The Scientist's Definition of Work

The term *work* has many different meanings in our society. We go to work in the morning, we work at our studies, a machine works—each of these uses of the word implies something different. The scientist's definition of work is related to the concept of the work done in lifting an object.

Suppose you wish to lift the object of weight W shown in Figure 4.1 through a height h, as indicated. To lift it, you must pull upward with a force large enough to overcome the downward pull of gravity on it. If you lift it slowly, its acceleration will be nearly zero. As a result, the upward force F which you exert on the object will be essentially equal to the downward pull of gravity on it, W. We will approximate situations such as this by setting $F = W$.

It is apparent that the work you do in lifting the object depends on (1) how heavy the object is, and (2) how far you lift it. Therefore, we define the work done in lifting a weight W through a height h to be the product of W and h; that is,

$$\text{Work done} = Wh$$

Let us check this in an example:

If you lift a 2-lb object upward a distance of 3 ft, our definition tells us that the work you do is

$$\text{Work} = Wh = (2 \text{ lb})(3 \text{ ft}) = 6 \text{ ft-lb}$$

As you see, the unit of work in the British system is the foot-pound.* If the object had weighed 8 lb instead of 2 lb, you would have done four times as much work since the load would have been four

---

* It is customary to call it a foot-pound, not a pound-foot.

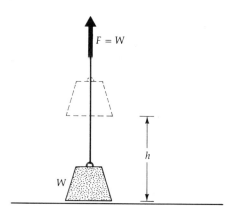

4.1 How much work is done in lifting the weight W to a height h?

times heavier. Our definition of work agrees with this, since, in this case,

Work = (8 lb)(3 ft) = 24 ft-lb

which is four times larger than before. Similarly, if you had lifted the weight only one-third as far, intuition as well as the definition tell us that only one-third as much work would have been done. Let us generalize the definition of work to situations other than the simple lifting of an object.

The scientist's definition for the work done by an applied force F is as follows.

> When a force F moves an object through a distance s in the direction of the force, *the work done is Fs*. In equation form,
>
> $$\text{Work done} = Fs \tag{4.1}$$

Notice the important stipulation about s; it is the *distance moved in the direction of the force*. To clarify this, refer to Figure 4.2. In Figure 4.2a a person is lifting a bucket of water. The work he is doing is the weight of the bucket times the distance through which he lifts it. In Figure 4.2b he is still exerting a lifting force on the bucket, but just to hold it in place. He is no longer moving the bucket upward, so the distance moved in the direction of the force is zero and no work is being done. In Figure 4.2c he still holds the bucket the same distance from the ground and has started to walk. He is exerting an upward force on the bucket in order to hold it up but he is moving it in a horizontal direction.

Even though the bucket is moved to the right a distance d, it requires only a very small *sideways* force to do this. As a result, only a negligible amount of work is done in moving the bucket horizontally.

As you can see, the scientific definition of work does not always match the layman's. For example, in Figure 4.3 we see a circus strong man supporting six friends. Though he will soon grow tired, he is doing no work according to the scientist's definition. He must exert a large force to counteract gravity, but the force will cause no motion, and so s, and thereby Fs, will be zero. He is doing no work because no motion in the direction of the force is occurring.

As we have pointed out, the unit of work in the British system is the foot-pound, the combined unit obtained from work = Fs. In the metric system the unit of work is the newton meter, since F is measured in newtons and s in meters. The unit newton meter occurs often, so we call it a *joule*, pronounced "jewel." In summary, when using the British system, F in pounds and s in feet, the unit of work is the foot-pound. When using the mks metric system, F in newtons and s in meters, the unit of work is the joule.

EXAMPLE 4.1

How much work is done in lifting a 60-kg man upward through a distance of 2 m?

Method: The weight of a 60-kg man is (mass)(g) or 60(9.8) newtons, and that large a force is needed to lift him. Therefore,

Work done = Fs = [60(9.8) newtons](2 m) = 1,180 joules

4.2 How much work is being done in each of these situations?

(a)

(b)

(c)

4.3 This circus strong man is doing no work according to the scientific definition of work. (Courtesy of The Bettmann Archive.)

WORK, ENERGY, ORBITAL MOTION 71

4.4 *While doing work Fs, the force accelerates the cart from v_0 to v.*

Work and Kinetic Energy

Work can be done in other ways than by lifting an object. For example, the horizontal force F in Figure 4.4 is accelerating the cart along a horizontal surface. The force is doing work since it displaces the cart a distance s in the direction of the force. If we ignore all friction forces, the effect of the work done by F is to increase the cart's speed. You might guess that some definite relation must exist between the work done, Fs, and the change in the cart's speed from v_0 to v. Such a relation can be found in the following way.

> Since we wish to relate the work done, Fs, to v and v_0, we must somehow express both F and s in terms of these quantities. To do this, we note in Figure 4.4 that the unbalanced force acting on the cart is F. As a result, the cart will accelerate in conformity with Newton's second law, and so we can write
>
> $$F = ma$$
>
> We know from Equation (3.2) that $a = (v - v_0)/t$, and therefore

$$F = m\frac{v - v_0}{t}$$

where t is the time taken for the cart to move through the distance s. We can then write the work done by F as follows:

$$\text{Work done} = Fs = m\frac{v - v_0}{t}s$$

Now we know that $s = \bar{v}t$, where $\bar{v} = \tfrac{1}{2}(v + v_0)$. After replacing s by this value in the work equation, we find

$$\text{Work done} = m\left(\frac{v - v_0}{t}\right)\left(\frac{(v + v_0)t}{2}\right)$$

Luckily, t cancels out, and the equation simplifies greatly.

When the arithmetic is done, we find that the work done by the force F causes the cart's speed to change from v_0 to v according to the equation

$$\text{Work done during acceleration} = \tfrac{1}{2}mv^2 - \tfrac{1}{2}mv_0^2 \qquad (4.2)$$

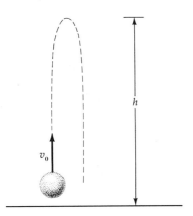

4.5 *The moving object can lift itself a height h by virtue of its kinetic energy.*

The quantity $\tfrac{1}{2}mv^2$ is of great importance; it is called <u>kinetic energy</u>. In terms of kinetic energy, we can restate Equation (4.2) in words:

The work required to accelerate an object of mass m from speed v_0 to v is equal to the change in kinetic energy of the object.

This relation is also true if the object is being decelerated rather than accelerated. In either case, the work done is equal to the change in kinetic energy.

As an example of the use of the concept of kinetic energy, suppose a 3,200-lb car moving at 80 ft/sec slams on its brakes, traveling a distance of 300 ft before it comes to a stop. How large must the friction force on the skidding tires of the car have been in order to stop it in this distance? Obviously, the car lost all its kinetic energy because of the work done by the retarding friction force between wheels and pavement. We can therefore write:

$$\text{Work by friction force} = \text{loss in KE}$$

where KE is kinetic energy.

If we call the friction force F and the distance the car slides s, this equation can be written in symbols as follows:

$$\text{Work done} = \text{change in KE}$$
$$Fs = \tfrac{1}{2}mv^2 - \tfrac{1}{2}mv_0^2$$

where m is the mass of the car, W/g. The final speed $v = 0$, while the initial speed $v_0 = 80$ ft/sec. If we place these and $s = 300$ ft in the equation, we find that $F = -1{,}070$ lb. (What does the minus sign mean?) This is about the maximum friction force you can achieve in such a situation (i.e., about one-third the car's weight); and so a car going at 80 ft/sec (55 mi/hr) cannot stop in less than about 300 ft unless it hits something.

Gravitational Potential Energy

The word *energy* comes from the Greek word for work, and so we say something has energy if that thing has the ability to do work. The word *kinetic* is derived from the Greek verb "to move," so kinetic energy is energy resulting from motion. Thus we can define <u>kinetic energy</u> as the ability of a system to do work by virtue of the system's motion.

A simple example of an object's ability to do work by virtue of its motion is illustrated in Figure 4.5. The upward-moving object can *lift itself* a distance h. In so doing, it does work equal to Wh (or mgh since $W = mg$). If you wish, you can think of this in a different way. The force of gravity W does work on the moving object and changes its kinetic energy. Since we know that the work done equals the change in kinetic energy, we can write at once for the situation shown in Figure 4.5 that

$$\Delta KE = \text{work done}$$

where ΔKE is change in kinetic energy. But the original kinetic energy of the object is $\tfrac{1}{2}mv_0^2$ and it loses all of this kinetic energy by the time it reaches the top of its path because it stops for an instant there. Hence its kinetic energy changes from $\tfrac{1}{2}mv_0^2$ to zero, a total change of just $\tfrac{1}{2}mv_0^2$. Since the work done in lifting the object is mgh, we find that

$$\Delta KE = \text{work done}$$

becomes

$$\tfrac{1}{2}mv_0^2 = mgh$$

We might notice in passing that this equation gives us a simple means for finding how high a thrown object will go. Solving for h, we see that m cancels out, and so

$$h = \frac{v_0^2}{2g}$$

This tells us that the height to which a thrown object will rise depends only upon the speed with which it is thrown and the acceleration due to gravity; the mass of the object is unimportant. Of course, this assumes that friction forces are negligible.

As we know, once the object in Figure 4.5 reaches the top of its path, it stops and then starts falling back down. Since it is stopped at the top, its kinetic energy there is zero. However, as the earth pulls down on the object with force W (or mg), the object accelerates and acquires kinetic energy. By the time the object has fallen a distance h, and just before it touches the ground, the force mg acting through a distance h has done work in the amount mgh on it. At the same time, the kinetic energy of the object has changed from its starting value of zero to $\tfrac{1}{2}mv^2$ where v is its speed just before it strikes the ground. We can therefore write

$$\text{Work done} = \Delta KE$$

and this gives

$$mgh = \tfrac{1}{2}mv^2$$

Notice that this is exactly the same equation we had for the upward progress of the object. We therefore conclude that in the absence of friction, *an object thrown upward will return with the same speed it had when thrown.* Moreover, since the object's mass m cancels from this equation as it did for the upward-moving object, we see that *all objects fall in the same way.* This is exactly what Galileo discovered about freely falling objects.

You will notice that the previous illustration implies that an object which can fall has the ability to do work. For example, in Figure 4.6a, the weight can fall, acquiring kinetic energy as it falls. After falling a height h, it has gained a kinetic energy

$$\tfrac{1}{2}mv^2 = mgh$$

Since the moving object is able to do work in the amount $\tfrac{1}{2}mv^2$, we conclude that before falling, the object already had the ability to do work in the amount mgh or Wh.

Figure 4.6b shows another example of the ability of an object to do work by virtue of its ability to fall. If we give the upper weight a slight downward push, the upper object will fall at a slow, constant speed. In doing so, it will lift the other weight and do work equal to Wh or mgh.

These examples show that an object of mass m at height h has the ability to do an amount mgh of work. The object therefore possesses energy, the ability to do work. We call this type of energy *gravitational potential energy* since the ability to fall under gravity is basic to this form of energy. To summarize:

> An object of weight $W = mg$ at a height h has a gravitational potential energy of Wh or, alternatively, mgh.

We have discussed two kinds of energy, kinetic and gravitational potential. In later chapters we shall consider several other forms of energy. They all have one property in common: *Since energy is the ability to do work, this ability is never lost until work is done.* This leads us to a conservation law for energy.

4.6 *An object of mass m has the ability to do an amount of work mgh in falling a distance h.*

(a)

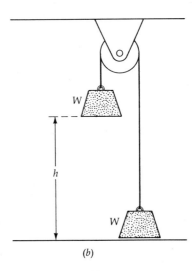

(b)

Conservation of Energy

We have seen that kinetic energy can be lost in two ways: it can do an equivalent amount of work against a force F, or it can produce an equivalent amount of potential energy. This latter alternative was simply work done against the gravitational force and so it could be included with the first alternative. However, work done against the gravitational force is so important that we separate this effect and refer to it as gravitational potential energy.

We have also seen that gravitational potential energy can be lost in two ways: it can be changed to an equivalent amount of kinetic energy, or it can be used to do an equal amount of work. Finally, when a force does work it can (1) increase an object's kinetic energy, (2) increase an object's potential energy, or (3) do work against friction forces. The total work done by a force is always equal to the sum of the increased kinetic and potential energies together with the work done against friction.

These diverse facts can be summarized in a fundamental law of nature, the *law of conservation of energy*. At this time we shall only state a portion of the law; the remainder of the law involves other forms of energy.

The total energy of an object, kinetic plus potential, is constant provided that no unbalanced force except gravity acts on the object.

As pointed out previously, the gravitational force is special since we make use of gravitational potential energy to express its effects.

Even if friction and other forces are acting on the object, we can still state something of value. If someone is pushing on the object so as to increase its speed, then the work done by the pushing agent will cause an equal increase in the energy of the object. Stopping forces on the other hand will cause the object to lose energy equal to the work done by them. The following statement summarizes these observations and actually encompasses the restricted law given above:

The original energy of an object plus the work done by accelerating forces minus the work done by stopping forces equals the final energy of the object.

This is sometimes called the *work-energy theorem*. We will now illustrate its use by a few examples.

EXAMPLE 4.2

An automobile weighing 3,000 lb hits a tree head-on at a speed of 40 ft/sec (about 27 mi/hr). It stops after the motor and radiator have been pushed back 3 ft into the car. Estimate the force the tree exerted against the car.

Method: The car originally had kinetic energy $\frac{1}{2}mv^2$. It lost all this because of the stopping force F exerted by the tree against the car, and this force was exerted on the car for a distance of 3 ft as it was coming to rest. Hence, the work done by the stopping force of the tree was $F(3\text{ ft})$. There was no accelerating force, and so the work-energy theorem says

Original KE − work by stopping force = 0

or

$$\tfrac{1}{2}mv^2 - F(3\text{ ft}) = 0$$

The right side of the equation is zero because the car stopped and therefore its final energy is zero. We can solve this equation for F by recalling that $v = 40$ ft/sec and $m = W/g = (3{,}000\text{ lb})/(32\text{ ft/sec}^2)$. The result is

$$F = 25{,}000 \text{ lbs} = 12.5 \text{ tons}$$

Even at this relatively low speed, such a large force would cause great damage.

EXAMPLE 4.3

A 10,000-kg space capsule reenters the earth's atmosphere at a height of about 100 km with a speed of 5,000 m/sec. Several minutes later, it drops slowly into the ocean, supported by its parachute. Estimate how much energy was lost to friction forces during its descent.

Method: The capsule had negligible kinetic energy as it drifted into the ocean. Therefore, it must have lost all its original kinetic and potential energy doing work against friction forces as it streaked through the atmosphere. As a result, the work-energy theorem tells us

Original KE + original PE − friction work = 0

which is

$$\tfrac{1}{2}mv_0^2 + mgh - \text{friction work} = 0$$

The friction work done is therefore

$$\text{Friction work} = \tfrac{1}{2}mv_0^2 + mgh$$

Since $m = 10{,}000$ kg, $v_0 = 5{,}000$ m/sec, $g = 9.8$ m/sec², and $h = 100{,}000$ meters, we find

$$\text{Friction work} = 13.5 \times 10^{10} \text{ joules}$$

Remember that 13.5×10^{10} means to multiply the 13.5 by 10 billion—that is, one with ten zeros after it. This large loss of energy to friction work results in a great deal of heat, causing the surface of the reentry capsule to become white-hot. Why aren't the men in the capsule burned up by the extreme heat?

EXAMPLE 4.4

The car shown in Figure 4.7 is moving with a speed of 90 ft/sec (61 mi/hr) when the motor fails. Assuming the car coasts without friction, how fast will it be going when it reaches the top of the hill?

Method: Since we are assuming the absence of all forces except gravity, the total energy of the car will remain unchanged. However, some of the kinetic energy will be changed to potential energy. The work-energy theorem gives

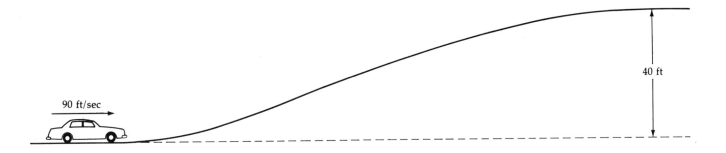

4.7 *If the car is coasting, how fast will it be going when it reaches the top of the hill?*

Original KE + original PE = final KE + final PE

Or in symbols,

$$\tfrac{1}{2}mv_0^2 + 0 = \tfrac{1}{2}mv^2 + mgh$$

Each term of this equation contains the car's mass m, and, after dividing each term of the equation by m, multiplying by 2, and rearranging, we have

$$v^2 = v_0^2 - 2gh$$

Substituting $v_0 = 90$ ft/sec, $g = 32$ ft/sec², and $h = 40$ ft gives the following result:

$$v = \sqrt{5{,}540}\ \text{ft/sec}$$

Since $70 \times 70 = 4{,}900$ and $80 \times 80 = 6{,}400$, the square root of 5,540 must lie about half-way between 70 and 80. The correct result is

$$v = 74.4\ \text{ft/sec}$$

Terminal Speed

As we have seen, when one form of energy is lost, either work must be done or another form of energy must appear. From this principle we have found how fast an object is going after falling a distance h. The object loses potential energy mgh and, if friction forces can be ignored, gains an equal amount of kinetic energy, $\tfrac{1}{2}mv^2$. How fast should an object be going after it drops 9 ft?

Equating the potential energy lost, mgh, to the kinetic energy gained, $\tfrac{1}{2}mv^2$, gives

$$\cancel{m}gh = \tfrac{1}{2}\cancel{m}v^2$$

or

$$v = \sqrt{2gh}$$

In our case, $g = 32$ ft/sec² and $h = 9$ ft, and so we find $v = 24$ ft/sec. This is the speed just before the object strikes the ground. We can compute how long it has taken to reach the ground by noting $\bar{v} = \tfrac{1}{2}(v + v_0) = 12$ ft/sec and using $s = \bar{v}t$, where $s = h = 9$ ft. The time is only $\tfrac{9}{12}$ or $\tfrac{3}{4}$ sec.

WORK, ENERGY, ORBITAL MOTION

If you let a tiny piece of paper fall from a height of about 9 ft, you will notice two curious facts. First, the slip of paper takes far longer than $\frac{3}{4}$ sec to fall; second, the paper seems to fall at almost constant speed once it has fallen a small distance. Clearly, this object does not fall in the manner we have been predicting.

The clue we need to interpret this behavior comes from the observation that the bit of paper falls with constant speed after the first instant; it no longer accelerates. Newton's second law tells us that since $a = 0$, no unbalanced force is acting on the falling object. Certainly the earth pulls down on it with a force equal to its weight W. Since the paper is no longer accelerating, the gravitational force must be balanced by an equal upward force. This balancing force is the friction of the air through which the paper is falling.

All objects falling through the air experience a friction force because of the air, and this force is opposite to the direction of motion of the object. The faster the object falls, the larger this retarding friction force is. At first, when an object is falling slowly, the upward-directed friction force is very small. As the object's speed increases, the friction force also increases. At a certain speed, called the _terminal speed_, the friction force becomes as large as the object's weight. When this speed is reached, the gravitational force is balanced by the friction force and the falling object stops accelerating. The object then falls with constant speed, its terminal speed.

A slip of paper or a feather has very little weight but offers a large surface for air friction. As a result, such objects soon reach their terminal speeds. Much heavier objects exhibit the same behavior if they fall far enough to reach their terminal speed. Typical terminal speeds are 25 ft/sec for a raindrop and about 200 ft/sec for a man. You might estimate how far a man would have to fall to reach his terminal speed; remember that g, the acceleration due to gravity, is 32 ft/sec².

Power

Mechanical devices such as machines, engines, motors, etc., use various forms of energy to do work. The _rate_ at which the work is done is often as important as the total work done. For example, a man with a shovel and a loading machine both do the same amount of work in loading a ton of gravel onto a truck. However, the machine can do the work much more quickly. The scientific quantity which measures the rate at which work is being done is called _power_. It is defined in the following way:

$$\text{Power} = \frac{\text{work done}}{\text{time taken}}$$

The units of power come directly from the definition. In the British system, work is measured in foot-pounds, and so power is measured in foot-pounds per second. A unit more frequently used is the _horsepower_ (hp), where

$$1 \text{ hp} = 550 \text{ ft-lb/sec}$$

You have also heard of the metric unit for power; it is the _watt_. When a machine is doing work at the rate of one joule per second, its power output is one watt. We will illustrate these points by a few examples.

EXAMPLE 4.5

A motor lifts a 2,000-lb elevator a distance of 60 ft in a time of 30 sec. Estimate the power output of the motor.

Method: In a time of 30 sec, the motor lifts 2,000 lb a distance of 60 ft. Therefore the work it does is

$$\text{Work done} = Wh = 120,000 \text{ ft-lb}$$

and the time taken is 30 sec. Hence the power is

$$\text{Power} = \frac{120{,}000 \text{ ft-lb}}{30 \text{ sec}} = 4{,}000 \text{ ft-lb/sec}$$

Since 1 hp = 550 ft lb/sec, the motor is capable of a power output of 4,000/550, or 7.3 hp. This is a large motor. In practice, the motor would have to have a greater capability than 7.3 hp, since extra work is needed to speed the elevator up and to overcome friction.

EXAMPLE 4.6

What minimum horsepower would a 3,200-lb car need to accelerate from rest to 40 ft/sec (about 27 mi/hr) in a time of 5 sec?

Method: If we assume the car to be on level ground and ignore friction forces, the work done will be equal to the increased kinetic energy of the car. Therefore,

$$\text{Work} = \Delta KE$$
$$= \frac{1}{2}\left(\frac{3{,}200}{32}\right)(40)^2 \text{ ft-lb}$$
$$= 80{,}000 \text{ ft-lb}$$

The power is then

$$\text{Power} = \frac{80{,}000 \text{ ft-lb}}{5 \text{ sec}} = 16{,}000 \text{ ft-lb/sec} = 29 \text{ hp}$$

Since even a Volkswagen is rated at twice this power, this seems unreasonable. Where does the error lie?

Energy in Vibrating Systems

Unlike the systems we have studied thus far, some mechanical systems vibrate back and forth. Typical examples are the vibrating pendulum and the back-and-forth motion of a mass at the end of a spring. The spring system occurs frequently in nature; even the atoms within molecules can be considered held to each other by springlike bonds. In spite of their diversity, all vibratory devices have certain features in common and we shall now discuss them.

The vibrating pendulum shown in Figure 4.8 moves back and forth between two limiting positions at the two ends of its swing. As you see, the pendulum moves most rapidly when it passes through its center position. When it is at the end of its swing, it stands still for a brief instant. Clearly, the pendulum ball loses gravitational potential energy as it swings down toward the center position. This lost potential energy changes to kinetic energy as the work-energy theorem says it must. Clearly, a swinging pendulum has a continual interchange of kinetic and potential energy.

The mass at the end of the spring shown in Figure 4.9 will also oscillate back and forth between the two limits on each side of the central position of the undistorted spring. When the mass is to the left of the central position, the spring is compressed; when it is to the right, the spring is stretched. Like the pendulum, the mass-spring system oscillates back and forth, having maximum kinetic energy when the mass is passing through the central position. At the two ends of its path, the mass has lost all its kinetic energy. The energy of the system is then stored in the spring and will be converted back to kinetic energy as the spring accelerates the mass. We call this new form of stored energy *spring potential energy*.

The vibrating mass-spring system has much in common with the vibrating pendulum. Both undergo constant interchange of kinetic and potential energy. However, unlike the pendulum, which stores energy in its gravitational potential form, the spring stores energy in the alternately stretched and compressed spring.

It is often convenient to draw a diagram showing the potential energy of the mass at the end of a pendulum or the energy stored in a spring. Figure 4.10a gives such a diagram for the pendulum above

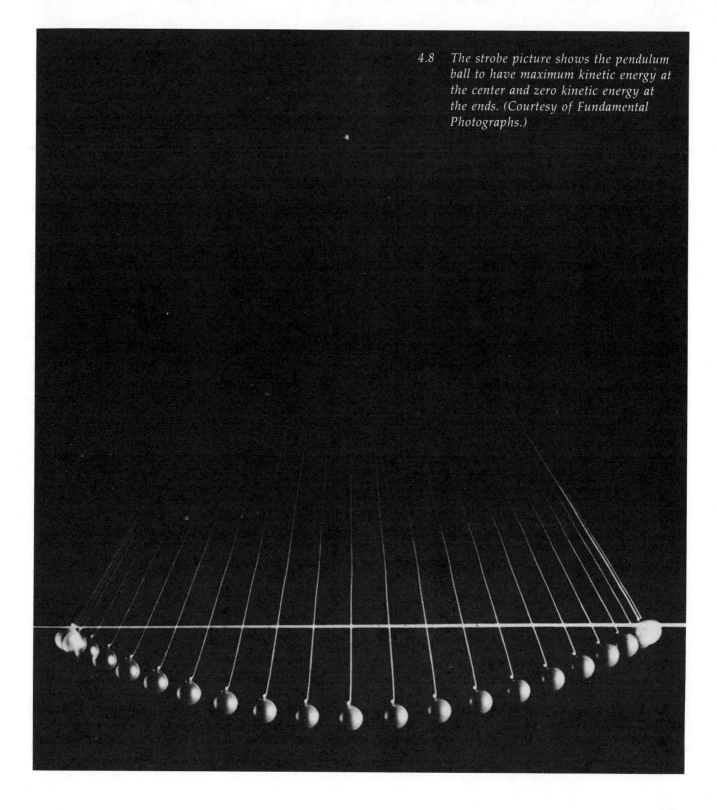

4.8 The strobe picture shows the pendulum ball to have maximum kinetic energy at the center and zero kinetic energy at the ends. (Courtesy of Fundamental Photographs.)

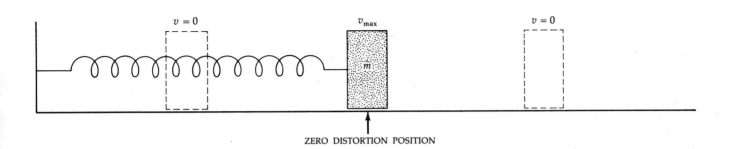

4.9 As the mass oscillates back and forth, the energy changes from potential to kinetic to potential, over and over.

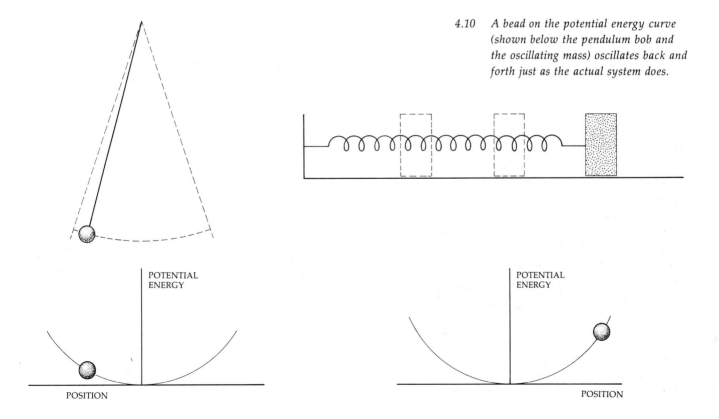

4.10 A bead on the potential energy curve (shown below the pendulum bob and the oscillating mass) oscillates back and forth just as the actual system does.

WORK, ENERGY, ORBITAL MOTION

it. Since the potential energy of the mass is proportional to the height of the mass, the potential energy diagram resembles the path of the pendulum. Similarly, in Figure 4.10b, the potential energy diagram for the spring system is shown. When the spring is compressed or stretched, the potential energy is large. At the center position, the undistorted spring has no energy stored in it.

If we mentally replace the curve in Figure 4.10 by a wire with a bead threaded on it, the bead will oscillate back and forth just as the actual system does. For example, suppose the bead is placed at one end of the curved wire. At that point its energy is all potential, just as it is for the pendulum. As the bead slides down the wire, the potential energy changes to kinetic. After passing the center point, the bead slides up the wire until all its energy is potential once again. Then the process repeats itself.

The concept of a bead sliding on the potential energy curve tells us how the actual system will behave. This conceptual device will be helpful in later chapters.

Newton's Law of Gravitation

We have discussed linear motion and vibrational motion. Another important type of motion is exhibited by spacecraft orbiting the earth and by the planets circling the sun. Newton was the first to explain the orbital paths of the planets satisfactorily.

According to Newton's first law, in the absence of an unbalanced force, an object will always move along a straight line. A force, such as that shown in Figure 4.11, for example, is needed to bend the path of an object into a circle. The force in the figure is a <u>radial force</u>, i.e., a force directed toward the circle's center. Moreover, the force is always perpendicular to the motion of the object and so the force does no work. It neither speeds up nor slows down the motion; *it simply changes the direction of the velocity.* This is a situation where a force causes a change in

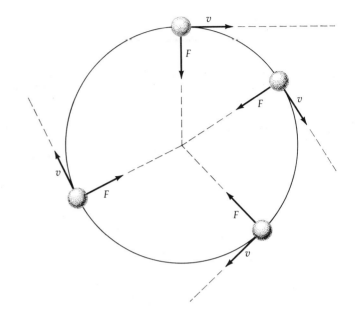

4.11 *A radial force is needed to bend the object out of a straight-line path into a circular path.*

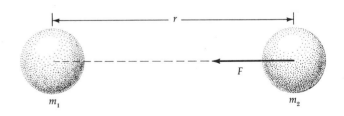

4.12 *The mass m_1 attracts the mass m_2 with the force F given by Newton's law of gravitation.*

velocity without changing the speed. Nevertheless, our definition of acceleration refers to any change in velocity, whether it is a change in direction or magnitude, and so the radial force causes an acceleration even though it does not alter the speed.

Newton analyzed the nearly circular motions of the planets around the sun in an effort to find the nature of gravitational force. (To do it, he invented a whole new branch of mathematics, which we call calculus.) The basic idea is reasonably simple: The sun must exert an attractive force on each of the planets to pull them into circular orbits. Since the moon orbits the earth, the moon must be similarly attracted by the earth. Moreover, the law of action and reaction tells us the earth must be attracted by the moon with an equal and opposite force.

From these considerations, Newton was able to show that the gravitational attraction of one object of mass m_1 for another object of mass m_2 is given by the equation

$$F = G \frac{m_1 m_2}{r^2} \quad (4.3)$$

where G is a constant of nature. The situation is shown in Figure 4.12. Notice that r is the distance between the centers of the spherical masses. This law, Newton's *law of universal gravitation*, shows that the gravitational force is an *inverse square* force; i.e., it varies as $1/r^2$. In other words, the gravitational force between two bodies decreases as the distance between them increases.

In order to evaluate the constant of nature G, we must measure the force of attraction F between two known masses, m_1 and m_2. Newton could not do this because the force between two ordinary-size masses is too small to be measured easily. In 1798, Henry Cavendish first managed to perform the delicate measurement, and found that

$$G = 3.4 \times 10^{-8} \quad \text{(British system)}$$

or

$$G = 6.67 \times 10^{-11} \quad \text{(mks system)}$$

for our two sets of units. (The symbolism "$\times 10^{-8}$" means to *divide* by 10 eight times. For example, 5×10^{-2} is simply 0.05.) We can appreciate the difficulty of such a measurement by computing F from Equation (4.3) for identical 2-lb spheres separated by 0.5 ft. The force turns out to be 0.000,000,000,5 lb (that is 5×10^{-10} lb), an extremely small force.

However, when one of the masses is as large as the earth, the force given by Equation (4.3) becomes sizable. For example, if m_1 is the mass of the earth and m_2 is the mass of your body with r equal to the earth's radius (since r is the distance from your center to the earth's center), then the force F is your weight. As you know, the weight of an object on the earth is the force of gravity pulling the object to the earth. In this case, your weight is the force with which the earth is attracting you.

According to Equation (4.3), the gravitational attraction force varies inversely as r^2. Hence, the farther an object is from the earth's center, the less strongly the earth will pull upon it. This results in a decrease in acceleration due to gravity as an object is taken to higher altitudes. Table 3.1 (page 50) shows that g generally decreases as an object is taken from sea level to the top of Pike's Peak.

Although the change in weight of an object is small from place to place on the earth's surface, the graph in Figure 4.13 shows that an object's weight decreases substantially as it is taken farther from the earth. When an object is on the moon, the earth's attraction for the object has decreased by a factor of about 1/270. If the object weighs W_e on the earth, the earth's attraction force for it will be only $W_e/270$ when the object is on the moon. However, when the object is on the moon, the moon attracts it with a force $W_e/6$, and so we can usually ignore the much smaller force exerted by the earth.

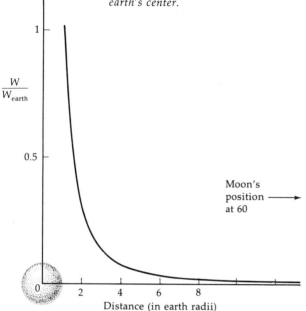

4.13 *The pull of the earth W on an object decreases with distance from the earth's center.*

Although the gravitational attraction force decreases with separation distance, the force of gravity never becomes zero. As r becomes exceedingly large, the force will still exist, although its value will become very small. This is why every part of the universe continues to attract every other part no matter how distant it may be. For most purposes, however, the forces exerted by individual distant objects in the universe are negligibly small. Except for the tides, the moon's gravitational force on earth objects has no noticeable effect. However, when a spaceship comes close to the moon or a planet, the gravitational effects of these bodies will predominate.

Spaceships and Weightlessness

When a spaceship orbits the earth, it is freely falling in space. To see why, consider the motion of a ball thrown horizontally on the earth. As we see from Figure 4.14, the gravitational force pulls the ball downward toward the earth's center and so the ball accelerates downward at 32 ft/sec each second it is in the air. Clearly, the thrown ball is falling freely.

At the same time, however, the ball is moving parallel to the earth. If we ignore the effects of air friction, no force is acting on the ball to change its horizontal velocity. Consequently the ball moves parallel to the earth with constant velocity as shown in Figure 4.14. The motion of the ball is the combined effect of its constant-speed motion parallel to the earth's surface and its falling motion toward the earth's center.

If the ball can be thrown hard enough, the curvature of the earth will make it possible for the ball to miss hitting the earth entirely, as Figure 4.15 shows. If a ball could be thrown with a speed close to 18,000 mi/hr, the curvature of the ball's path would exactly match the earth's curvature, and so the ball would circle the earth. Clearly, to put an object in orbit around the earth, we must shoot it parallel to the earth's surface at just the right speed. This is what happens when a spaceship is put in orbit. It is shot straight up until it gets above the earth's atmosphere, and then it is turned so it can travel parallel to the earth's surface at just the proper speed. Thereafter, it continues to circle the earth at this speed.

Even at such high altitudes, however, a few air molecules exist. As a result, the ship slows over a period of years and comes closer and closer to the earth. Eventually, it reaches such a low altitude that air friction increases and eventually the ship is burned up. Of course we may bring the ship safely back to earth by slowing it suddenly with a reverse burst from its rockets. In that case, it passes through the atmosphere quickly enough so that only its outer protective shell is destroyed by heat.

Just as a certain velocity and mass determine a spaceship's orbit around the earth, so the velocity and mass of a planet determine its orbit around and distance from the sun.

You have no doubt heard that objects in an orbiting spaceship appear to be weightless. Clearly, they are quite close to the earth, and so there is still a large force of gravity acting on them. Therefore, if

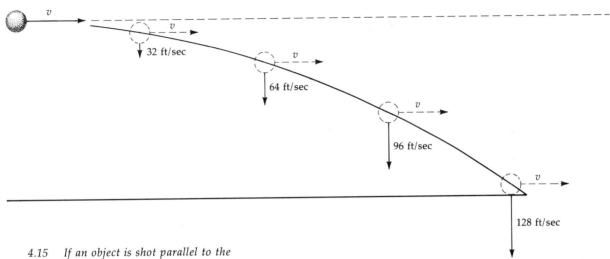

4.14 The ball follows the solid line. Its horizontal and vertical velocities are given at 1-sec intervals.

4.15 If an object is shot parallel to the earth's surface and fast enough, it will go into orbit around the earth. [Part (b) is taken directly from Newton's Principia.]

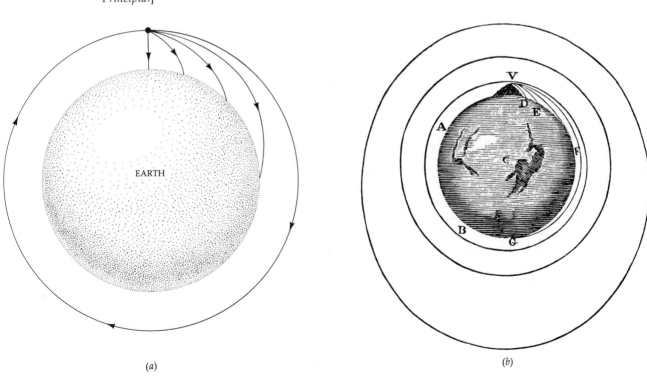

WORK, ENERGY, ORBITAL MOTION 85

$a = 0$
Scale reads W
(a)

a down
Scale reads $< W$
(b)

4.16 *The apparent weight of the object depends upon the acceleration of the elevator.*

we define weight as the force of gravity, objects in the spaceship do have weight. However, they *appear* to a man in the spaceship to have lost their weight. This is a property of all freely falling objects.

We usually judge the weight of an object by measuring how much force is required to hold it in place. However, this force, the supporting force which holds it, is not always equal to the weight of the object. For example, suppose a weight hangs in an elevator, as shown in Figure 4.16. When the elevator is at rest or moving with constant speed, the acceleration of the object is zero and Newton's second law tells us no unbalanced force acts on the object. Hence, when $a = 0$, the upward force of the support must equal the downward force of gravity W. This is the situation in Figure 4.16a where the scale acts as the support. Since the scale reads the force needed to support the object, the scale reads the weight W.

However, if the elevator and the object in it are accelerating downward, $F = ma$ tells us that an unbalanced downward force must be causing this acceleration. Only two forces act on the object, the downward pull of gravity W and the supporting force exerted by the scale. Since the object is accel-

erating downward, W must be larger than the supporting force. Hence, in Figure 4.16b, the scale reading will be less than W. The object will have an apparent weight less than W.

If there were no supporting force at all on the object, the object would fall with acceleration g under its own weight, of course. Now suppose that the cable holding the elevator has broken and the elevator is falling freely. Its acceleration, together with everything in it, would be the free-fall acceleration g. The scale would read zero in this circumstance; the apparent weight of the object would be zero.

An object can appear weightless whenever it and its surroundings are freely falling. In particular, a spaceship in orbit around the earth is falling freely toward the center of the earth. As a result, everything within the space ship appears to be weightless, like the contents of the freely falling elevator. The contents will require no form of support, and will float around inside the cabin. Because the ship is falling freely, the astronauts within it are subjected to the same sensation you may have noticed in a rapid downward start of an elevator, or during a roller coaster ride as the coaster starts down a steep hill.

When the spaceship is coasting toward the moon or back toward the earth, it is freely falling also. Under these conditions, objects within the ship appear weightless too. However, when the ship's rocket engines fire, the ship is accelerated forward. During that time, objects floating in the ship will fall toward the rear of the ship. Can you explain why?

BIBLIOGRAPHY

1. NEWTON: *Principia* (paperback), University of California Press, Berkeley, 1962. Contains Newton's original discussion of planetary motion.

2. ANDRADE, E. N.: *Sir Isaac Newton* (paperback), Anchor Science, Doubleday, Garden City, N.Y., 1954. A readable account of the life and discoveries of Newton.

3. GAMOW, G.: *Gravity* (paperback), Anchor Science, Doubleday, Garden City, N.Y., 1962. A famous physicist noted for his ability to popularize physics discusses our understanding of gravity through the the work of Galileo, Newton, and Einstein.

4. HOLTON, G., and ROLLER, D.: *Foundations of Modern Physical Science,* chapters 11 to 18, Addison-Wesley, Cambridge, Mass., 1958. Chapters 9 to 18 contain material pertinent to this chapter for the student who wishes to pursue these topics further.

5. WENT, F. W.: "The Size of Man," *American Scientist*, vol. 56, no. 400, 1968. An interesting discussion of how size has influenced evolution on the earth.

SUMMARY

When a force F pushes an object through a distance s in the direction of the force, the force does an amount of work $W = Fs$. In the British system F is in pounds, s is in feet, and so W is in foot-pounds. When mks units of the metric system are used, F must be in newtons, s in meters, and W in joules.

A moving object has the ability to do work and therefore has kinetic energy. A mass m slowing from speed v_0 to speed v can do an amount of work equal to $\frac{1}{2}mv_0^2 - \frac{1}{2}mv^2$. The kinetic energy of a moving object is $\frac{1}{2}mv^2$. In order to accelerate a mass m from a speed v_0 to a speed v, the accelerating force must do an amount of work $\frac{1}{2}mv^2 - \frac{1}{2}mv_0^2$.

An object of mass m which has been lifted to a height h required an amount of work mgh to lift it. When it falls back down, it has the capability of doing this same amount of work. Therefore, we say the object has a gravitational potential energy equal to mgh.

The law of conservation of mechanical energy states that the sum of the kinetic and potential energies of an object is a constant as long as no unbalanced forces except gravity act upon it. The effect of forces may be combined with this law to give the work-energy theorem: The original energy of an object plus the work done by accelerating forces minus the work done by stopping forces equals the final energy of the object.

Objects falling through the air are retarded in their motion by the friction force of the air. Since this force increases with speed, it can become large enough to balance the pull of gravity on the object. Hence a body falling through the air will eventually reach a speed large enough so that no unbalanced force exists upon it. The object then ceases to accelerate and continues to fall at a constant speed called the terminal speed of the object.

Power is defined to be the rate of doing work, i.e., power = (work done)/(time taken). Its units are foot-pounds per second or joules per second. The joule per second unit is called a watt. A common unit, the horsepower, is equivalent to 550 ft-lb/sec.

Any system which vibrates undergoes a continual interchange of energy between kinetic and potential. Typical examples are a pendulum and a mass vibrating on a spring.

Newton's law of gravitation states that objects attract each other according to an inverse square law. He used this law to explain the motion of the planets around the sun and the weight of objects.

He showed that an object such as the moon or a satellite circling the earth is actually freely falling toward the center of the earth at all times.

Since a spaceship circling the earth is a freely falling object, everything within it is also freely falling. Hence objects within the ship require no support and appear to be weightless. A similar situation exists in a ship coasting through space.

TOPICAL CHECKLIST

1. What is the scientist's definition of work?
2. What restriction is placed on s in work = Fs?
3. What are the British and metric units for work, for energy, and for power?
4. What is meant by kinetic energy?
5. What is meant by gravitational potential energy?
6. State the law of conservation of energy as given in this chapter.
7. Paraphrase the work-energy theorem.
8. Why do falling objects reach a terminal speed if they fall far enough through the air?
9. Define power and give the units in which it is measured.
10. Discuss what happens to the system's energy as a mass vibrates on the end of a spring.
11. What is the bead-on-a-wire analogy?
12. What sort of force is required to make an object move on a circular path?
13. State Newton's law of universal gravitation.
14. What is an inverse square law force?
15. Why does an object weigh less on the moon than it does on earth?
16. How is an orbiting spaceship similar to a thrown rock?
17. What is meant by weightlessness and how does it arise?

QUESTIONS

1. When an object loses kinetic energy, the energy can go (*a*) to potential energy, (*b*) to doing work on another object, and (*c*) into work against friction forces. Give an example of each case.
2. Give an example for each of the following ways in which potential energy can be lost: (*a*) lost to work against friction; (*b*) changed to kinetic energy; (*c*) lost to doing useful work.
3. A block of wood sits on a flat, but not friction-free, tabletop. If a bullet is shot horizontally into the block, what happens to the bullet's kinetic energy?
4. (*a*) It is possible to push on an object without doing work on it. Give an example to show this. (*b*) It is possible to move an object from place to place without doing appreciable work on it. Explain how.
5. Objects weigh only one-sixth as much on the moon as they do on the earth. Explain why g there is about 5 ft/sec^2.
6. What data would you need in order to estimate the power output of a boy climbing a rope?
7. It has been proposed that the high tides in the Bay of Fundy could be utilized as an energy source. Describe a system that might be used to capture this energy.
8. Reasoning from the interchange of kinetic and potential energy, explain why the speed of a satellite in a noncircular orbit about the earth keeps changing. Is its speed largest when it is at apogee (farthest point from earth) or when it is at perigee (the closest point)?
9. A pendulum has a hollow ball at its end. It is drawn aside and allowed to swing back and forth. On successive trials, the experiment is altered by putting just a little water in the ball, by filling the ball half full of water, and finally

by filling the ball completely with water. In which case will the swinging die down most rapidly? Most slowly?

10 Just for fun consider the following impossible situation. Two men stand at opposite ends of a railroad car and play catch with a baseball. If the car is weightless and if there is no friction in the wheels, describe the motion of the car as the game goes on.

11 A person standing on a scale in an elevator notices that the scale reads more than his actual weight. Assuming the scale to be accurate, is the elevator accelerating upward, going upward at constant speed, going downward at constant speed, or accelerating downward? Justify your answer.

12 Draw a diagram showing a spaceship circling the earth. Show on the diagram what happens if the rocket engines are ignited to speed up the ship in its orbit. Repeat for the case in which the engines are used to slow down the ship. When are each of these two operations performed in the journey of a ship to the moon and its return?

13 A scientist has pointed out that man could not be much taller than he is and survive on earth. For example, if he were 50 ft tall and fell down, he would certainly break a bone. What implication does this have for life on a much more massive planet which has about the same temperature and atmosphere as ours? (See reference 5 in the Bibliography.)

14 What data would you need to determine the mass of the sun? Assume the earth-to-sun distance is known but all other data must be obtained by measurements on the earth.

15 A friend of mine claims that a roller coaster was going down a steep hill so swiftly that she had to hold herself down in the seat. Explain why she had that sensation. Was she correct?

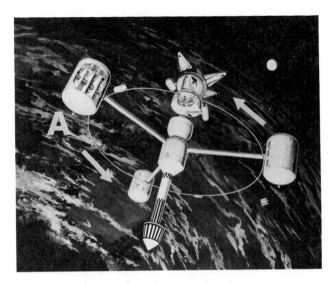

4.17 A rotating space station. (Courtesy of Grumman Aerospace Corp.)

16 It is possible to orbit a satellite above the earth in such a way that it stays above a fixed point on the earth at all times. In what direction must the satellite be shot? Explain.

17 A comet that wanders into our solar system can do one of three things if we consider only it and the sun: (a) it can hit the sun; (b) it can go into orbit about the sun; (c) it can leave the solar system, probably never to return. Qualitatively, what must be different in the three cases to account for these different behaviors?

18 One tentative plan for a space station is shown in Figure 4.17. The spacemen would live and work in the three capsules on the edges of the ring. To simulate gravity, the station would rotate in the direction shown. (a) How could it be set in rotation? (b) Why would this rotation eliminate weightlessness? (c) Which direction would be "down" for a man at point A in the station?

PROBLEMS

1 How much work must you do in carrying a 40-lb child from one end of a football field to

the other? To carry him up a 15-ft flight of stairs?

2. A liter of water has a mass of 1 kg (1 liter ≅ 1.06 qt). How much work must a pump do to fill a 10,000-liter tank with river water if the tank is 20 m above the river surface?

3. How much work is done in accelerating a 2,400-lb car from rest to a speed of 50 ft/sec provided friction forces can be neglected?

4. A 1,500-kg car going at 20 m/sec slams on its brakes and stops. How much work is done against friction forces during the stopping process?

5. How much work do you do in lifting your body up several flights of stairs if the vertical distance from top to bottom is 50 ft? Estimate the shortest time it would take you to run up a set of stairs like this and then compute the horsepower output capabilities of your body.

6. The friction force between a 3,200-lb car's skidding wheels and the level roadway is about 1,000 lb. If the car is going at a speed of 60 ft/sec (about 40 mi/hr), how far will it skid before stopping? (Use the energy method.)

7. If the car in Problem 6 starts from rest and the maximum total force at the rear (traction) wheels is 500 lb, what is the absolutely shortest possible distance in which it can attain a speed of 60 ft/sec?

8. Victoria Falls on the Zambezi River in Africa has a height of 355 ft and about 200,000 lbs of water falls down it each second. If all this water could be used to drive turbines, how much work could it do in 1 sec? What horsepower could the falls produce?

9. Using energy considerations, determine how high a stone will rise if it is thrown straight up with a speed of 50 ft/sec.

10. A 2,400-lb car parked on a hill starts to coast down the hill. If the hill is 300 ft long and 20 ft high, what is the fastest speed possible for the car when it reaches the bottom of the hill?

11. When a 50-ft-long pendulum is pulled aside to an angle of 8° to the vertical, the ball at its end is lifted 0.5 ft above its lowest position. If the pendulum is then released, how fast will the ball be moving when it reaches the lowest position?

12. What is the approximate magnitude of the gravitational force of attraction between you and another person when you are 10 yd apart?

13. If the car in Problem 10 is actually moving with a speed of 10 ft/sec when it reaches the bottom, how much work was done against the brakes and other friction forces as it rolled down? How large was the average friction force?

14. An average person consumes enough food each day so that the food energy he or she derives from it is about 2,000 kilocalories (usually the nutritionists simply call this 2,000 calories). This is equivalent to about 6×10^6 ft-lb of energy. (1 calorie = 4.185 joules = 3 ft-lb.) About how close do you come to doing this much work each day? Where does most of the rest of the energy go?

15. A 20-lb object should weigh slightly more on a spring scale at midnight than it does at noon because of the earth's change in position relative to the sun. Estimate this variation in apparent weight, using the fact that the sun is about 3×10^5 times as massive as the earth and the earth-to-sun distance is about 20,000 times larger than the earth's radius.

16. Write the following in decimal form: 2×10^3; 2×10^{-3}; 1.6×10^{-2}; 500×10^{-2}.

17. The newspapers reported some time ago that when a 20-lb child fell from a window about 60 ft from the ground, a man below caught the child in his arms. Estimate the force the man had to hold to do this. Why would someone making such a catch not stand rigidly?

Chapter 5

HEAT ENERGY AND MOLECULAR MOTION

We have discussed the well-defined motions of one or two easily observed objects. But if we seek to understand the motions of the billions upon billions of molecules of air in a bottle or of the molecules in a drop of liquid, we must devise entirely new techniques. Instead of measuring the velocity and position of each particle separately, we turn to quantities that measure average properties—quantities such as temperature, pressure, and viscosity. We will see that heat energy is molecular kinetic energy and that temperature is one of its measures. Once we understand the meaning of heat, we shall be able to discuss such phenomena as boiling, freezing, and evaporation.

Rumford's Discovery

Everyone knows that heat flows from hot to cold, that heat is necessary to melt or boil a substance, that heat is given off by fire. Indeed, the effects of heat are so widely known that there is no need to discuss them in a text. Instead, we will show why heat leads to these effects. In so doing, we will see once again how science progresses from observation to understanding with the aid of a few fundamental laws of nature.

For a long period before about 1800, people commonly believed that heat was a fluid called *caloric* which all bodies contained. It was supposed that hot objects contained more of this fluid than cold objects, and that when a hot object was placed in contact with a cold object, caloric flowed from the hot to the cold object, cooling one and warming the other. A system became uniform in temperature when each part of the system contained the same amount of caloric. It was not until about 1800 that this concept was disproved by the experiments of Count Rumford.

Rumford was intrigued with the fact that metals became hot when a hole was being bored in them. He noticed that when a cannon barrel is being bored, enough heat is generated to boil water. Upon investigation he found that even if the drill was so dull that it simply rubbed against the metal without making a hole, heat was produced as long as the rubbing action was continued. He concluded that "... any thing which ... can continue to be furnished without limitation, cannot possibly be a material substance [such as caloric was supposed to

HISTORICAL PICTURES SERVICE

COUNT RUMFORD (BENJAMIN THOMPSON)
(1753–1814)

Born in Concord, New Hampshire (then called Rumford), Thompson was a young man when the American Revolution began. After serving briefly in the American army, he grew dissatisfied and deserted to England. For a time, he was subsidized by an English nobleman so that he could carry out scientific research. A few years later he went to the kingdom of Bavaria (now the southern part of Germany), where he became Minister of War and was granted the title by which he is best known. His duties gave him access to the Bavarian arms factory and enabled him to use the heavy machinery and equipment there for experiments involving the nature and properties of heat. Today's military research frequently leads to peaceful applications (nuclear energy, radar, jet travel, and communications satellites are recent examples), and it is interesting to note that such was the case even in Rumford's time.

be]: and it appears to me to be extremely difficult, if not quite impossible, to form any distinct idea of any thing, capable of being excited and communicated, in the manner the heat was excited and communicated in these experiments, except it be MOTION.'"*

Since Rumford's discovery that heat energy is simply molecular kinetic energy, others have extended his understanding immensely, and the existence of molecules and atoms has been proved, a fact disputed until the late 1800s. We shall not attempt to give a historical summary of our acquisition of this knowledge. Suffice it to say that progress was often frustratingly slow and was marred by passionate disagreements among scientists of the time. Instead of an historical approach, then, we shall develop the molecular concept of heat in a more logical way. Our first step will be to learn the relation between temperature and the molecular motion within gases.

Temperature and Its Measurement

There are many forms of thermometers, but, unfortunately, most of them do not agree with each other. Although slipshod construction leads to most of the variations, there is a fundamental reason why even a precisely made mercury thermometer will not always agree exactly with a thermometer which uses alcohol in its tube. This difference goes back to the way in which temperature scales are defined. In essence, we have agreed on two universal temperatures and then defined all others in relation to these.

For many years, it was agreed that the freezing and boiling points of water should be the calibration points for temperature scales. As shown in Figure 5.1, two marks are placed on a thermometer column to show the level to which the thermometer liquid

---
* W. Magie, *Source Book in Physics,* Harvard University Press, Cambridge, Mass., 1962, p. 151.

rises when placed in ice water and in boiling water. These are the fixed points on the scale. In the United States and many English-speaking countries, these two points are labeled on the Fahrenheit scale. On this scale, freezing water is assigned a temperature of 32°F, and water boiling at sea level is assigned the value 212°F. There are 180 Fahrenheit degrees between these two temperatures, and the thermometer tube is marked off in 180 divisions between the 32- and 212-degree marks. Equal-size divisions are extended above and below these temperatures as well.

Most people, however, do not use the Fahrenheit thermometer; they make use of the centigrade (or Celsius) scale. On this scale, the boiling point of water is 100°C, and the freezing point is 0°C. As a result, the region that is divided into 180 degrees on the Fahrenheit thermometer is divided into 100 degrees on the centigrade scale. Hence, a centigrade degree is a factor nine-fifths larger than a Fahrenheit degree. The method for conversion between these two scales is given in Appendix 2. Unfortunately, when different liquids are used in thermometers, the thermometers will not agree when reading most temperatures other than those two calibration temperatures. Most liquids do not expand with temperature* in exactly the same way or even in exact proportion to the temperature. For example, suppose a thermometer with alcohol in it is placed in a 0°C bath and a 100°C bath and calibrated at those points. If you now compare it with a similarly calibrated mercury thermometer by placing both in a bath which reads 50°C on the mercury thermometer, the alcohol will not rise exactly to the height of the 50° calibration on the alcohol thermometer. Although this slight disagreement is not important for most purposes, it prompted scientists to seek a more precise and fundamental way to measure temperature. As a result, the gas thermometer was developed.

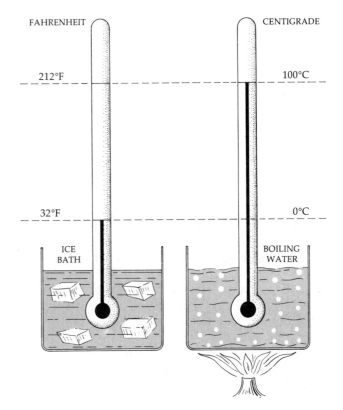

5.1 A possible way to calibrate a thermometer.

---

* The reason why liquids expand as temperature increases will be discussed later.

Gas Pressure and Absolute Temperature

The gas thermometer works because the pressure of a confined gas increases as its temperature is increased. Figure 5.2 shows a diagram of a gas thermometer. The thermometer bulb is a gas-filled container attached to a device to read the pressure of the gas. (We will discuss how pressures are measured when we describe the operation of the barometer.) The volume of the gas is kept constant by proper design of the pressure-measuring device.

If we calibrate the gas thermometer by measuring the gas pressure at the temperature of melting ice and at the temperature of boiling water, we can draw the two-point graph shown in Figure 5.3. From this graph we can define any other temperature of the thermometer. For example, if the pressure is found to be P_1 as shown on the graph, then the temperature will be T_1. As with the mercury thermometer, we are simply subdividing the interval between 0 and 100°C into 100 equally spaced degrees. Temperatures outside this range are obtained by extending the scale.

As we have seen, because of the way thermometers are calibrated, they all agree at 0°C and at 100°C, but not at intermediate temperatures. The advantage of the gas thermometer is that *all gas thermometers agree with each other*, provided the gas pressure is not too high and the gas is not close to condensing. When three different gases, or three different amounts of gas, are used in the thermometer of Figure 5.2, the calibration graphs appear as indicated in Figure 5.4. Notice that the pressure of the gas appears as though it would be zero if the gas could be cooled to −273°C. This temperature is called *absolute zero*. In practice, all gases condense to liquids before they can be cooled that far.

A temperature scale that has zero degrees at absolute zero would appear to be more fundamental than either the centigrade or Fahrenheit scale. This

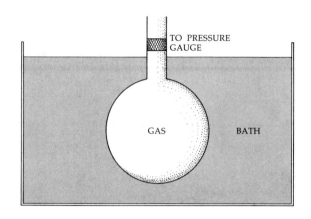

5.2 The temperature of the gas is measured by noting the gas pressure.

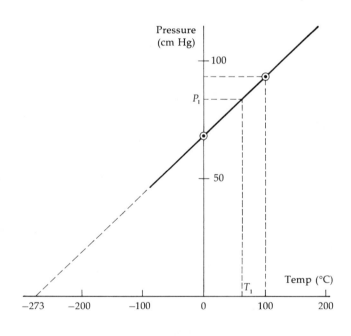

5.3 The pressure of the gas in Figure 5.2 varies linearly with temperature.

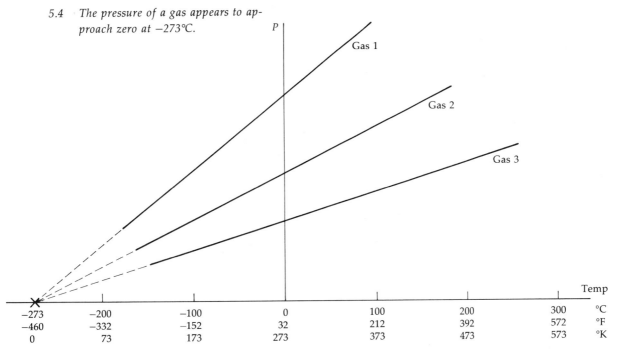

5.4 *The pressure of a gas appears to approach zero at −273°C.*

turns out to be true. The scale which uses absolute zero as one of its calibration points* is called the <u>absolute</u> or <u>Kelvin</u> temperature scale (measured in degrees Kelvin, or °K). Its degrees are the same size as those used for the centigrade scale, and so it is simply shifted by 273 degrees from that scale. It is shown in Figure 5.4 along with the centigrade and Fahrenheit scales.

In terms of the Kelvin temperature scale, the graph of Figure 5.4 for the relation between pressure P and temperature T for a gas can be expressed in an equation. Since P is a linear function of T and since both go to zero together (*if* we use degrees Kelvin), we have

$$P = (\text{const})T$$

Notice that this equation is true only if T is measured in degrees Kelvin because only on this scale is $P = 0$ when $T = 0$. The constant can be evaluated by measuring P at a known temperature. Experiment shows that the constant depends upon the number of molecules N in the volume of the container V, and that

$$PV = NkT \qquad (5.1)$$

where $k = 1.38 \times 10^{-23}$ joule/°K. This important constant of nature is called *Boltzmann's constant*. Equation (5.1) is one form of the <u>ideal gas law</u>. In using it, T must be in degrees Kelvin.

In the next section our discussion of the gas law, Equation (5.1), will give us an understanding of heat and temperature. In addition, this relationship has great practical importance since it tells us how a gas behaves. For example, suppose we wish to find how the pressure of the gas in a closed container will change as its temperature is changed from T_1 to T_2. If P_1 is its pressure at temperature T_1 and P_2 its pressure at T_2, we can write Equation (5.1) twice, once for each temperature:

$$P_1 V = NkT_1$$
$$P_2 V = NkT_2$$

---

* The other calibration point is the unique temperature at which water, ice, and water vapor can coexist. This complication need not concern us here. However, we should mention that the Celsius temperature scale uses these calibration points. For our purposes, the Celsius and centigrade scales are identical.

HEAT ENERGY, MOLECULAR MOTION

Since the container is closed, the number N of molecules in the fixed volume V does not change. Because two equal quantities are still equal after dividing by equal quantities, we can divide one equation by the other to give

$$\frac{P_1 \cancel{V}}{P_2 \cancel{V}} = \frac{\cancel{Nk}T_1}{\cancel{Nk}T_2}$$

or

$$\frac{P_1}{P_2} = \frac{T_1}{T_2}$$

which tells us that the pressure varies proportionately to the Kelvin temperature in this situation. Other examples of the gas law are given below.

EXAMPLE 5.1

What happens to the pressure of the air in a tire as the temperature of the tire rises from 0 to 100°C.

Method: If we assume that the tire's volume remains constant, the relation obtained above applies. $T_1 = 0 + 273 = 273°K$ and $T_2 = 100 + 273 = 373°K$. Then, after cross-multiplication,

$$P_2 T_1 = P_1 T_2$$

and so

$$P_2 = \tfrac{373}{273} P_1 = 1.37 P_1$$

The tire's pressure rises by nearly 40 percent. In view of this, can you give a reason why blowouts occur when tires get too hot? (The strength of rubber decreases at high temperatures and this also is a factor of importance.)

EXAMPLE 5.2

If we decrease the volume of gas in a cylinder to 0.10 its original volume by pushing in a piston, by how much does the pressure of the gas change providing the temperature remains constant and no gas escapes?

Method: Again using $PV = NkT$, the original P_1 and V_1 change to P_2 and $0.10V_1$ after compression. Hence,

$$P_1 V_1 = NkT \quad \text{and} \quad P_2(0.10 V_1) = NkT$$

Dividing one equation by the other gives

$$\frac{P_1 \cancel{V_1}}{P_2(0.10 \cancel{V_1})} = \frac{\cancel{NkT}}{\cancel{NkT}}$$

Or, after cross-multiplying, $P_2 = 10 P_1$. The pressure of the gas becomes 10 times larger.

Molecular Interpretation of Gas Pressure

A gas consists of many tiny molecules shooting at high speeds in random directions. We know from experience that the molecules are in motion. For example, if a bottle of perfume is opened in a room, the perfume molecules soon reach all corners of the room, as we can tell from the scent of the perfume. Since the perfume molecules do not settle to the floor, they must have much more kinetic energy than that required to lift them from floor to ceiling.

We can infer something about the size of gas molecules from the observation that they fill a room from floor to ceiling. We know that smoke and dust particles eventually fall out of the air and form a layer on surfaces to which they fall. Although the smaller the particles, the less this tendency, even the smallest particles we can see eventually fall out

5.5 *When the molecules hit the container wall, they exert a force upon it.*

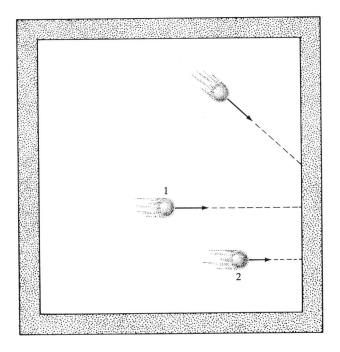

of the air. Since the air molecules themselves do not deposit on the ground, they must be smaller than the smallest dust particles.

From these observations and from more scientific experimental results, we picture gas molecules as exceedingly tiny, ball-like entities which move about rapidly. If we have a container filled with many gas molecules flitting about, we can easily see why the gas exerts a pressure on the walls of the container. As shown in Figure 5.5, the molecules will strike the wall of the container, and thus exert a force on it. Since *pressure* is defined to be *the force on the wall divided by the area of the wall (i.e., pressure = force/area)*, The force on the wall gives rise to an equivalent pressure on the wall.

A thimble full of air has about 10^{18} (a billion billion) air molecules, each moving at a speed of close to 500 m/sec (or 500 yd/sec). Therefore, to be realistic, Figure 5.5 would have to show billions of molecules hitting the container wall each second. This explains why we do not detect the individual impacts of the molecules with the wall. The pressure is the cumulative effect of billions of small impacts.

Clearly, the pressure exerted on the wall by the gas will be proportional to three quantities:

1 The number of molecules hitting the wall in a unit of time
2 The average speed of the molecules, v
3 The mass of a molecule, m

We would expect the pressure to double if the number N of molecules in the container doubled, since there would be twice as many impacts and therefore twice the force at any given time. However, if the volume V were doubled, the number of molecules close to the wall would be reduced by half and so would the number of impacts on the wall. Hence, the number of impacts with the wall is proportional to the number of molecules per unit volume, N/V, not just to N alone.

In addition, the number of impacts each second depends upon how fast the molecules are moving. For example, if the average speed of the molecules in Figure 5.5 were small, perhaps only the molecule labeled 2 would strike the wall in a time of 1 sec. Molecule 1 would be going too slowly to reach the wall. But if the molecular speed were high, both 1 and 2 would strike the wall in a second. Therefore, the number of molecules hitting the wall each second is proportional to the average molecular speed v as well as to N/V. Quantity 1 listed above will therefore be proportional to Nv/V.

Since the gas pressure P is proportional to the three quantities listed above, namely Nv/V, v, and m, we can write

$$P = (\text{const})\frac{Nv}{V}vm$$

After multiplying through the equation by the container volume V, we have

$$PV = (\text{const})Nmv^2$$

HEAT ENERGY, MOLECULAR MOTION

More advanced texts show (see reference 1, page 108) that the constant in this equation is simply $\frac{1}{3}$. Therefore

$$PV = \tfrac{1}{3}Nmv^2 \qquad (5.2)$$

This tells us how the pressure of a gas is related to the behavior of the molecules within the container. Thus the pressure of a gas increases in proportion to the number of molecules per unit volume, N/V, the mass of a molecule, m, and the square of its speed, v^2.

You may already have noticed that Equation (5.2) is very similar to the gas law, Equation (5.1). The gas law came directly from experimental observations and our definition of temperature, whereas Equation (5.2) came from our deductions of what a gas must be. Thus, Equation (5.1) is an *experimental* relation based directly on observation, and Equation (5.2) is a *theoretical* relation, based on theoretical considerations of the behavior of a gas composed of ball-like molecules. Let us compare them:

$$PV = NkT \qquad (5.1)$$
$$PV = \tfrac{1}{3}Nmv^2 \qquad (5.2)$$

Since two things equal to the same thing are equal to each other, we can equate the right sides of these equations. We then find

$$NkT = \tfrac{1}{3}Nmv^2$$

which can be rearranged to give

$$T = \frac{1}{3k}(mv^2) \times 2 \times \tfrac{1}{2}$$
$$T = \frac{2}{3k}(\tfrac{1}{2}mv^2) \qquad (5.3)$$

We multiplied the equation by $(2) \times (\tfrac{1}{2})$, which is unity, so that we could introduce the factor $\tfrac{1}{2}$ to yield $\tfrac{1}{2}mv^2$, the kinetic energy of a molecule. Moreover, k is Boltzmann's constant, simply a number.

Hence we can conclude:

Kelvin (or absolute) temperature is a direct measure of molecular kinetic energy.

We see from this that temperature has a fundamental molecular meaning. When a gas is hot, the gas molecules have large kinetic energies, so their speeds are high. When a gas is cold, the gas molecules move slowly; they have very little kinetic energy. When two containers of gas have the same temperature, the average kinetic energies of the molecules are the same in both containers. Even if the gases are different chemically, say nitrogen and oxygen, the average kinetic energies* are identical. At $T°K$ the average kinetic energy $\tfrac{1}{2}mv^2$ is equal to $3kT/2$ for a gas molecule, independent of the chemical composition of the molecule.

Now we can see the importance of $T = 0°K$, absolute zero. When $T = 0°K$, the molecular kinetic energy must be zero according to Equation (5.3). Therefore all molecular motion should cease at absolute zero. For this reason the pressure of a gas decreases toward zero as T approaches absolute zero. Before that temperature is reached, however, the molecules of a gas will have too little energy to pull away from each other. Thus, long before absolute zero is reached, all gases condense into liquids. As we will see in Chapter 10, a law of nature discovered in the mid 1920s tells us that molecular motion must be present even at absolute zero. In spite of this complication at extremely low temperatures, we can say for all temperatures with which we shall be concerned that *temperature is a measure of molecular kinetic energy.*

EXAMPLE 5.3

Find the average speed of the nitrogen molecules in air at 27°C. The mass of a nitrogen molecule is $28 \times 1.67 \times 10^{-27}$ kg. (Recall that 10^{-27} means to divide by 10 twenty-seven times.)

---

* Strictly speaking, these are the energies of translational motion, or straight-line motion, of the particle.

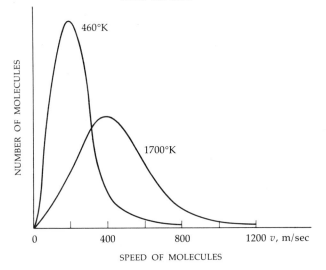

5.6 Although these curves for the distribution of speeds in a gas composed of mercury atoms appear to go to zero at large speeds, they do not quite touch the axis.

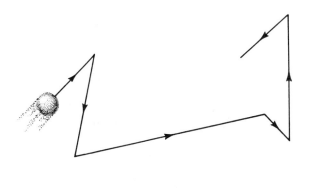

5.7 The small smoke particle is seen to follow a zigzag path when it is observed with a microscope; this type of motion is called Brownian motion.

Method: Our previous discussion shows that

$$\tfrac{1}{2}mv^2 = \tfrac{3}{2}kT$$

where $T = 27 + 273 = 300°K$ and $k = 1.38 \times 10^{-23}$ joule/°K. We then have

$$\tfrac{1}{2}(45 \times 10^{-27} \text{ kg})v^2 = \tfrac{3}{2}(1.38 \times 10^{-23} \text{ joule/°K})(300°K)$$

Doing the arithmetic, we find $v \cong 540$ m/sec.

The Kinetic Theory

Thus far we have been concerned only with the average speed of gas molecules. If you make a mental picture of molecules bumping into each other as they shoot through space, however, you can see that it is unrealistic to think that they will retain their original speeds. Instead, we would expect them to have a wide range of speeds. Long before it was technically feasible to measure the speeds of molecules, the exact way the speeds should vary had been predicted. This prediction is shown in graphical form in Figure 5.6.

We see from Figure 5.6 that a gas has a wide range of molecular speeds. For a gas of mercury at 460°K (187°C), most of the molecules have speeds near 200 m/sec. A few molecules have speeds near zero, and these are nearly standing still. (Typically, such a molecule will be struck and set in motion by another molecule in a fraction of a second.) Other molecules are moving much faster than 200 m/sec. The curve seems to go to zero at large speeds, but it does not quite touch the axis, and so a few particles will predictably have exceedingly high speeds. Since we know that increased temperature leads to increased kinetic energy, the difference between the 460 and 1700°K curves shown in the figure is to be expected. Notice that most of the molecules at 1700°K have speeds of about 400 m/sec.

We can make other predictions for our model of a gas. For example, suppose a smoke particle large enough to be seen in a microscope (far larger than an air molecule) is floating through the air. A few air molecules will strike it each second. Under the impact of such collisions, the smoke particle will be pushed here and there. Its path as seen in a microscope should be a zigzag as shown in Figure 5.7.

HEAT ENERGY, MOLECULAR MOTION

THE BETTMANN ARCHIVE

JAMES PRESCOTT JOULE (1818–1889)

Joule inherited a prosperous brewery in Manchester, England; he pursued science as an avocation. He carried out many experiments showing the relations between electrical, mechanical, and chemical effects. Scientist friends, recognizing the young brewer's abilities, urged him to apply for a professorship at St. Andrews, Scotland. But one of the university's electors held that Joule's slight physical deformity disqualified him, and so he was rejected. Undaunted, he continued his experiments and in 1847 gave a lecture describing his work. Largely because of his amateur status, his discoveries were ignored. Later that year, he was allowed a few minutes to present a paper before a scientific meeting at Oxford. Just when it seemed he would be ignored once again, a well-known scientist prolonged the discussion and soon it became evident that the paper had great merit. In the end, the paper caused a great sensation and, almost overnight, Joule became a scientist of note. He is most well known today for his determination of the mechanical equivalent of heat energy. In addition, he was one of the founders of the principle of conservation of energy.

Robert Brown, a botanist, first observed this type of motion in 1827; it is called *Brownian motion* in his honor.*

Our model of molecular motion in gases is known as the *kinetic theory of gases*. Many investigators during the middle and late 1800s worked on its development. As early as 1851, James Prescott Joule (1818–1889) showed that the pressure of the air could be interpreted in terms of nitrogen molecules (since the air is mostly nitrogen) moving with speeds of a few thousand feet per second. However, Ludwig Boltzmann (1844–1906) and the great physicist James Clerk Maxwell (1831–1879) were the first to formulate the motions and behavior of gas molecules in a precise mathematical theory.

In those days, no one could be sure that the kinetic theory was anything more than the product of Boltzmann's imagination. In 1898, criticized on all sides for a theory based upon the unproved existence of molecules, Boltzmann wrote: "I am conscious of being only an individual struggling weakly against the stream of time." His critics maintained that a theory involving hypothetical, invisible molecules was pure fantasy. (They had a point, because at the time there was no direct experimental proof for the existence of molecules, even though chemists had been remarkably successful in explaining chemical reactions in terms of molecules.)

Not until 1908 did scientists develop experiments that would allow them to test the detailed predictions of the kinetic theory. (These experiments came too late to console Boltzmann, who committed suicide in 1906.) Boltzmann's theory predicted exactly how the number of gas particles should change with height above the earth, but he could not confirm this prediction with experiments.

---

* Brown actually observed the effect in droplets of water that contained pollen dust particles. The water molecules in the liquid act like gas molecules. Brown originally believed the motion of the pollen particles was a characteristic of the "life" carried by the genetic material in the pollen. He later discarded this idea when he noticed that even powdered stone showed evidence of this supposed "life."

LUDWIG BOLTZMANN (1844–1906)

Boltzmann was born in Vienna, Austria, and lived there until he graduated from the University in 1866. Thereafter, he held professorships at Graz, Munich, Leipzig, and Vienna. The molecular picture of gases is a monument to his life's work, and much of our present understanding of molecular motion grows out of his theories. But in spite of his achievements, Boltzmann had the misfortune to live in a time when the molecular picture of matter was under heavy attack. A popular school of thought, led by the great scientist-philosopher Ernst Mach, held that atoms and molecules were imaginary constructs and as such had no place in a philosophically sound science. Boltzmann was unable to rally support to his views, especially since experimental evidence of the existence of molecules was not available. He became increasingly subject to depression and ultimately killed himself. Less than three years after his death, researchers achieved the first of a series of direct experimental proofs of his molecular theories.

However, the theory also predicted that if very fine particles were dispersed in a column of liquid the number of particles should be less at increasing heights in the column. In 1908 Jean Perrin confirmed this prediction by measuring the Brownian movements of particles dispersed in such a manner.

Other measurements confirmed the molecular picture of matter shortly thereafter. The most elegant proof of the kinetic theory came in 1926 when the creation and use of vacuums had developed sufficiently for Otto Stern to measure the speeds of gas molecules directly. Stern showed that the speeds of the molecules in a gas agreed precisely with the kinetic theory curves shown in Figure 5.6.

The Earth's Atmosphere

Although we seldom think about it, our bodies are continuously subjected to the pressure of the air molecules around us. You may protest that this pressure is too small to concern us, but we only need to perform the experiment in Figure 5.8 to appreciate how large this pressure is. As we see there, when the air is removed from inside the metal can, the unbalanced force of the air on the outside causes the can to collapse. The pressure of the air is obviously large. Indeed, the force of the air pressing upon your chest is about 1,000 lb.

We do not ordinarily notice this huge pressure on our bodies for the same reason that the metal can did not collapse until after air was removed from it. Before, the pressure of the air within the can was the same as that outside. Hence, each surface had equal and opposite forces pushing on it; the force of the air inside balanced the opposite force exerted by the outside air. Similarly, the air pressure inside our lungs and other body cavities balances the air pressure from outside. What do you think happens to a child's gas-filled balloon when it is released, rising into the reduced air pressure high in the atmosphere?

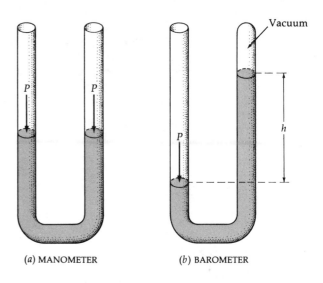

(a) MANOMETER (b) BAROMETER

5.9 *In a barometer, the height h of mercury balances the atmospheric pressure and is a measure thereof.*

5.8 *When the air is removed from the can, it collapses.*

As you know, the air becomes thinner as we go higher above the earth. At a height of 5 mi, the air is only about 0.2 as dense as it is on the surface of the earth; at 100 mi, the density is only about one billionth as large as at sea level. We can understand this variation in density if we recall that a molecule must do work against gravity to lift itself above the earth's surface. The kinetic energies of the air molecules are changed to potential energies as the molecules move upward. If all the molecules in the air had the same energy, the atmosphere would end abruptly at the maximum height to which they could ascend with that energy. However, as we saw in Figure 5.6, molecules have a wide range of energies. As a result, some are capable of rising much higher above the earth than others.

Even at sea level, the pressure of the air varies. When the barometric pressure is given in a weather report, the forecaster is telling you the pressure of the air. As you may know, rapidly falling barometric pressures presage bad weather. The tool used to measure air pressure is the <u>barometer</u>. When a liquid such as mercury is placed in a U-shaped tube, as shown in Figure 5.9a, the mercury stands at equal levels in the two sides. The air pressure on the surface of one column of mercury balances the pressure on the surface of the other column. However, if the air is pumped out of one side, as shown in (b), the mercury will rise until the extra height h of mercury on one side balances the air pressure on the other side. The larger the air pressure P, the larger the height h will have to be to balance it.

At sea level, this balancing height is close to 76 cm, or 29.9 in. The air pressure (or barometric pressure) is then said to be 76 cm of mercury (or 76 cm Hg, where Hg is the chemical symbol for mercury). Atmospheric pressure is lower in the higher altitudes, of course, since the air is less dense. For example, in the "mile high" city of Denver, atmospheric pressure is close to 60 cm of mercury. This lower pressure affects cooking (water boils at a lower temperature at high altitude) and can cause angina heart pains (the air is less dense at high altitude and so less oxygen reaches the lungs with each breath).

EXAMPLE 5.4

If the average air molecule at 27°C could go straight upward without striking other molecules, how high would it go? Take the mass of an air molecule to be $28 \times 1.67 \times 10^{-27}$ kg.

Method: In moving upward, the molecule will lose all its thermal kinetic energy (energy due to heat), $\frac{3}{2}kT$, to potential energy. We can write

$$\text{Thermal KE at bottom} = \text{PE at top}$$
$$\tfrac{3}{2}kT = mgh$$

Since $T = 27 + 273 = 300°$K, we find

$$\tfrac{3}{2}(1.38 \times 10^{-23} \text{ joule/°K})(300°\text{K})$$
$$= (28 \times 1.67 \times 10^{-27} \text{ kg})(9.8 \text{ m/sec}^2)h$$

Solving for the height h to which the molecule will ascend yields $h = 14,000$ m, which is about 9 mi.

Heat Energy in Liquids and Solids

Now that we understand something of the thermal behavior of gases, let us turn our attention to liquids and solids. As in gases, the temperature of liquids and solids measures the kinetic energies of the molecules. In these cases, the motions of the molecules are not as simple as those in a gas. Even so, the fact that molecules in hot substances have higher kinetic energies than those in cold substances enables us to understand much of the thermal behavior of these substances. Let us see what happens in a metal rod as one end is heated.

We know that the heat will travel along the rod from the hot to the cold end; eventually the whole rod becomes hot. What has happened is this: The hot (i.e., high-energy) molecules in the flame that is heating the rod strike and impart kinetic energy to the molecules of the rod. These molecules* then strike adjacent molecules and pass energy on to them. They, in turn, strike the cold, low-energy molecules to which they are adjacent, and so on. As a result, the kinetic energy, called thermal or heat energy in this context, is transmitted along the rod and even the distant end becomes warm. This process is called _heat conduction._

Another phenomenon is the fact that liquids and solids expand when heated; this is known as _thermal expansion._ We have seen that when substances are at a temperature higher than absolute zero, the molecules within them are in motion. This motion is primarily a vibration of the molecule as it bounces back and forth between the neighbor molecules pressing against it. The greater the energy of the molecule, the larger the space it can preserve for itself against the intrusion of the neighbor molecules. When a liquid or solid is heated, the increased

---

* In metals, a large part of the energy is carried along the rod by the free electrons rather than by the molecules themselves. The process is the same in either case.

5.10 The liquid on the left has a much higher viscosity than that of the liquid on the right.

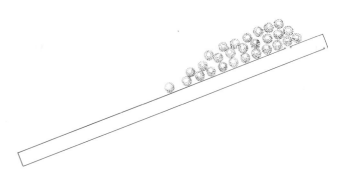

5.11 The molecules of the liquid will flow over each other most readily when the temperature of the liquid is high.

energy of the molecules allows them to keep their neighbors farther away. Hence the solid or liquid expands as it is heated.

A similar phenomenon involves a liquid's *viscosity*, which characterizes how it flows. For example, the liquid shown on the left in Figure 5.10 flows much less easily than that on the right; we say that the liquid on the left has a *high viscosity*; it is very viscous; it flows with great difficulty. To understand this phenomenon, let us refer to the idealized model of a flowing liquid shown in Figure 5.11. As we shall see in Chapter 13, the molecules of a liquid are attracted to each other. For this reason, the molecules shown in the figure do not easily slide over each other, and so the liquid does not flow instantaneously down the incline. However, the molecules do have kinetic energy. They periodically pull loose from each other as a result of their thermal motions, and so the liquid flows. Clearly, the hotter the liquid, the more fluid it will be. This is why oils and syrups, for example, flow more readily when they are hot than when they are cool. For most substances, *viscosity decreases with increasing temperature.*

Evaporation and Boiling

As pointed out in the last section, most molecules attract molecules like themselves. Nitrogen molecules attract each other, as do water molecules. This mutual attraction causes *condensation*. When the molecules in a gas are cooled sufficiently, their kinetic energies become too low to allow the molecules to escape from each other. As a result, the attraction forces between several water molecules in the air, for example, bind them together in a droplet. Fog forms by this process when the temperature drops rapidly on a humid day. All vapors (i.e., gases) condense into droplets of liquid if they are cooled enough.

A liquid, then, is a collection of molecules whose kinetic energies are too small to allow them to break loose from each other. But we know that all liquids tend to evaporate; therefore some of their molecules must actually have enough energy to break loose from the others. From Figure 5.6 we know that a few of the molecules possess speeds (and therefore energies) far above the average molecular speed and energy. These highly energetic molecules are able to escape from the surface of a liquid and thus cause the phenomenon known as *evaporation.*

As you know, evaporation causes cooling. For example, when perspiration evaporates from your body, you cool down. Refrigerators work because of the cooling action of evaporation. Let us see why this cooling action occurs. Since only the most energetic molecules in a substance evaporate, the molecules left behind are those with the lowest energies, and so the average energy of the substance decreases. At the same time the temperature decreases, since the temperature of a gas or liquid is measured by the kinetic energy of its molecules. The liquid is therefore cooled as evaporation takes place.

In summary, then, a condensed gas (a liquid) is simply a collection of molecules each of whose kinetic energy is too small to allow it to break loose from the others. The molecules still possess kinetic energy in proportion to the temperature of the liquid. Since they cannot now free themselves, they vibrate back and forth between their neighbor molecules; they rotate and wiggle and move from place to place within the liquid, but by erratic and complex paths. Only a few of the most energetic ones escape from the surface when they happen to wander to it.

It is impossible to say exactly how much energy a particular molecule in a liquid will have at any given instant because the motions of the molecules are far too erratic and complicated to analyze in specific detail. But we can state the *average* energy of a molecule at a given temperature. We can also estimate the fraction of molecules which have energies within a certain range, but we are unable to say exactly which molecules will possess these energies. As a result, we can predict that certain things will happen in a liquid but we can't say where or when they will happen.

For this reason, boiling is also unpredictable. You know that a liquid is boiling when bubbles within it rapidly form and grow continuously. These bubbles are filled with a gas of the molecules composing the liquid. When a liquid boils, a number of molecules with high energies congregate by chance in the same small region of the liquid. Their energies are large enough so that they break loose from each other, forming a tiny gas bubble. Clearly, the gas inside the bubble must have a pressure large enough to keep the bubble from collapsing under the pressure of the liquid and the atmosphere above it. (For this reason, boiling occurs less easily in a pressure cooker, where the pressure is high, than in an open pot. Consequently, liquids can be heated to a higher temperature in a pressure cooker before they boil, and so the food cooks more rapidly. An egg in an open pot will cook faster at sea level than in Denver, which is high in the mountains. Can you explain why?)

Not only must the pressure inside the bubble be large enough to keep the bubble from collapsing, but the bubble must grow. More molecules must evaporate into the bubble and stay there than are colliding with the surface of the bubble and returning to the liquid. Thus the pressure of the evaporating molecules (the *vapor pressure* of the liquid) must exceed the pressure of the atmosphere and surroundings, which in turn determine the pressure inside the bubble. Clearly, *boiling occurs only if the vapor pressure of the liquid (the pressure of the evaporating molecules) exceeds the pressure of the atmosphere upon the liquid.* Since the vapor pressure increases with temperature, the boiling temperature of a liquid will increase as the outside pressure on the liquid increases.

If you cook, you are familiar with a phenomenon known to chemists as "bumping" of a liquid. Frequently, when liquids are heated, boiling occurs with almost explosive violence. In such cases the soup, for example, suddenly starts to boil and throws the lid off the pot. What has happened is this. In order for a liquid to boil, some of the more energetic molecules must cooperate to form a small bubble; this is a random process, occurring only by chance. In a very pure or uniform liquid, this process sometimes does not occur until the temperature slightly exceeds the usual boiling temperature. Then, when a bubble does form, it grows explosively—forming many other bubbles in the process—and the liquid boils extremely vigorously until evaporation lowers the temperature to the normal boiling temperature of the liquid. This is a spectacular example of <u>*super-heating*</u> a liquid, that is, heating it above its boiling temperature. Boiling, therefore, takes place when molecular energies are so high that the liquid literally tears itself apart. As a liquid is cooled below its boiling point, its molecules lose energy and their motions slow. The viscosity of the liquid increases, since the molecules are no longer as capable of breaking away from their neighbors. Some materials that are markedly viscous near room temperature are molasses, asphalt, and grease. Materials such as window glass are already so viscous at room temperature that they appear solid. If you examine windows in very old buildings, you may find panes that have thinned at the top and thickened at the bottom; the glass has flowed slowly through the centuries. Many transparent plastics (Lucite, Plexiglass, polystyrene) are also glasslike liquids.

Crystalline Solids

Before cooling enough to appear glasslike, many liquids <u>*crystallize*</u>. Whereas the molecules in liquids are thrown together at random, the atoms within a crystal are arranged in a precise, ordered pattern. At low temperatures, the forces between individual molecules can hold the molecules in definite positions relative to one another. For example, in a crystal of table salt (NaCl is its chemical formula), the sodium (Na) and chlorine (Cl) atoms are arranged in the pattern (or lattice) shown in Figure 5.12*a*. The light balls represent the chlorine atoms and the dark balls represent the sodium atoms. (Could it be the other way around?) Actually, the atoms pack tightly against each other, unlike the balls in the figure.

The sodium chloride lattice is called a *cubic lattice*; that is, the atoms are arranged in cubes throughout the crystal. Of course, the figure shows only a tiny portion of the crystal. An ordinary grain of table salt will have perhaps a billion billion little cubes, whereas the one shown has only 27 cubes in it. It is useful to remember that in solids and liquids, the atoms are about 1 or 2 Å (angstroms) apart. An *angstrom* is a unit of length 0.000,000,01 (or 10^{-8}) centimeters long; we use it when speaking of atoms because atoms have diameters of about 1 Å.

There are many other crystal forms besides the simple cubic structure of sodium chloride. For example, sodium metal exists in cubic crystals of this same type (all balls being sodium atoms) but with an additional atom at the center of each cube. This is called a body-centered cubic lattice. Other crystal forms are also shown in Figure 5.12.

Because of this ordered array of atoms in a crystal, the crystal itself will frequently take on beautiful, sometimes highly intricate, shapes. Two examples are given in Figure 5.13. Sometimes the shape of the crystal is similar to the shape of the atomic groupings within the crystal. But, as you can see from the crystal of ice, this need not be the case.

Even in a crystal, the atoms undergo thermal motion. However, they are constrained to vibrate back and forth around their average lattice position. The distance through which they vibrate is only a fraction of the normal distance between atoms, and the frequency with which they vibrate is extremely high, on the order of 10^{13} complete vibrations each second.

If the temperature of a crystal is raised high enough, the atomic vibrations become so energetic that the crystal breaks up. The material melts, i.e., changes to a liquid. As you already know, this happens at a temperature of 32°F (0°C) for ice melting to water. The crystal must melt completely before its temperature rises above the melting temperature. This is why well-stirred ice water will remain at 0°C until all the ice melts.

5.12 *Typical crystal lattices.*

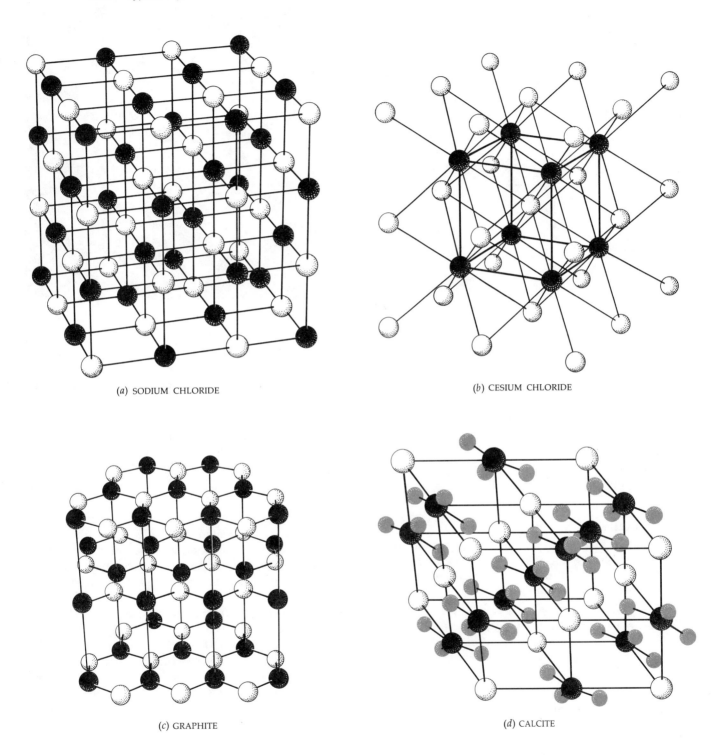

(*a*) SODIUM CHLORIDE

(*b*) CESIUM CHLORIDE

(*c*) GRAPHITE

(*d*) CALCITE

HEAT ENERGY, MOLECULAR MOTION 107

5.13 A calcite crystal (a) and an ice crystal (b); each crystal contains more than 10^{18} atoms. [Part (b) courtesy of W. A. Bentley, NOAA.]

(a)

(b)

In this chapter we have examined the molecular behavior of gases, liquids, and solids with respect to heat energy. Although we have invoked the presence of forces between the molecules, we have said little about the nature of these forces. The forces between atoms and molecules are far stronger than the force of gravitation; they are basically electrical in nature. We shall begin our study of electricity in the next chapter; then we can better appreciate the interactions of molecules and atoms.

BIBLIOGRAPHY

1. BUECHE, F.: *Principles of Physics,* 2d ed., McGraw-Hill, New York, 1972. Chapters 10 to 12 have a more detailed treatment of the material considered in this chapter.

2. MAGIE, W. F.: *Source Book in Physics,* Harvard University Press, Cambridge, Mass., 1962. This contains the nonmathematical writings of Brown, Rumford, etc.

3. BROWN, S. C.: *Count Rumford* (paperback), Anchor Science, Doubleday, Garden City, N.Y., 1962. A readable account of the life and work of the pioneer in the field of heat.

4. HOLTON, G., and ROLLER, D. H.: *Foundations of Modern Physical Science,* Addison-Wesley, Cambridge, Mass., 1958. Chapters 19 to 21 and 25 contain a semihistorical discussion of the material in this chapter at a somewhat higher level.

5. MACDONALD, D. K. C.: *Near Zero* (paperback), Anchor Science, Doubleday, Garden City, N.Y., 1962. The fascinating story of the peculiar things that happen as matter is cooled close to absolute zero.

SUMMARY

Heat energy is molecular kinetic energy. When an object or gas is heated, the molecules within it are given more kinetic energy. When it is cooled its molecules lose kinetic energy. The temperature of an object or gas is a measure of the kinetic energy of its molecules.

Three common temperature scales exist. The Fahrenheit scale has the normal freezing and boiling points of water at 32°F and 212°F. On the centigrade or Celsius scale, these points are 0°C and 100°C. Although the absolute or Kelvin scale has the same size degrees as the Celsius scale, the numbers on the scale are shifted by 273 degrees, so that these two points are at 273°K and 373°K.

When a confined gas is heated, its pressure is proportional to its Kelvin (or absolute) temperature. If N molecules of gas exist in a volume V at a temperature of $T°K$, the pressure P of the gas is given by the gas law $PV = NkT$, where k is a constant of nature called Boltzmann's constant. This indicates that at absolute zero, where $T = 0°K$ (or $-273°C$), the pressure of a gas becomes zero. All gases condense, however, before this temperature is reached.

The gas law can be derived from a model that pictures the pressure to be the result of ball-like molecules colliding with the walls of the container. From this model we learn that the absolute temperature is a measure of the kinetic energy of the molecules. Precisely, $T = \frac{2}{3}k(\frac{1}{2}mv^2)$. This predicts that all molecular motion must cease at absolute zero. Gases show a wide variation in molecular speeds. At any instant, some molecules are at rest while some have extremely high speeds. The average speed of nitrogen molecules in the air at room temperature is about 500 m/sec.

Tiny particles large enough to be seen in the microscope undergo a zigzag motion when suspended in a gas or liquid. This is called Brownian motion, and it results from the unbalanced collisions of the surrounding molecules upon the particle.

When heat is conducted through a solid, such as a rod, the heat energy (molecular kinetic energy) is passed down the rod by collisions of the hotter molecules with those having less energy.

Since energy is required for a particle to rise above the earth's surface, only the most energetic molecules are able to reach high altitudes. As a result, the air is much less dense at high altitudes than it is at the lower levels. About 100 mi above the earth, the air is only one billionth as dense as it is at the earth's surface. We can measure the pressure of the atmosphere by use of a barometer.

The viscosity of a liquid is large when the molecules within the liquid move over each other only with great difficulty. Since molecular motion is greater at higher temperatures, the viscosity of liquids decreases as the liquid is heated.

Boiling occurs within a liquid when the pressure due to the molecules escaping from the liquid surface equals the pressure of the atmosphere upon the liquid. Under these conditions, vapor bubbles can grow within the liquid. Even at lower temperatures evaporation can occur as the most energetic molecules escape from the liquid surface. Since they carry away more energy than the average molecule possesses, evaporation leads to cooling.

Within liquids and glasses, the molecules do not arrange themselves in precise patterns. In crystalline solids, however, the molecules exist in a definite pattern (or lattice) which is repeated throughout the crystal. Heat energy must be added to a crystal in order to break the atoms loose from the lattice and thus melt a crystalline solid.

TOPICAL CHECKLIST

1. How did Rumford's experiments negate the "caloric" concept?
2. What is the primary difference between a hot substance and the same substance when cool?
3. Why is it that thermometers based on the expansion of liquids cannot be made to agree perfectly?
4. What are the designations of the freezing point of water on the centigrade, Fahrenheit, and Kelvin scales?
5. How does the pressure of a gas behave as a function of temperature?
6. Describe an experiment by which absolute zero can be defined.
7. What is the ideal gas law and what does it imply for a graph of P versus T when V is kept constant?

8. As the temperature of a gas increases, V being constant, what happens to P? If V is decreased with T held constant, what happens to P?

9. In terms of molecules, what is the physical mechanism by which a gas exerts a pressure?

10. Of what is T a measure and what is the relation between the two quantities?

11. About how fast is the average molecule moving in air at room temperature? Why doesn't a perfume molecule move across a room this fast?

12. Sketch a graph showing how the speeds of the molecules in a gas vary. How does the graph change as the temperature is varied?

13. In a container of gas, what is the speed of the slowest molecule? What can you say about the speed of the fastest?

14. What is the Brownian motion?

15. Why is it reasonable that the atmosphere should grow thin as we go higher above the earth?

16. What is a mercury barometer and how does it work? About how large is atmospheric pressure where you are?

17. When a solid or liquid is heated, what is happening within the substance?

18. What is the molecular process involved in the process of heat conduction? In thermal expansion?

19. What is meant by viscosity? How and why does it change with temperature?

20. Explain, using molecular concepts, what is happening during evaporation. Why does this process lead to cooling?

21. What determines whether or not a liquid will boil at a given temperature? Why does the boiling temperature change with atmospheric pressure? What is meant by vapor pressure?

22. What is superheating and why does it occur?

23. How are glasslike plastics as well as glass itself related to liquids?

24. What is the essential difference between a liquid or glass and a crystalline substance?

25. What happens when a crystalline substance melts? Are the atoms in a crystal motionless?

QUESTIONS

1. In terms of the kinetic theory and its molecular model, explain the cause of the following: (a) Pressure increases with T; (b) pressure increases as more molecules are added to a volume; (c) temperature increases as a gas is suddenly compressed.

2. In a diesel engine, the vaporized oil is not ignited by a spark from a spark plug. Instead, the vapor ignites when the piston in the cylinder suddenly compresses the vapor. Explain.

3. Explain why a 60 mi/hr wind high in the mountains exerts far less force than a similar wind at sea level.

4. Explain how we carry out the process of breathing. Shouldn't it be sufficient to simply leave one's mouth open so the air can get in and out?

5. Dalton's law of partial pressures states that "the pressure of a mixture of gases is equal to the sum of the pressures of the individual gases." Justify the law.

6. Since gas molecules possess kinetic energy, they should be capable of pushing objects about and hence of doing work. Give some examples of cases in which this occurs.

7. The higher you go above the earth, the larger the fraction of hydrogen gas is in comparison to nitrogen. Explain why the composition of the atmosphere should vary in this way.

8   Why can a jet airliner operate more efficiently at 30,000 ft than at 5,000 ft if the wind speeds are the same at the two levels?

9   An empty tank which is used to hold gasoline is being inspected. Since it does not slosh when shaken, the can is empty. Yet when a match is dropped into it in an effort to examine its interior, the can explodes. Explain what probably happened.

10 Consider two liquids, A and B, which have different viscosities when they are at the same temperature. Liquid A is much more viscous than liquid B. (*a*) All other things being equal, which has the larger attractive force between molecules, A or B? (*b*) All other things being equal, in which liquid are the molecules largest? (*c*) Why would an increase in temperature cause A's viscosity to decrease? (*d*) Which liquid would you expect to boil at the lower temperature and why? (*e*) Which liquid would you expect to evaporate faster and why?

11 The molecules of skin and flesh within the region of a burn are found to be torn into fragments. Explain how touching a very hot object could lead to this result. Why is a "floor burn" (obtained by an arm sliding on a gym floor) similar to a burn from a stove? How is this related to the fact that potatoes take longer to cook in Denver than at sea level?

12 One way to prevent a liquid from "bumping" when it is heated is to place a *boiling chip* in it, which is a chip of porous rocklike material. Why should this be of any aid in this situation? (*Hint:* The pores are filled with air at the start and the chip cannot be used over unless it is dried.)

13 Water boils at 100°C, and benzene, a molecule about four times as massive as water, boils at 80°C. What can you conclude about the forces between the two types of molecules in the pure liquids?

14 A flask is partly filled with water at 60°C. When a vacuum pump is attached to the flask and the air is removed, the water begins to boil. Why?

15 Solid carbon dioxide, dry ice, does not melt when laid on a table in the air. Instead, the dry ice simply disappears slowly. What is happening?

16 A machinist friend of mine maintains the following. He often heats one end of a metal rod to red-hot in order to temper it. When he thrusts the red-hot end in oil to cool it quickly, he says the heat moves up the rod from the end in the oil and the upper end becomes much hotter than the lower end. Can he be right?

17 As a piece of metal is heated, it expands. Does its weight increase also?

18 Even though acetone molecules are about three times more massive than those of ethyl alcohol, the alcohol is more than three times as viscous as the acetone. What can you conclude from this?

19 Which would be harder to heat from 0°C to room temperature, a glass full of water or a glass full of ice and water?

20 A bullet is shot into a piece of wood. Upon examination, it is found that the bullet has partly melted during the impact. Explain how this could have happened.

PROBLEMS

1   What are the following temperatures on the centigrade scale? (*a*) 72°F (room temperature); (*b*) 95°F; (*c*) 10°F; (*d*) −13°F.

2   Change the following to Fahrenheit temperatures: (*a*) 30°C; (*b*) 90°C; (*c*) −40°C.

3   If the temperature of a gas in a tight enclosure is changed from 0 to 100°C, by what factor will the gas pressure increase?

4. A can of air is closed at 27°C and cooled. At what temperature will the gas pressure within the can be reduced to three-fourths its original value?

5. The pressure within a 1-liter metal tank filled with oxygen gas is 100 lb/in.² (pounds per square inch). How much volume would this gas occupy if it expanded enough for its pressure to drop to atmospheric pressure, about 15 lb/in.²?

6. In air at atmospheric pressure, 76 cm of mercury, a nitrogen molecule moves only about 10^{-6} cm before hitting another molecule. (We say its *mean free path* is 10^{-6} cm.) About how low must the pressure be if the mean free path is to be about 1 cm?

7. The fastest speed any material object can have is the speed of light, 3×10^8 m/sec. How hot would a gas have to be if the average speed predicted by the kinetic theory were this large? Use $m = 20 \times 10^{-27}$ kg for the molecular mass.

8. Helium gas is the most difficult of all gases to condense; it does so at −269°C. What is the average speed of helium molecules in a gas at this temperature? ($m = 4 \times 1.67 \times 10^{-27}$ kg.)

9. The energy unit commonly used when discussing heat is the calorie, where 1 cal ≅ 4.2 joules. Nutritionists use the "calorie" in categorizing food energies, but their calorie is really a kilocalorie. Normally we acquire about 2,000 cal of food energy daily in our diets. Show that this is equivalent to about 8.4×10^6 joules of energy. How many times could a 50-kg woman ascend a 3-m-high flight of stairs before she uses up this much energy? Assume no energy is lost to other uses (an unrealistic assumption).

10. The unit of heat energy, the calorie, is defined to be the amount of energy one need add or remove from 1 gm of water to change its temperature by 1°C. Also 80 cal of heat energy are needed to change 1 gm of ice at 0°C to 1 gm of water at 0°C. Suppose 1.0 gm of 0°C ice is added to a glass of water (200 gm) at 20°C. What will be the temperature of the glass of water after the ice has melted?

Chapter 6

ELECTRICAL CHARGES AND CURRENTS

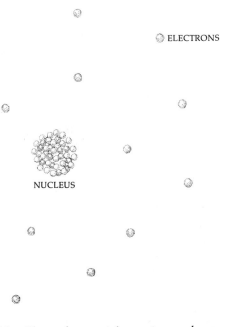

6.1 *The nucleus contains protons and neutrons; the nucleus and the electrons are much farther apart than this picture indicates.*

In previous chapters we considered only one fundamental force, the gravitational force. A much stronger force, electrical in nature, is responsible for the structure of atoms and molecules. In this chapter we will consider the electrical force and its practical uses. Later we will see how this force holds atoms, molecules, and solids together.

The Quantum of Charge

Atoms are composed of three types of particles—the electron, the proton, and the neutron. Although we shall discuss atomic structure in detail in Chapter 12, it will simplify our study of electricity if we learn the basic facts about atomic structure, as shown in Figure 6.1. An atom consists of a central ball-like nucleus composed of protons and neutrons. Orbiting the nucleus are equally tiny particles called electrons. They are held to the nucleus by electrical forces, just as the planets are held to the sun by gravitational forces.

Long before the existence of the electron, proton, and neutron had been established, two types of electricity, or <u>electrical charge</u>, were known. One of

6.2 *Like charges repel while unlike charges attract.*

UNCHARGED
(NEUTRAL)
(a)

LIKE
CHARGES
(b)

UNLIKE
CHARGES
(c)

these types, <u>negative charge</u>, can be placed on a hard rubber (or ebonite) rod by rubbing the rod vigorously with a piece of fur. The other type of electrical charge, designated <u>positive</u>, can be placed on a glass rod by vigorous rubbing with a silk cloth. It is a simple matter to show that these two crude defining experiments do, indeed, result in different kinds of charge. You probably performed the experiments yourself in elementary school. However, we will review them here because of their fundamental importance in all electrical phenomena.

Suppose two small, light balls are hung from threads as shown in Figure 6.2a. They hang straight down, of course. But if some of the positive electricity from a charged glass rod is placed on each ball, the two balls repel each other as shown in Figure 6.2b. When the two balls are charged negatively by contact with the charged ebonite rod, they also repel each other, as in (b). However, if the originally uncharged balls are charged separately, one by positive charge from the glass rod and the other by negative charge from the ebonite rod, the two balls attract each other as in (c). We can conclude from this experiment that there is a difference between the charges on the glass and ebonite rods; in addition, the experiment shows the following:

Like charges repel each other whereas unlike charges attract each other.

Many other objects produce a charge when rubbed together. If you vigorously rub your plastic pen against your clothes, the pen will become charged. Or, as you have surely noticed, in dry weather your whole body can become charged if you walk across certain types of carpet or slide across a seat in an automobile. You know many other examples. But no matter how a charge is generated, it always corresponds to one of the two charges described above. We conclude from this that two, and only two, kinds of charge exist. One of these we call positive, the other is designated negative.

We know now that these charges can be explained by the particles in an atom: The electron carries a negative charge and the proton carries a positive charge. As its name indicates, the neutron is neutral; it carries no net charge at all. We define the unit of charge as the <u>coulomb</u> (abbreviated coul). The electron's charge is -1.602×10^{-19} coul; the proton's charge is identical except for sign, $+1.602 \times 10^{-19}$ coul.* *No smaller charge has ever been found.* More-

---

* Recall that 10^{-19} indicates the number should be divided by 10 nineteen times.

over, all charges found in nature are integer multiples of 1.602×10^{-19} coul. In other words, charged objects always have 1, 2, 3, 4, etc., times the charge on the electron or proton; intermediate-size charges do not exist.

From this we see that charge comes in little pieces (called *charge quanta*) whose size is 1.602×10^{-19} coul. The quantum of charge is designated by the symbol e, the proton's charge being $+e$ and the electron's being $-e$. You might guess from this that charged objects always acquire charges by virtue of an excess of protons or electrons. Indeed, this is the usual case. However, there are other tiny particles similar to the electron and proton which carry charges of $\pm e$ even though they do not contain protons or electrons. These other particles are not important in practical electricity and so we shall discuss them later. Let us now investigate the nature of the forces that charges exert upon each other.

Coulomb's Force Law

Although it is sufficient in many situations simply to know that like charges repel and unlike charges attract, the exact dependence of the force upon charge magnitude q and separation distance r is also of value. Augustin Coulomb (1736–1806) first discovered this dependence; it is called Coulomb's law in his honor. He found that if the ball on the left in Figure 6.3 carries a positive charge of magnitude q_1 coul, it will repel the ball on the right, which carries a like charge of magnitude q_2, with a force of F newtons. Moreover, Newton's law of action and reaction tells us that an equal and opposite force will push q_1 toward the left. If the distance between centers of the balls is r meters, Coulomb's law for F is

$$F = 9 \times 10^9 \frac{q_1 q_2}{r^2} \text{ newtons} \qquad (6.1)$$

6.3   *The force of repulsion F exerted on q_2 by q_1 is given by Coulomb's law.*

We see that the electrical force law is an inverse square law. Like gravitational force, electrical force decreases inversely with the square of the separation distance. However, the electrical force is much stronger than the gravitational force, as we can see from the following example.

EXAMPLE 6.1

Compare the gravitational force between two protons to the electrical repulsion between them. The mass of a proton is 1.67×10^{-27} kg.

Method: Newton's law of gravitation tells us the gravitational attraction force on a particle of mass m due to an identical particle a distance D away is [see Equation (4.3)]

$$F_G = 6.7 \times 10^{-11} \frac{mm}{D^2}$$

The Coulomb electrical force for the same two particles, each of charge q, is

$$F_e = 9 \times 10^9 \frac{qq}{D^2}$$

Dividing the first equation by the second gives

$$\frac{F_G}{F_e} = \frac{6.7 \times 10^{-11}}{9 \times 10^9} \frac{mm}{qq}$$

But $m = 1.67 \times 10^{-27}$ kg and $q = +e = 1.6 \times 10^{-19}$ coul, and so the ratio of the two forces is

$$\frac{F_G}{F_e} = 8 \times 10^{-37}$$

This almost inconceivably small number indicates that most often the gravitational force between charged particles is totally negligible in comparison to the electrical force.

Conductors and Insulators

According to Coulomb's law, when one charge is brought close to another, each will experience a force because of the presence of the other charge. Therefore a charge near an atom will exert forces on the charges within the atom. Since there are as many positive protons as negative electrons within the atom, the oppositely directed forces on protons and electrons will cancel out. The atom by itself experiences no net force because it is electrically neutral.

When the atoms are packed together to form a solid or liquid, two types of electrical behavior are possible depending upon the particular atoms involved. In certain solids and liquids, called _insulators_, the atoms remain distinct entities. Each retains a firm hold on the electrons associated with it. As a result, the electrons are not free to move away from the atom to which they are attached. The protons, of course, are held tightly in the nucleus and connot escape from the atom. Since the charges in an insulator are bound in place, it is impossible for charge to move about in the solid or liquid.

Typical examples of insulators are wood, plastics, glass, and paper.

Substances in which charge can move around are called _conductors_; metals (copper, iron, aluminum, and so on) are the most familiar of this group. In these substances, a few of the outermost electrons in each atom come loose from their parent atom when the atoms are packed together to form a solid or liquid. As a result, the metal has a large number of free electrons, and these negative charges can move from place to place under even the slightest force.

Materials in a third, intermediate, grouping are called _semiconductors_. They are used widely in solid-state electronic devices, particularly in transistors, and are gaining an increasingly important place in scientific technology.

Induced Charges

In this section we will see how electrical charges behave in several situations. As our first example, let us consider the ebonite rod again. It is an insulator, yet when the rod is rubbed with fur, it acquires a negative charge. How does this happen?

Apparently, some electrons from the atoms in the fur move over to the ebonite rod during the rubbing. Thereby, the rod acquires extra electrons and becomes _negatively charged_. Since the rod is an insulator, these excess negative charges remain on that portion of the rod which was in contact with the fur; they cannot move from place to place.

When a glass rod is rubbed with silk, electrons rub off the surface of glass and so the remaining electrons in the glass can no longer balance the positive charges of the protons. The glass rod is _positively charged_, since it has more protons than electrons.

Suppose the negative ebonite rod shown in Figure 6.4 is brought close to a neutral metal object. Since the free electrons within the metal can move under the slightest force, some of them move to the distant

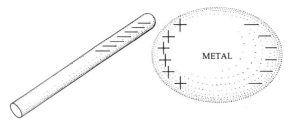

6.4 *The charged ebonite rod gives rise to charge separation within the metal.*

side of the metal object because they are repelled by the negative charge on the rod. But then there are more protons on the left side of the object, and so that region has an equal amount of positive charge. In this way, the negative rod separates some of the previously balanced charges within the metal. We say that the rod <u>*induces*</u> a positive charge on the left side of the object and a negative charge on the opposite side. Charges which have been separated this way are called <u>*induced charges*</u>. Before proceeding further, you might wish to test your understanding by explaining what will happen in an analogous situation with a positively charged glass rod.

Now, suppose the ebonite rod in Figure 6.4 is touched to the metal surface and then removed. Since the negative charges on the rod are strongly attracted to the nearby positive charges on the metal, some of them will jump from the rod to the metal when the two touch. The metal then has extra electrons and is negatively charged. (Before touching the rod, the metal had equal numbers of protons and electrons even though some were separated from each other.) When the ebonite rod is removed, the charges in the metal no longer experience an outside force. They are then free to redistribute on the metal object as shown in Figure 6.5. Under the force of their mutual repulsions, the excess negative charges tend to move to distant ends of the object.

Figure 6.6 shows another case of charge induction. The two metal plates shown in (*a*) have been charged oppositely with equal quantities of excess positive and negative charge. The excess positive charge on the left plate moves to the side nearest to the negative plate. In the same way, the left plate induces the negative charge on the other plate to move to the inside surface. By mutual induction, the excess charges are held on the inner surfaces of the two parallel plates. We will refer frequently to this situation. For convenience, the plates will usually be drawn end-on, as shown in part (*b*) of the figure. Notice that the two plates will attract each other because of their opposite charges.

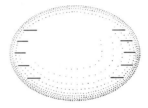

6.5 *Why do the excess negative charges localize near the ends of the metal object?*

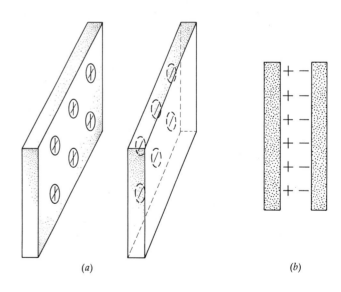

6.6 *The oppositely charged metal plates induce each other's charges to the inner surfaces.*

ELECTRICAL CHARGES, CURRENTS 117

The Electric Field

When a man pushes a car or pulls a rope attached to a post, the forces involved are easily pictured and understood. Forces such as these are exerted by material objects, one acting directly on the other. However, the two fundamental forces we have encountered thus far, gravitational and electrical, are not like this. The sun pulls on the earth with no visible connection between the two; a positively charged ball attracts one with a negative charge without touching it. These are examples of *forces at a distance*. We will represent such forces in terms of <u>*fields*</u>.

You may already be aware of the concept known as the *earth's gravitational field*. In fact, as you read this, you are immersed in the gravitational field of the earth. The direction of the field is downward, since we say the direction of a field is the same as the direction of the force it represents. Frequently we draw pictures of gravitational fields using lines that point in the direction of the force and spacing them closer together where the force is stronger. Figure 6.7 shows an appropriate field picture for a room. We can see at a glance that the gravitational force in the room is downward (the field line arrows point down) and the force is uniform and does not depend upon location (the force lines are everywhere equally spaced).

The earth's field is not exactly uniform since the earth is a sphere, not a flat plane. The gravitational field of the earth as a whole is shown in Figure 6.8. Again the field lines tell us that the gravitational force is directed toward the earth's center. Moreover, the larger spacing between lines as the distance from the earth increases tells us that the earth's field (i.e., the gravitational force) decreases in strength as we go away from the earth. In summary, the direction of the gravitational force is indicated by the arrows on the field lines; the relative magnitude of the force is represented by the relative density (or closeness together) of the field lines.

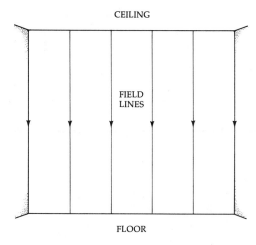

6.7   *The gravitational field inside a room.*

We can also use fields to describe electrical forces. The Coulomb electric force is also a force at a distance. However, unlike gravitational force, which involves only one type of mass,* electrical force involves two types of charge, positive and negative. Therefore we must specify the direction of the electrical force field further. We customarily *take the electric field direction to be the direction of the force on a positive charge*. Let us apply this to a particular situation.

Suppose we consider the region between the two charged parallel plates shown in Figure 6.9a. Assume that the plates are larger than shown so that the effects at the ends can be neglected. If we place a small positive charge called the *test charge* at point A, it will experience a force toward the right since it is repelled by the positive charge on the left and attracted by the negative charge on the right. Therefore the electric field at A is directed toward the right as shown. Similar reasoning shows us that the electric field everywhere between the plates is directed toward the right. Hence the electric field can be drawn as shown in Figure 6.9b; it is uniform throughout the interior region.

---

* Another type of mass does exist, called *antimatter*. However, it is annihilated by ordinary matter and does not exist in stable form in our galaxy. We will mention it again in Chapter 11.

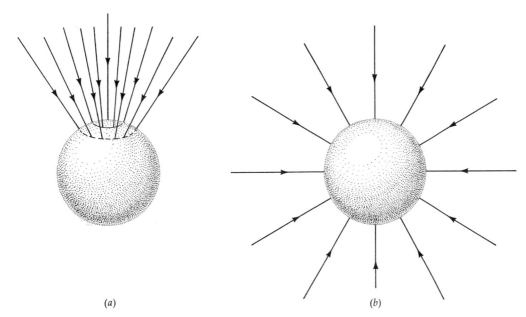

(a)   (b)

6.8 Although the earth's field should be drawn in three dimensions as in (a), we frequently use the convenient representation shown in (b).

We can define a quantity E, the *electric field strength or intensity*. It is the force a one-coulomb test charge would experience if it were placed in the field. For example, the field strength E at point A in Figure 6.9a is equal to the force a one-coulomb (i.e., one unit) test charge would experience if it were placed at A. The units of E are those of force per unit charge, namely, newtons per coulomb.

As a second example of an electric field diagram, consider the field around the positively charged sphere shown in Figure 6.10a. To find the direction of the field at A, we need only ask what direction the force on a positive test charge would be if placed at A. Since like charges repel, the force, and therefore the field in the region close to a positive charge, would be directed radially outward. The fields for other charge situations are shown in the other portions of Figure 6.10. You should be able to justify the field direction shown at point A in each case. Notice also that the lines of force always originate on positive charges and terminate on negative charges. Why should this be?

Like gravitational field lines, electric field lines tell us that the force is strongest where they are closest together. Thus we can see from Figure 6.10 that the field (and force) is strongest close to the charges that cause the field. This is exactly what we would expect.

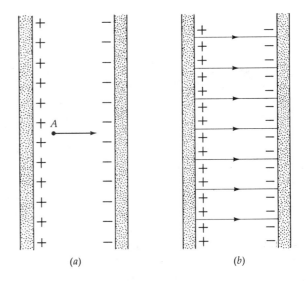

(a)   (b)

6.9 The electric field lines originate on positive charges and end on negative charges.

ELECTRICAL CHARGES, CURRENTS 119

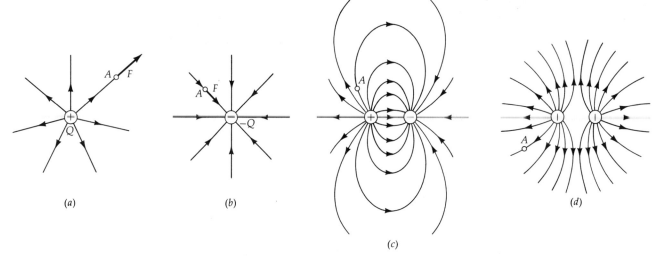

6.10 As in Figure 6.9, the lines of force originate on positive charges and end on negative charges.

Electric Current—The Ampere

Let us now turn our attention to currents and voltages. As you already know, metals have large numbers of free electrons. For example, a 1-m length of copper wire has about 10^{20} free electrons. This number is so large that we can liken the free electrons in a metal wire to the water molecules filling a water pipe. In using this analogy, we shall consider metal wires to be the "pipes" through which a "fluid" composed of electric charges flows. Unfortunately, however, there is a complication.

Only a few difficult experiments can show whether the charges that move within a metal are positive or negative. None of these experiments was even considered in the early days of electricity, and hence the early investigators simply guessed which moved, the positive or negative charges. With a 50-50 chance of being correct, they picked the wrong one and assumed that metals have movable positive charges. By the time it became clear that this assumption was wrong, the field of electricity was already extensively developed. The concept of positive charge motion in metals had become strongly entrenched, and, even today, almost everyone uses terminology based on that mental picture of charge flow. For this reason we, too, will usually ignore the fact that negatively charged electrons move in metals and, instead, assume positive charge motion. You must forgive us this inaccuracy; we are simply conforming with the rest of the world since it is impossible to tell without special techniques which charge is moving.

We can define the rate of charge flow (or electrical current) in a wire by comparing it to water flow rate (or current) in a pipe. If 2 gal of water flow past a given point in a pipe each second (out the spout for example), the water current is said to be 2 gal/sec. In the electrical case, we have a pipe filled with positive charges instead of water molecules. As shown in Figure 6.11, these charges flow through the wire like water flowing through a pipe. If a total charge of 5 coul flows through the pipe cross section at A each second, the electric charge flow, the current, is simply 5 coul/sec.

To make this definition of electric current more concise, we can write

$$\text{Electric current} = \frac{\text{charge passing point in wire}}{\text{time taken in seconds}}$$

If we call the current I, and write q for the charge passing through the cross section A in Figure 6.11 in a time t, the definition becomes

$$I = \frac{q}{t} \qquad (6.2)$$

Since the unit of charge q is the coulomb, the electric current unit is coulombs per second. Electric currents are so important that we designate a coulomb per second as an *ampere* (amp). To summarize, if a wire is carrying a current of 10 amp, then 10 couls of charge flow through the cross section of the wire each second.

A current of 10 amp represents a flow of a tremendous number of electrons. The number of electrons needed to make 10 coul of charge is $10 \div 1.6 \times 10^{-19}$ since the charge on each electron is 1.6×10^{-19} coul. Therefore, 6×10^{19} electrons must pass through a cross section each second when the current is 10 amp. Since this is 6 with 19 zeros after it, there are obviously too many electrons moving through the wire for us to be able to see the effect of each individual one. Like water molecules flowing in a pipe, the particles flow abundantly.

Batteries and Electromotive Force Sources

Since charges are not perfectly free to move, they need an energy source to push them through the wire. In the same way water, because of its viscosity, needs an energy source to push it through a pipe. A simplified water-circulating system is shown in Figure 6.12; the water has gravitational potential energy when it is in the reservoir. This energy is lost as the water flows downhill (through many different paths rather than the single one shown, depending on how the consumer uses the water).

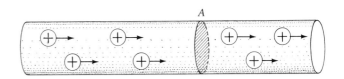

6.11 *The current in the wire in amperes is defined to be the quantity of positive charge in coulombs flowing through a cross section such as A in one second.*

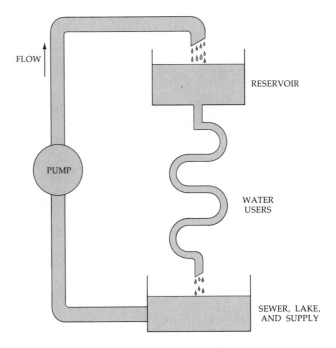

6.12 *Schematic diagram of a water system. It is to be hoped that the sewer to water supply section is vastly oversimplified.*

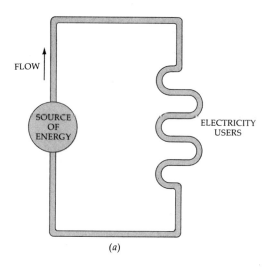
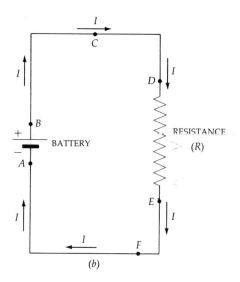

6.13 It is assumed in the diagram at the right that the total fall in electrical level occurs in the small region labeled Resistance.

Eventually the water flows through a sewer and ends up in a lake, river, or sea. Aside from the fact that the water is no longer clean, it is useless in a more fundamental way; it has lost its potential energy.

To make the water useful once again, it must be lifted to the reservoir (after purification, of course). The lifting agent, whether it be a pump as shown or the evaporation process using the energy of the sun, once again gives the water potential energy. The entire process repeats over and over again.

Compare the water system with Figure 6.13a, which shows the flow path of electric charge. In this case, a battery or some other energy source (all given the name *source of electromotive force*, or *emf*) raises the positive charge to a high *electric potential energy* level. It is customary to represent the high-energy side of a battery by a plus sign, as shown in Figure 6.13b. Also, notice the symbol for a battery, ———, where the longer, thinner line represents the + or high-energy side of the battery.

Like the pump in Figure 6.12, the battery raises positive charge (through itself) from point A to point B. Positive charge at B has a large amount of electric potential energy. Usually, circuits are made with good conducting wire large enough so that the charge loses a negligible amount of energy as it moves through the wire from B to C to D. Since the charge has the same energy at all these points, we say that points B, C, and D are all at the same *electric level* or *electric potential*.

The charges meet considerable resistance to flow in the section of the circuit between D and E. Hence we say that section has *resistance*. This section might be the white-hot wire inside an incandescent light bulb. Obviously the light bulb is using up electric energy to produce its heat and light. The energy comes from the potential energy lost by the positive charges as they flow from D to E through the resistance. By the time the charges reach E they have lost all their electric potential energy. Since E, F, and A all are at the same low electrical level, the charges' potential energy remain zero until the charges reach the battery at A. The battery now gives the charges more energy by lifting them to the high electrical level of B, and the process repeats itself.

In summary, the battery (or any other source of emf) supplies energy to the charge. The charge carries the energy to the portion of the circuit between D and E. There the energy is used to light a

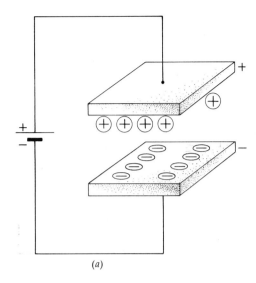
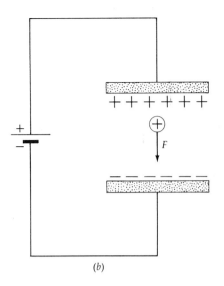

(a) (b)

6.14 *A potential difference exists between the two plates; the upper plate is the high potential side.*

bulb, run a motor, operate a radio, or for some other similar purpose. Its energy now gone, the charge returns from E to the battery where it receives more energy.

Electric Potential Difference—Voltage

The term "voltage" of a battery is familiar; we commonly refer to a 12-volt car battery or a 1.5-volt flashlight battery. It is our way of measuring how much energy a charge has. Let us see exactly what voltage means.

Suppose we connect two metal plates to a battery by means of metal wires as shown in Figure 6.14. The plate attached to the positive side of the battery acquires a positive charge and the other plate becomes negatively charged. We can think of the positive plate as being at the top of an electrical hill and the negative plate as being at the bottom of the hill. This is in line with our previous thinking that a positive charge at the positive side of the battery has a higher electric potential energy than when it is at the negative side.

To make this point more plausible, consider what happens if we place a positive charge between the two plates as in Figure 6.14b. The positive charge is pushed toward the negative plate by the repulsion of the charges on the positive plate and is pulled by the attraction of the charges on the negative plate. If released, the positive charge will "fall downhill" from the positive to the negative plate. It should be clear from this that the positive charge does indeed have potential energy when it is at the level of the positive side of the battery. We shall therefore picture the positive side (or terminal) of a battery as being at the top of an electrical hill.

Suppose, however, we replace the positive charge between the plates in Figure 6.14b by a negative charge. This negative charge, if released, will fall upward to the positive plate. We see, therefore, that what was uphill for a positive charge is downhill for a negative charge, and vice versa. Since we have agreed to pretend that the positive charges move to produce currents, we will also agree to use the terms "uphill" and "downhill" to refer to positive charges. In other words, the positive terminal of a battery or other source of emf will be called the *high side* of the battery. Current, being a flow of positive charge, will always flow from points of high potential energy to points of low potential energy.

We still need a quantitative measure for the height of the electrical hill furnished by a battery or source of emf. It is conventional to express this electrical height in terms of the work done in carrying a positive one-coulomb (+1-coul) charge from the low to the high side of the battery, that is, from the negative to the positive terminal of the battery. The height of the electrical hill from negative to positive terminal is the same as the height of the electrical hill from the negative to positive plate in Figure 6.14. Clearly, work must be done to carry a positive charge from the negative plate to the positive plate, since the charges on the plates exert a downward force on a positive charge. We specify the height of this electrical hill, the so-called *potential difference* between the two plates, in the following way:

> *The potential difference between two points (or plates) is the work done in carrying a +1-coul charge from one of the points to the other. The unit of potential difference is the volt.*

For example, if we say that the battery in Figure 6.14 is a 6-volt battery, we mean that the work done in carrying a +1-coul charge from the minus to the plus side of the battery is 6 volts. Notice that the volt, being work per unit charge, is a work unit divided by a charge unit. Hence the volt is a joule per coulomb.

Let us consider a few numerical uses of the concept of electric potential difference. Suppose we connect a 12-volt battery to two metal plates in an evacuated tube* as shown in Figure 6.15. If a proton at rest ($q = 1.6 \times 10^{-19}$ coul and $m = 1.67 \times 10^{-27}$ kg) is released at the positive plate within the tube, it will fall toward the right, to the negative plate. How much kinetic energy will it acquire as it falls downhill to the plate on the right?

---
* We shall usually consider the motion of charges in vacuum rather than in air; in that way, collisions with air molecules can be avoided.

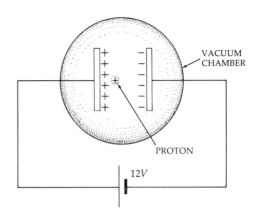

6.15 *The proton will fall downhill from the positive to the negative plate. In so doing it will acquire a kinetic energy qV where q is the proton's charge and V is the voltage of the battery.*

To find how much kinetic energy the proton acquires, we need to know how much work is required to lift it from the negative to the positive plate. It gains this work back as kinetic energy when it falls from the positive to the negative plate. But we know from the definition of potential difference that to carry a +1-coul charge from one plate to the other requires work equal to V, where V is the potential difference between the plates (12 volts in this case). To carry a charge of +3 coul would require an amount of work $3V$ since the charge being carried is now three times as large. Similarly, to carry a charge q would require an amount of work equal to qV. Therefore:

> *The work done in carrying a charge q through a potential difference V is qV. When a charge q falls through a potential difference V, it acquires an energy qV.*

It is now a simple matter to find the kinetic energy and speed of the proton just before it strikes the negative plate. We know that in its fall from the positive plate the proton acquires an energy qV which will be in the form of kinetic energy. Therefore

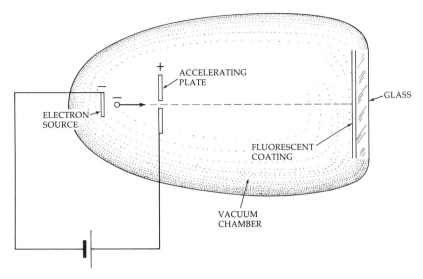

6.16 *A greatly oversimplified diagram of a TV tube.*

$$\tfrac{1}{2}mv^2 = qV$$

or

$$\tfrac{1}{2}(1.67 \times 10^{-27} \text{ kg})v^2 = (1.6 \times 10^{-19} \text{ coul})(12 \text{ volts})$$

which, when solved for v, gives

$$v \cong 4.8 \times 10^4 \text{ m/sec}$$

This speed, 48,000 m/sec or about 100,000 mi/hr, is high in comparison to the speeds we usually encounter, but low compared to the speed of light (3×10^8 m/sec).

As another example, let us consider what happens inside a TV tube. Figure 6.16 is a schematic diagram of such a tube. Basically it has a source of electrons, a potential difference to accelerate the electrons, and a fluorescent screen that gives off a spot of light where the electrons hit it. All of this is inside a glass tube from which the air has been evacuated.

The electron source, or cathode, is made of a material that will emit (i.e., give off) electrons if it is heated red-hot.* (We do not show the heater for the cathode, since that would further complicate the diagram.) A metal plate with a small hole in it is placed a short distance away from the cathode. From the diagram you can see that the cathode and plate are connected to the minus and plus sides of a voltage source. Notice that the plate is positive because it is connected to the positive side of the battery. The battery's potential difference V is between the cathode and the plate.

When an electron (a negatively charged particle) is emitted by the cathode, it accelerates toward the positively charged plate. As it moves from cathode to plate, it falls through the potential difference V and acquires a kinetic energy equal to qV. (Recall that what is uphill for a positive charge is downhill for a negative charge.) Consequently, the electron has a high speed when it reaches the plate. A few of the electrons shoot through the hole in the plate and continue at high speed across the tube until they strike the fluorescent material at the opposite end. For reasons we shall discuss in later chapters,

---

* You can picture this process as the evaporation of electrons from the very hot material of the cathode.

the collision of the high-energy electrons with the fluorescent coating causes the coating to give off the light we see on the TV tube. In an actual tube, the electron beam is deflected back and forth and up and down so that it traces out the picture on the fluorescent screen in a small fraction of a second.

To find the speeds of the electrons as they pass through the small hole in the plate, we equate qV to $\frac{1}{2}mv^2$, since the electrons acquire their kinetic energy by falling through the potential difference V. In an actual tube V is about 10,000 volts and for an electron, $q = -1.6 \times 10^{-19}$ coul and $m = 9 \times 10^{-31}$ kg, and so

$$qV = \tfrac{1}{2}mv^2$$

becomes

$$(1.6 \times 10^{-19} \text{ coul})(10{,}000 \text{ volts}) = \tfrac{1}{2}(9 \times 10^{-31} \text{ kg})v^2$$

which gives

$$v \cong 60 \times 10^6 \text{ m/sec} = 6 \times 10^7 \text{ m/sec}$$

The speed of light is 30×10^7 m/sec, and so we see that the electron in the TV tube moves with a speed about one-fifth as great as the speed of light. In cases like this, the computed speed is not quite right. You may already know that objects cannot move faster than the speed of light. When an object moves with a speed close to that of light, the mass of the object begins to increase. We will see why when we discuss Einstein's theory of relativity in Chapter 10.

Ohm's Law

Now that we understand current, voltage, and the purpose of an emf source, we can discuss the relation on which all practical electricity is based. If we turn our attention to the circuit shown in Figure 6.17, we notice that the battery of voltage V furnishes energy

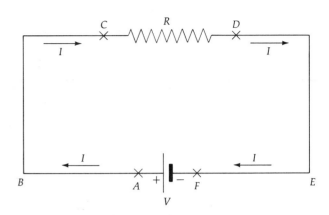

6.17 Ohm's law tells us that $V = IR$.

to the circuit. Since current always flows out of the positive terminal of a single battery, the current must flow clockwise as indicated. Since we have agreed to neglect the small resistance to flow caused by the connecting wires, the only resistance to flow occurs in the portion of the circuit from C to D, that is, within the resistance R.

Because no energy is lost in the portion of the circuit ABC, these three points must be at the same high electric potential as the positive terminal of the battery (or at the top of an electrical hill). Similarly, the portion of the circuit DEF is at the same potential energy level as the low (negative) terminal of the battery. Clearly, charge from the battery at A flows downhill through the resistance from C to D and is then lifted from F to A, up to the top of the hill, by the battery. The charge loses energy as it falls downhill through the resistor, and gains energy when the battery lifts it from F to A.

Let us try to formulate the equation which will relate the current I to the resistance R trying to stop the flow and to the potential difference (or height of the electric hill) V. Certainly, the larger the resistance to flow, the smaller will be the current that flows. We might guess that if the resistance were made 10 times larger, the current would become only one-tenth as large. Indeed, experiment shows that this is exactly what happens. To make an educated guess about the effect of V on I, we recall that for a "dead" battery, $V = 0$ and no current flows; that is, $I = 0$. Clearly, the larger the voltage of the battery, the larger the current that flows. We might therefore guess that I should increase in proportion to V. This, too, is confirmed by experiment.

The experiments to determine the relation between I, V, and R were first carried out by George Simon Ohm (1789–1854). He summarized his findings in what is now termed <u>Ohm's law</u>, namely,

$$I = \frac{V}{R} \quad \text{or} \quad V = IR \qquad (6.3)$$

Notice that if V is tripled, say, then I also triples. But if R is increased by a factor of 5, then I decreases by this same factor. Therefore, Ohm's law confirms our guess regarding how I, V, and R should be related.

Ohm's law also forms the basis for our definition of the unit of electrical resistance R. Solving Equation (6.3) for R, we find

$$R = \frac{V}{I}$$

and so its unit is volt per ampere. For convenience, we call this unit the <u>ohm</u>. For example, if a resistance element is connected from the top to the bottom of an electrical hill of height 20 volts (V), and a current of 5 amp (I) then flows through it, its resistance is

$$R = \frac{V}{I} = \frac{20 \text{ volts}}{5 \text{ amps}} = 4 \text{ ohms}$$

As another example of the use of Ohm's law, consider the circuit shown in Figure 6.18. Notice the two meters in the circuit; for practical purposes they do

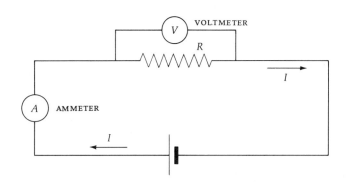

6.18   *How can the readings of the ammeter and voltmeter be used to find R, the resistance of the light bulb?*

not disturb the circuit. The ammeter, A, placed so that all the current must flow through it, measures the current in amperes. The voltmeter, V, is connected to both the top and the bottom of the electrical hill. It reads the height of the electrical hill, the potential difference between its two ends. Let us use these two meters to measure the resistance of the household light bulb shown as R in the circuit. The meter readings are as follows:

Ammeter: Current = 0.0078 amp

Voltmeter: Potential difference across R = 1.56 volts

Ohm's law tells us that if the potential drop across a resistor is V and if the current through the resistor is I, then

$$V = IR \quad \text{or} \quad R = \frac{V}{I}$$

In the present case,

$$R = \frac{1.56 \text{ volts}}{0.0078 \text{ amp}}$$

or

$$R = 200 \text{ ohms}$$

Actually, the potential difference assumed here is too small to light a household light bulb. As we saw earlier, energy is lost as the charge falls downhill through a resistor, i.e., as the current flows through the resistor. This energy is lost as heat, and it is this heat energy that causes an ordinary bulb to glow white-hot. In the U.S., household bulbs are designed to operate on a potential difference of 120 volts, far more than is available in this circuit. Actually, neither I nor V alone determines how hot the bulb becomes. The important quantity is the power, VI.

Electrical Power

When an amount of charge q falls downhill through a potential difference V in a time t, it loses electrical energy equal to qV. Therefore, the energy which is lost each second, or the *power loss*, will be qV divided by t. As an equation,

$$\text{Power} = V\frac{q}{t}$$

But in Equation (6.2) we defined q/t as the current I. As a result, we can rewrite the definition of power as

$$\text{Power} = VI \qquad (6.4)$$

If I is in amperes and V is in volts, the unit of power is called the <u>watt</u>. It is the same unit used in describing mechanical power: one watt is equal to one joule per second. In summary:

The power lost as a current I flows downhill through a potential difference V is simply VI. It is measured in watts.

As an example, consider the power loss in a common flashlight bulb. When white-hot, the flashlight bulb will have a resistance of perhaps 2 ohms and will use two 1.5-volt batteries as a voltage source. The flashlight circuit is shown in Figure 6.19.

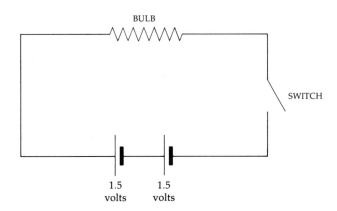

6.19 *The two flashlight batteries are said to be in series aiding and are equivalent to a single 3.0-volt battery.*

When we close the switch to light the flashlight, a current I flows clockwise in the circuit. Since the two batteries aid each other (both want current to flow clockwise) they act like a single 3.0-volt battery. As a result, the voltage drop across the resistor (the bulb) is 3.0 volts. Since the resistance of the bulb is 2 ohms, we find from $V = IR$ that the current is

$$I = \frac{V}{R} = \frac{3.0 \text{ volts}}{2 \text{ ohms}} = 1.5 \text{ amp}$$

Of course the brightly lit flashlight bulb is using electrical energy and converting it to heat and light energy. The electrical power (i.e., energy per second) lost in the bulb is

$$\begin{aligned}\text{Power} &= VI \\ &= (3.0 \text{ volts})(1.5 \text{ amps}) \\ &= 4.5 \text{ watts}\end{aligned}$$

We say that such a bulb is a 4.5-watt bulb. Notice that it is designed to operate on a certain voltage or potential difference. If the two batteries were old and weak so that their voltages were lower than 1.5 volts, the current would be less and the bulb would

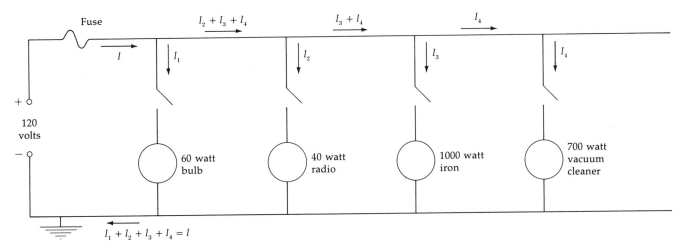

6.20 *A schematic diagram of a house circuit. The symbol ⏚ represents the ground wire.*

not glow brightly; you certainly have seen this effect. If you used three batteries instead of two with this bulb, the current would be much larger. The filament inside the bulb would become so hot that it would melt and the bulb would burn out.

House Lighting Circuits

When you buy a light bulb you need to know whether it is an incandescent or fluorescent type and what power it will use. Typical incandescent bulb sizes are 20, 40, 60, 75, 90, 100 watts, etc. The higher the power of the bulb, the more energy it uses and gives off as light; a 100-watt bulb gives off more light than a 40-watt bulb. In the United States, bulbs are usually designed to operate with a voltage source of 120 volts. Thus, a 60-watt bulb will have stamped on it "60 watts, 120 volts."

To provide electrical energy, power companies commonly furnish two wires to each house. The potential difference between them is about 120 volts. One of the wires is grounded (i.e., attached in some way to the earth), and we say it is at zero electrical level. The second wire coming to the house has a potential about 120 volts higher than the first. (Actually, these are alternating voltages, but for the present purposes, we will consider them to be steady or *direct* voltages.) Both wires run to the main power box in each house; several pairs of wires then run from this box to various sections of the house.

Figure 6.20 shows a set of such wires. Notice that the top and bottom wires have a potential difference of 120 volts. Current flows from the top wire to the bottom through the electrical device being used when the switch is closed.

The circuit shown in Figure 6.20 is called a parallel circuit: each electrical device is parallel to the others. The 60-watt bulb might represent a table lamp plugged into an electric outlet. The plug's two metal prongs connect the bulb to the 120-volt supply. One prong connects one end of the bulb's resistance wire to the top wire, and the second prong connects the other end to the lower wire. If that particular plug is controlled by a switch, no current can flow from the top wire to the bottom one through the bulb until the switch is closed. In the parallel circuit, the bulb's circuit is unaffected when switches are opened or closed in the radio and iron circuits. The current flowing through the bulb depends only upon the bulb itself and the potential difference between the upper and lower wire.

Suppose now that all the switches are closed in Figure 6.20. We can find the current flowing through each device because power in watts is simply VI. Therefore the currents are:

60-watt bulb: $I_1 = \dfrac{\text{Power}}{V} = \dfrac{60}{120} = 0.50$ amp
40-watt radio: $I_2 = = 0.33$ amp
1,000-watt iron: $I_3 = = 8.33$ amp
700-watt cleaner: $I_4 = = 5.83$ amp
Total current $= I_1 + I_2 + I_3 + I_4 = 14.99$ amp

All the current flowing from the top wire to the bottom wire in Figure 6.20 comes through the left end of the top wire. As a result, about 15 amp flows through the fuse. Of this, 0.50 amp is diverted and flows through the bulb while the rest, 14.5 amp, continues on. In the next-to-last section of the upper wire, a current $I_4 = 5.83$ amp flows through the vacuum cleaner and back to the source by way of the lower wire. No current flows in the extreme right ends of the upper and lower wires because there is no place for it to go.

When too much current flows in a wire, the wire becomes hot. Depending upon the house, house wires may safely carry up to 15 or, in some houses, 30 amp. Since excessively hot wires can set a house on fire, all house circuits contain a fuse or a circuit breaker to limit the amount of current that can flow.

A fuse is simply a piece of wire (in a replaceable plug) that melts if the current becomes too large. When it melts, the wire breaks and all current stops flowing. In newer houses, the fuse is replaced by a circuit breaker, an automatic switch that opens when the current becomes too large. When a house line becomes overloaded, the fuse or circuit breaker shuts off the whole line and all the devices on it.

When the cause of the overload has been removed, it is safe to replace the fuse or close the circuit breaker.

The line in the circuit of Figure 6.20 is already heavily loaded. Most houses have several such lines leading from the power box to various parts of the house, with a separate fuse for each line. Thus a blown fuse or circuit breaker can shut off the electricity in one part of a house without affecting other parts.

Electrical appliances such as large air conditioners and clothes driers require large amounts of power. For example, an electric stove might require 2,400 watts. If this were operated on 120 volts, a current of 20 amp would have to flow through it. However, if it were operated on 220 volts, only 10.9 amp would be needed, since power is VI. Since excessive current overheats the wires, we reduce the current by using a higher voltage. Thus large appliances are usually run from a line that gives a 220-volt potential difference; the power company runs a third wire to each house for this purpose. Many countries of the world use 220 volts instead of 120 volts as their common house voltage. Can you think of any disadvantage to using 220 volts?

Electrical Safety

Since we use electrical apparatus daily, we should understand the elements of electrical safety. Electricity can kill a person in two ways: it can cause the muscles of the heart and lungs (or other vital organs) to malfunction or it can cause fatal burns.

Even a small electric current can seriously disrupt cell functions in that portion of the body through which it flows. When the electric current is 0.001 amp or higher, a person can feel the sensation of shock. At currents 10 times larger, 0.01 amp, a person is unable to release the electric wire held in his hand because the current causes his hand muscles to contract violently. Currents larger than

6.21 *About how large could the voltage source potential be and still not injure the person shown?*

0.02 amp through the torso paralyze the respiratory muscles and stop breathing. Unless artificial respiration is started at once, the victim will suffocate. Of course, the victim must be freed from the voltage source before he can be touched safely; otherwise the rescuer, too, will be in great danger. A current of about 0.1 amp passing through the region of the heart will shock the heart muscles into rapid, erratic contractions (ventricular fibrillation) so the heart can no longer function. Finally, currents of 1 amp and higher through body tissue cause serious burns.

The important quantity to control in preventing injury is *electric current*. Voltage is important only because it can cause current to flow. Even though your body can be charged to a potential thousands of volts higher than the metal of an automobile by simply sliding across the car seat, you feel only a harmless shock as you touch the door handle. Your body cannot hold much charge on itself, and so the current flowing through your hand to the door handle is shortlived [recall from Equation (6.2) that $t = q/I$], and the effect on your body cells is negligible.

Now let us consider the situation shown in Figure 6.21 where a young man is grasping wires from the two terminals of a battery with his hands. A current will flow through his body, in through his right hand and out through his left. To find the magnitude of the current, we use Ohm's law,

$$I = \frac{V}{R}$$

where V is the battery voltage and R is the resistance of the man's body from hand to hand. R depends greatly upon the moisture on the person's hands and similar factors; however, it normally ranges from 20,000 to 500,000 ohms. If we use the lowest value to emphasize the hazard and assume that the battery has 12 volts, we have

$$I = \frac{V}{R} = \frac{12}{20,000} = 0.0006 \text{ amp}$$

The young man would probably not even feel a shock.

However, if a person touched two wires in a house lighting circuit of 120 volts, the current passing through him would be

$$I = \frac{V}{R} = \frac{120}{20,000} = 0.006 \text{ amp}$$

This current is high enough so that the person would certainly experience a severe jolt and possibly be in serious danger.

In some circumstances, the 120-volt house circuit is almost certain to cause death. One of the two wires of the circuit is always attached to the ground, so it is always at the same potential as the water pipes in a house. Suppose a person is soaking in a bathtub; his body is effectively connected to the ground through the water and piping. If his hand accidentally touches the high potential wire of the house circuit (by touching an exposed wire on a radio or heater, for example), current will flow through his body to the ground. Because of the large, efficient contact his body makes with the ground, the *resistance* of his body circuit is low. Consequently the current flowing through his body is so large that he will be electrocuted.

Similar situations exist elsewhere. For example, if you accidentally touch an exposed wire while standing on the ground with wet feet, you are in far greater danger than if you are on a dry, insulating surface. The electrical circuit through your body to the ground has a much higher resistance if your feet are dry. Similarly, if you sustain an electrical shock by touching a bare wire or a faulty appliance, the shock is greater if your other hand is touching the faucet on the sink or is in the dishwater.

As you can see from these examples, the danger from electrical shock can be eliminated by avoiding a current path through the body. When the voltage is greater than about 50 volts, avoid touching any exposed metal portion of the circuit. If a high-voltage wire must be touched (for example, in case of a power-line accident when help is not immediately available), use a dry stick or some other substantial piece of insulating material to move it. When in doubt about safety, avoid all contacts or close approaches to metal or to the wet earth. Above all, do not let your body become the connecting link between two objects that have widely different electric potentials.

BIBLIOGRAPHY

1. DART, F. E.: *Electricity and Electromagnetic Fields* (paperback), Merrill, Columbus, Ohio, 1966. A more advanced treatment than that given here, but one that should be understandable to you.

2. MORE, A. D.: "Electrostatics" (paperback), Anchor Science, Doubleday, Garden City, N.Y., 1968. A nonmathematical discussion of the behavior of charges and charged objects.

3. ANDERSON, D. L.: *Discovery of the Electron* (paperback), Momentum, Van Nostrand, Princeton, N.J., 1964. The story of how we became aware of the existence of the electron. Some mathematics used, but you may skip those parts if you wish.

4. FINK, D. G., and LUTYENS, D. M.: *Physics of Television* (paperback), Anchor Science, Doubleday, Garden City, N.Y., 1960. A completely nonmathematical discussion of television which tells how both color and black and white TV works. Clearly written and easily understood.

5. MAGIE, W. F.: *Source Book in Physics,* Harvard University Press, Cambridge, Mass., 1965. Contains excerpts from Ohm's writings.

SUMMARY

All charges found in nature have magnitudes which are integer multiples of the charge quantum, $e = 1.6 \times 10^{-19}$ coul. There are two kinds of charge, plus and minus. The electron charge is $-e$, and the charge on the proton is $+e$.

Like charges repel each other and unlike charges attract. According to Coulomb's law, the force upon charge q_1 because of a charge q_2, a distance r from it is $F = 9 \times 10^9 \ q_1 q_2 / r^2$. The force is in newtons. Like the gravitational force, the electric force obeys an inverse square law.

In materials called conductors, electrons are free to move from place to place. These are the metals and they can be used for conduits (wires) through which charges can flow. A second class of materials, called insulators, have no free charges and do not conduct electricity. An intermediate classification, semiconductors, also exists.

When a charged object is brought close to a metal, the charges within the metal redistribute under the force of the nearby charge. As a result, unbalanced charges, called induced charges, appear on various portions of the metal.

If a test charge experiences an electric force at a certain place, we say that there is an electric field at that point. We define the electric field strength E at a point in space as the force which a +1-coul test charge would experience at that point. The units of E are newtons per coulomb. It is customary to represent the field by drawing field lines; these lines are closest together where the field is strongest.

We define the electric current I in a wire to be the charge flowing past a given cross section of the wire each second. In symbols, $I = q/t$, where q is the charge which flows in time t. The unit of I is called the ampere.

For a current to flow, a potential difference V is required. If the potential difference between two points is V volts, then work in the amount qV joules is required to carry q couls from one point to the other. In addition, if a charge q falls freely through a potential difference V, it acquires an amount qV of kinetic energy in the process. Sources of electrical energy such as batteries are characterized by the potential difference that they can supply between their two terminals. This is called the emf of the battery.

When a potential difference of V volts exists across a resistance of R ohms, a current of I amp flows through the resistance. The relationship is Ohm's law, $V = IR$.

Devices that use electricity consume electrical energy at a certain rate. The rate of energy loss or production, as with mechanical systems, is called power. The power lost as a current I flows downhill through a potential difference V is VI watts.

A person experiences an electric shock when a current greater than about 0.001 amp flows through his body. The most dangerous situation arises when current flows through the portion of the body that includes the lungs or heart. Even a small current can disrupt the operation of these vital organs and cause death.

TOPICAL CHECKLIST

1. How large is the quantum of charge and what is its importance?
2. How many kinds of charge are there and what sort of forces do they exert on each other?
3. What was the original defining experiment for positive charge? Negative charge? How can you tell the type of charge on a charged object?
4. What is the unit used to measure charge?
5. Give the mathematical law that expresses the force exerted by one charge on another.
6. Compare the strengths of the electrical and gravitational forces.
7. On a molecular scale, what is the difference between a conductor and an insulator? Give examples of each category. What are the intermediate substances called?
8. How can you induce charges to appear on various portions of metals? What are induced charges?
9. When a charge is placed on a metal rod, where do the charges localize?
10. What is a gravitational field? What is its direction in this room? Draw the gravitational field of the earth.
11. What is an electric field? Draw the field near plus and minus point charges.
12. The direction of the electric field lines tells us what? The relative closeness of the lines has what significance?
13. What do we mean by "forces at a distance"?
14. What is the definition of E, the electric field intensity?
15. About how many electrons are there in a 1-ft length of wire?
16. Define electric current. What unit is it measured in? How is it analogous to water flow in a pipe?

17. Why is the current direction taken to be the direction of positive charge flow?
18. What is the mathematical relation between I, q, and t? Define each quantity.
19. Define an emf source and state what its function is in causing currents.
20. What happens as current flows through a resistance element?
21. Give the meaning of electrical potential difference. In what unit is it measured? How is it related to work?
22. What does the electric field look like between two parallel, oppositely charged metal plates?
23. As a charge q falls through a potential difference V, how much potential energy does it lose? If this happens in a vacuum, what happens to the energy?
24. How does the light originate on a TV screen?
25. What is Ohm's law and to what does it apply? How is it used to define the unit for resistance?
26. What is the definition of electric power? What is its unit? It's relation to V and I?
27. How does a house lighting circuit work? What is the purpose of a fuse or circuit breaker in it?

QUESTIONS

1. The girl touching the metal ball on the van de Graaff generator in Figure 6.22 is having her whole body charged. Why does her hair stand up? (The potential difference between the van de Graaff generator's sphere and ground is about 10^5 volts. The way it achieves such a high voltage is discussed in Chapter 11.)
2. You may have noticed that sparks sometimes jump from one piece of clothing to another when you are undressing in the winter. (This is best seen in complete darkness.) What is

6.22 *The girl is touching a van de Graaff generator; see Question 1.*

the origin of the sparks? How is this similar to the sparks you sometimes notice when combing your hair?
3. How can we be sure that the whole earth doesn't have a very large positive charge on it?
4. When a charged rod (or plastic pencil) is brought close to a metal ball (or even a tiny, slightly moist piece of paper), the rod attracts the object even though the object carries no net charge. Why?
5. Lightning is caused when charge jumps from (or to) a highly charged cloud. Suppose a highly positively charged cloud moves slowly parallel to the moist surface of the earth. What influence does it have on the earth's surface? On a tree over which it passes? On a person standing in the middle of a shallow lake? On a lightning rod?
6. In reference to Question 5, list a few of the best and worst things you could do in a severe lightning storm.
7. Phonograph records sometimes acquire an electrostatic charge, making it difficult to remove dust from them. Why is the situation

only made worse if you try to clean the record with a dry cloth? (Before answering this, rub a dry plastic pencil vigorously against wool or some similar cloth and then bring it close to a tiny slip of paper.)

8 Draw a typical house electrical system with a table lamp, a toaster, and a TV set operating on it. (You can show each as a resistor, but label each.) Show the position of the switch for each, assuming that the lamp is plugged into a wall outlet and is controlled by two switches, a wall switch and one on the lamp. Also show the fuse in the circuit and tell what its purpose is.

9 In Figure 6.9a, suppose an electron is shot to the right from point A. Describe its subsequent motion.

10 Suppose the ceiling of the room you are in were negatively charged and the floor were positive. What would the electric field diagram look like for the room? How could the existence of the field be detected?

11 During a storm, a high-voltage wire falls down and touches a man. He is knocked unconscious. (a) Discuss safe means for freeing him from the wire. (b) What first aid should be given to him?

12 A bird can stand on a high-voltage wire without being harmed. Why is this situation similar to an open garden hose lying on level ground with water passing through it?

13 One of two parallel wires in a house lighting system is at the same electrical level as the earth (provided the electricians wired it properly). If you touch this wire you will usually not be shocked. The other wire is 120 volts higher in electrical level than the earth. Why will you feel a shock if you touch it?

14 In Question 13, the intensity of the shock you will feel depends upon what you are doing at the time. If you are wearing dry, rubber-soled shoes and are not touching anything else but a dry floor, the shock will be slight. However, if your feet are wet and you are standing on a basement floor made of moist cement, or if you are holding onto a water pipe or are in a bathtub at the time, you will be hurt severely. Explain the reasons for these differences.

15 A person can have an arm burned off by a high-voltage line and still live while another person can be killed by a shock that leaves no visible damage to the body. Explain this in terms of the sensitivity of various parts of the body to electrical shock.

16 Christmas tree lights that stay on even if one bulb goes out are connected in parallel. Draw a diagram of the circuit from the wall plug through the lights.

17 Christmas tree lights that all go out when a single bulb burns out are connected in series. Draw a diagram of the circuit from the wall plug through the lights.

18 Suppose a child is playing with a lamp cord and is near a wall plug. Is he in any danger? What will happen if he cuts the lamp cord in two with a pair of cutting pliers?

19 Arrange the following in increasing order of power consumption: reading lamp, transistor radio, hair drier, clothes drier, refrigerator, electric stove. Why are some run from 220 volts rather than 120 volts?

20 Is it dangerous to use a battery transistor radio while in a tub or pool? How about a 120-volt line-powered radio? A hair drier?

PROBLEMS

1 Two identical balls are separated by a distance of 2 m. Each has a mass of 10 mg (i.e., a weight of 0.010×9.8 newtons). How large an excess charge must each ball possess if the repulsion force on one is to be equal to its weight?

2. Two identical balls, each with mass of 10 gm, carry equal but opposite charges of $\frac{1}{2}$ coul each. How far apart must they be placed if the electric attraction force of one on the other is to equal the earth's gravitational attraction for it?

3. A penny has about 10^{23} electrons in it which can move around relatively freely. Suppose two pennies, 2 m apart, have 1 percent of their free electrons removed. Then each would have a net charge of $0.01 \times 10^{23} \times 1.6 \times 10^{-19}$ or 160 coul. Find the approximate force of one penny on the other in newtons and in pounds. What do you conclude from this?

4. Through how large a potential difference must an electron be accelerated to achieve a speed of 3×10^7 m/sec, 0.10 the speed of light? ($m = 9.1 \times 10^{-31}$ kg, $q = -e$.)

5. An alpha particle is simply a helium nucleus and has $q = +2 \times 1.6 \times 10^{-19}$ coul and $m = 4 \times 1.67 \times 10^{-27}$ kg. Through how large a potential difference must it fall to be accelerated to a speed of 2×10^7 m/sec? Through how large a potential difference must a 2×10^7 m/sec alpha particle rise in order to be brought to rest?

6. In a particular position in space, a -2-μcoul charge (i.e., -2×10^{-6} coul) experiences a force of 3 newtons toward the left. What is the magnitude and direction of the electric field at that point?

7. How much work is done in carrying a 2-μcoul charge (i.e., 2×10^{-6} coul) from one terminal of a 500-volt power supply to the other?

8. For the circuit shown in Figure 6.23, find (a) current flowing through the wire at point 1; (b) the resistance of the device through which 4 amp flow; (c) the power being used by the 2-amp device.

9. In the circuit shown in Figure 6.24, suppose the current passing through the 6-ohm resistance is 3 amp. (a) Is the current flowing from B to A or from A to B? (b) Is the current through the 2-ohm resistor 1 amp or 9 amp? (c) How much current flows through the battery? (d) What is the battery voltage?

10. How much current does a bulb marked "60 watts, 120 volts" draw when properly lit?

11. A lighted 60-watt 120-volt bulb has a current of $\frac{1}{2}$ amp flowing through it. How much charge passes through it each second? How many electrons flow through it each second?

12. (a) For the circuit shown in Figure 6.25, what is the resistance of the appliance connected between AB? (b) There is a bulb at EF. What is its power rating in watts? (c) How much current is flowing through the wire at point 1? (d) What is the purpose of a fuse in a circuit such as that shown? Indicate on the diagram where it should be placed in the circuit.

13. Refer to the 120-volt lighting system shown in Figure 6.26. (a) When the switch for the 60-watt bulb is closed, how much current flows through it? (b) When the toaster is on, how much current flows through it? (c) Calling the current through the bulb I_b and that through the toaster I_t, how much current flows through the wire at point A? At point B?

14. How much current will a $\frac{1}{8}$-hp motor draw on a 120-volt circuit? (One hp = 746 watts.)

15. Estimate the electrical power (in watts) which a city of 100,000 people consumes at night (when we will assume all industry and stores to be closed). If the city obtains its power at 120 volts, how much current must flow to the city? In practice, the power is sent to the city on high-voltage towers with a potential difference of about 200,000 volts. How much current is needed at this voltage?

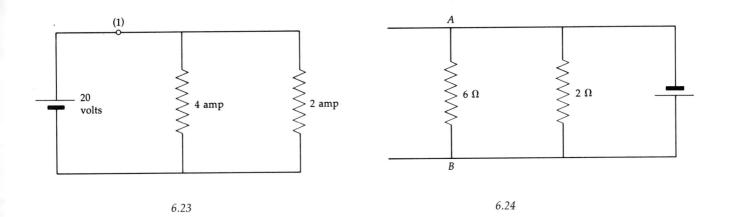

6.23

6.24

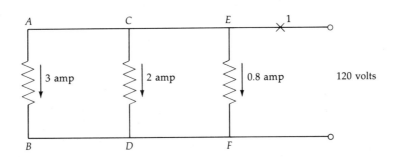

6.25

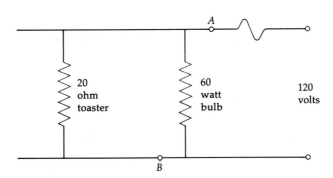

6.26

Chapter 7

ELECTROMAGNETISM

Magnetism played an important role in the development of civilization, for when man realized that needles made of magnetic material always pointed north, he was able to chart his path accurately on long journeys across the seas. Today magnetism is basic to such different aspects of technology as the transmission of television signals, the generation of electrical power, the operation of a telephone bell, and thousands of other applications. In this chapter we shall review and extend our knowledge of magnets so that we may better understand how magnetism is used in our world. (A small compass and magnet may help you in the following section.)

Magnets

You probably know that each magnet has a north and a south pole, and that the north pole of one magnet will attract the south pole of another, giving rise to the saying "unlike poles attract." The observation that two north poles repel each other, as do two south poles, leads us to state "like poles repel."

We also know that magnets attract originally unmagnetized pieces of iron. For example, in Figure 7.1, the nails hang together because the magnet's north pole induces a south pole on the nail, which is then attracted to the north pole of the magnet. The nail, now acting like a magnet, induces another nail to become magnetized, and so on. Although nails do not usually retain their induced magnetism once the inducing magnet is removed (we say they are made of "soft iron"), some alloys of iron with other metals retain their magnetism tenaciously (they are made of "hard iron") and are used in permanent magnets.

Most materials, however, have no obvious magnetic properties at all. Iron, steel, nickel, and their alloys are the only commonly encountered magnetic substances. Paper, wood, copper, aluminum, and so on, are not affected by a stationary magnet nearby. (We shall see later in this chapter, though, that all metals can be influenced by a *moving* magnet in their vicinity.)

One of the first uses we think of when discussing magnets is the compass. The compass needle is simply a bar magnet; you can construct one by suspending a bar magnet as shown in Figure 7.2. This experiment defines the north pole of a magnet as the pole that points northward. Of course, given a compass, the north pole of any other magnet can be found. Since unlike poles attract, the south pole of the compass points to the north pole of the magnet.

The primary purpose of a compass is to show direction on the earth's surface. We commonly say that the compass needle lines up in the earth's *magnetic field*. Unlike the gravitational field, which influences masses, or the electric field, which exerts forces on electric charges, the magnetic field exerts forces on magnetic poles.* Let us now discuss the magnetic fields caused by several objects.

Suppose that a compass is placed near the north pole of a bar magnet, as shown in Figure 7.3. Since the compass needle is just a tiny bar magnet mounted so that it can rotate, the needle's north pole (its pointed end) is repelled by the bar magnet's north pole. Hence it lines up pointing away from the magnet's north pole as shown. We define the magnetic field direction at the position of the compass needle to be the direction in which the compass needle points. If we represent the magnetic field by field lines as in Figure 7.4a, the field lines must include arrows to show the direction the compass needle will point when placed on the field line. Since the compass needle points away from the north pole of a bar magnet, the field lines come out of and away from north poles.

Similarly, we find that the compass needle must point *toward* the south pole of the bar magnet, as shown in Figure 7.3. Because the compass needle tip, a north pole, is attracted to the south pole of the bar magnet, the field lines must point toward the south pole as shown in Figure 7.4a. Clearly, the other compass needles in Figure 7.3 will point as shown, and so the field lines must be drawn as shown in Figure 7.4a. Similar reasoning can be used to draw the field lines in Figures 7.4b and c. Remember that *field lines always come out of north poles and enter into south poles.*

* We will see later that the magnetic field also exerts forces on moving electric charges and that it is related to the electric field.

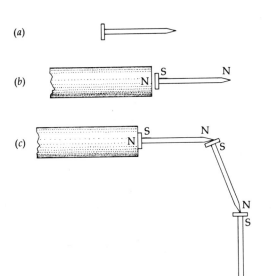

7.1 Magnets cause iron objects to become magnets by induction.

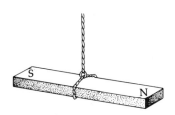

7.2 The north pole of a magnet is defined to be the pole which points north when the magnet is suspended as shown.

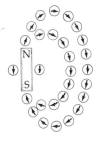

7.3 A compass needle points in the direction of the magnetic field.

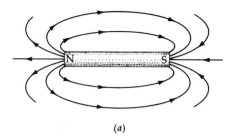

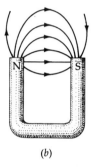 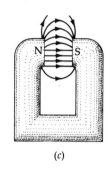

7.4 *Using the fact that a compass needle should line up along the field lines, you should be able to show that the lines drawn are reasonable.*

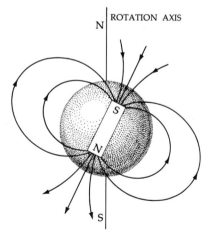

7.5 *The magnetic poles of the earth do not coincide with the poles defined by its axis of rotation. Moreover, the pole near the earth's north pole is actually a south magnetic pole.*

As we know, a compass needle points toward the north on the surface of the earth. This means that the field lines of the earth are directed from the Antartic to the Arctic. A picture showing the earth's field is given in Figure 7.5. Notice that the field is similar to that of a bar magnet. But, as we see, this bar magnet does not coincide with the earth's rotation axis. Instead, the magnet's axis comes out of the earth's surface in the vicinity of Labrador. Over a period of thousands of years, the earth's inner magnet shifts in direction. This helps us to chart the history of the earth, as we shall see in Chapter 16. It may surprise you at first to learn that the Arctic pole of the magnet, which is near the earth's geographic North Pole, is a *south magnetic pole*. This is obviously true because the pole in Labrador attracts the north pole of a compass, and we know that the compass tip is attracted by south magnetic poles. Although the causes of the earth's magnetic field are not well understood, we shall see in the following section that all magnetic fields appear to be the result of electric charge flow.

Sources of Magnetic Fields

The first clue to the origin of magnetic fields was discovered in 1820 by Hans Christian Oersted. He found that a compass needle deflects when a switch is thrown to turn on a current in a nearby circuit. Pursuing this observation further, he showed that a magnetic field always exists near a wire through which current is flowing. By use of a compass, the field can be traced; it is found that the magnetic field circles around a straight wire through which a current is flowing. The field is shown in Figure 7.6.

As you might guess, the strength of the magnetic field (which we represent by the symbol B) is directly proportional to the current in the wire. Furthermore, B decreases with increasing distance

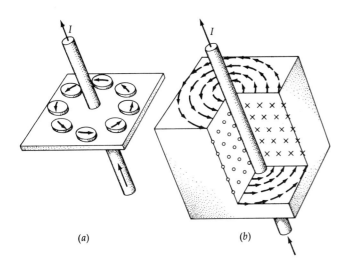

7.6 Notice that the magnetic field lines caused by a current have no ends; they circle back upon themselves.

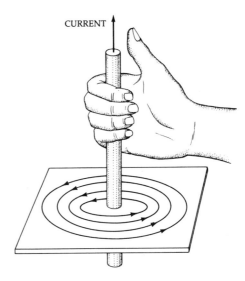

7.7 The right-hand rule for remembering the direction of a magnetic field caused by a current in a wire.

from the wire; B is proportional to $1/r$, where r is the distance from the axis of the wire. In the units commonly assigned to B, the exact relation is

$$B = \frac{2 \times 10^{-7} I}{r} \qquad (7.1)$$

(Recall that 10^{-7} means to divide by 10 seven times.) When I is measured in amperes and r in meters, the unit of B is the *tesla*. Another unit, called the *gauss*, is sometimes used; 10,000 gauss = 1 tesla.

To get an idea about the relative strengths of magnetic fields, the earth's field is about 1 gauss (10^{-4} tesla). The magnetic field at a distance of 10 cm (0.10 m) from a wire carrying a current of 5 amp is, from Equation (7.1),

$$B = \frac{(2 \times 10^{-7})(5 \text{ amp})}{0.10 \text{ m}} = 100 \times 10^{-7} \text{ tesla}$$

which is 0.10×10^{-4} tesla or 0.10 gauss. This is only about one-tenth as large as the earth's field. We will see, however, that the field may be greatly increased by winding the wire in the form of a coil on a bar of iron.

A simple rule for remembering which way the magnetic field circles a wire is illustrated in Figure 7.7. It is called the *right-hand rule*. If you grasp the wire with your right hand so that your right thumb points in the direction the current is flowing, the fingers of your hand circle the wire in the same direction as the magnetic field. Notice that the magnetic field lines have no beginning or end; they always circle back upon themselves. Thus they seem to be quite unlike electric field lines, which start on positive charges and end on negative charges. Even the magnetic lines that appear to stop and originate on the poles of magnets actually do not do so. Instead, they simply circle through the magnet and come out the other end.

When a wire is bent into the form of a circular loop, the magnetic field due to the current in the wire still circles the wire. However, it is distorted as shown in Figure 7.8. Using the right-hand rule, you should be able to show that the field directions are at least qualitatively correct. Notice that the field in part (*b*) of the figure is much like that you would expect from a short, fat bar magnet, as shown in part (*c*). The region above the loop in (*b*) looks like the north pole, while the lower region looks

ELECTROMAGNETISM 141

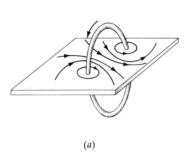
(a)

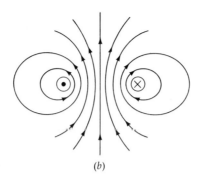
(b)

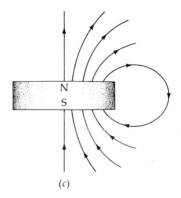
(c)

7.8 Two views of the magnetic field near a current-carrying loop are shown in (a) and (b). In (b), the current is going into the page at the cross and coming out at the dot. In (c) we see that a short, fat bar magnet has a field similar to that of a current loop.

like the south pole. This resemblance is even more striking in the following configuration.

Consider the coil of wire shown in Figure 7.9a. Each loop of the coil generates a magnetic field, and the combined effect is as illustrated. The more closely wound coil (called a _solenoid_) shown in cross section in Figure 7.9b is more typical. In this case the magnetic field is uniform inside the solenoid. This is, in fact, one of the best ways for obtaining a uniform magnetic field.

> The strength of the field inside a solenoid is given by a very simple formula:
>
> $$B \text{ inside solenoid} = 4\pi \times 10^{-7} \frac{NI}{L} \quad (7.2)$$
>
> where N is the total number of loops on the length L of the solenoid. Clearly, B is much larger than for a single wire, as we can see by comparing Equations (7.2) and (7.1).

The field inside a solenoid can be intensified by a factor of nearly a thousand by winding the solenoid on an iron rod. In the next section we will see what happens to iron in this situation.

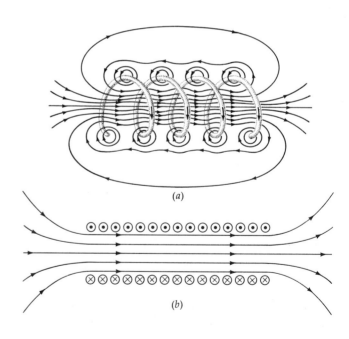

7.9 The magnetic field of a solenoid looks much like the field of a bar magnet.

Atomic Interpretation of Magnets

The magnetic properties of iron and similar substances which show large magnetic effects were first correctly explained by André Marie Ampère in 1820. He based his explanation upon the similarity between the field of a solenoid, shown in Figure 7.9, and the field of a bar magnet. The right-hand end of the solenoid in Figure 7.9 acts like the north pole of a magnet, since the field lines come out of that end. The opposite end behaves like a south pole, since the fields lines come back to it. Although we usually draw the field lines as though they ended on the poles of a bar magnet, they actually pass through the magnet like the lines in a solenoid.

Ampère suggested the model of a bar magnet shown in Figure 7.10. Unlike many of his contemporaries, Ampère believed that all solids are composed of small particles, the particles we now refer to as atoms. He suggested that atoms have charges circling around in them. As a result, an atom is like a tiny current loop. In a bar magnet, all the atoms line up as shown in the cross section in Figure 7.10. (We have pictured the atoms to be rectangular in cross section and vastly enlarged, in order to facilitate the discussion.)

We see that the currents due to adjacent atoms cancel each other in the interior of the magnet (i.e., they are side by side and opposite in direction). However, the atoms at the surface have no neighbor to cancel their currents and so a current flows around the magnet's surface as shown. This is equivalent to a current-carrying coil of wire wound around the bar magnet, and so the bar magnet should have a field exactly like that of a solenoid.

According to Ampère's picture, the atoms in a bar magnet act like current loops. Clearly, when we magnetize a bar of iron, we line up the atoms in the iron. In fact, to magnetize a piece of iron, we can place it inside a solenoid. The magnetic field inside

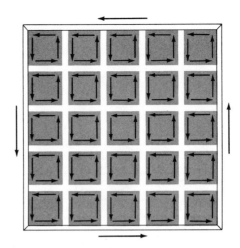

7.10 *Amperian currents within the oriented atoms cancel inside the bar magnet but give rise to a solenoid-like current flowing around the outside of the magnet.*

the solenoid forces the atoms of the iron to line up along the solenoid axis, and so it becomes a magnet. With few exceptions, the magnetic field caused by amperian currents in iron is much stronger than the field caused by the current in a solenoid.

The combination of the solenoid current and the amperian currents increases the field by a factor of hundreds or thousands, depending on the particular situation. This combination of an iron core inside a solenoid is called an <u>electromagnet</u>. In an electromagnet, the type of iron used is such that the atoms spontaneously disalign when the solenoid current (and thus the solenoid field) is reduced to zero. As a result, an electromagnet can be turned on and off. In contrast, the iron of permanent magnets is modified so that the atoms remain aligned.

Forces on Currents in Magnetic Fields

From a practical standpoint, magnetic fields are important because they exert forces on currents in wires. This feature forms the basis for operation of electric motors and similar devices. The physical

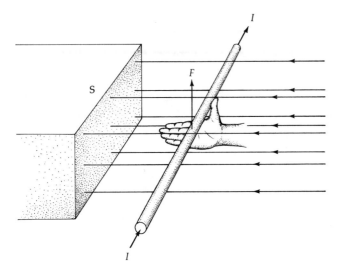

7.11 *When the right hand's fingers are directed along the field, the thumb in the direction of the current, the palm of the hand pushes in the direction of the force.*

phenomenon involved is shown by the experiment illustrated in Figure 7.11. In this figure, a wire is carrying a current I past the south pole of a magnet. The magnetic field flows into the south pole. If we perform this experiment, we find that the wire experiences a force F which is perpendicular to both the current and the field lines. Here too a right-hand rule can help us remember the direction of the force:

> *If you hold your opened right hand with fingers pointing in the direction of the field and thumb pointing along the wire in the direction of current flow, the palm of your hand will push in the direction of the force.*

As you might expect, the magnitude of the force is proportional to the strength of the magnetic field B and to the current in the wire I. However, the force also depends upon the angle between the magnetic field lines and the current line (the wire). If the angle is 90° as shown in the figure (i.e., the field lines are perpendicular to the wire), the force on the wire is maximum. However, if the field lines are parallel to the wire, there is no force at all on the wire. This is illustrated in the drawing at the right

in Figure 7.12. We shall be concerned only with cases where the field and wire are perpendicular (maximum force) or parallel (zero force).

To see how this effect makes an electric motor turn, refer to Figure 7.13, which shows the basic principle. The motor consists of a magnetic field (usually furnished by an electromagnet) and a coil mounted so that it can rotate about an axis through its center. When a current I flows through the coil, it activates the magnetic field of the electromagnet.

If you examine the forces on the vertical section of the coil in Figure 7.13, you will see how they can cause the coil to rotate about the axis as indicated. (Try this using your hand as the magnetic field and a book as the coil.) Notice that the rotation effect of the forces is at a maximum in (b). Can you show that the forces in (a) cause no rotation at all because the plane of the coil is perpendicular to the field lines?

In a motor, the axis of the coil is the rotating shaft of the motor and it rotates with the coil. The shaft, in turn, rotates whatever the motor is designed to operate. We see that electrical energy, which causes the magnetic field and the force on the coil, is transformed to mechanical energy as the motor rotates and performs work. We have oversimplified the

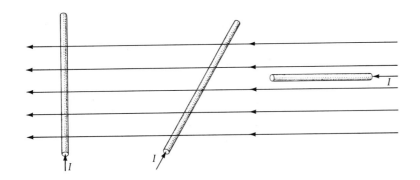

7.12 The force on the wire is largest in the situation at left. It is zero for the wire at right. Can you show that the force would rise perpendicular to the page?

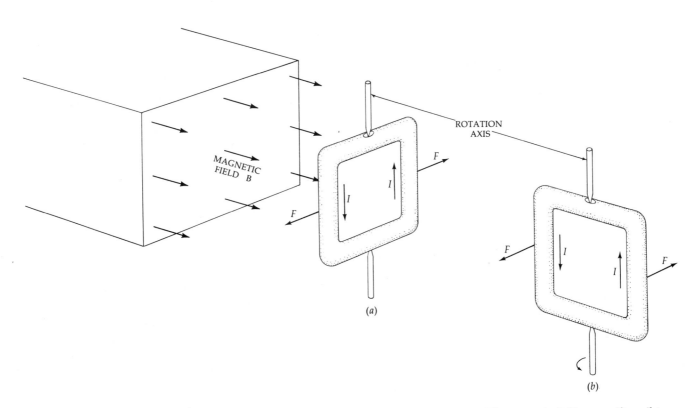

7.13 The magnetic field causes the coil to rotate about the axis.

ELECTROMAGNETISM 145

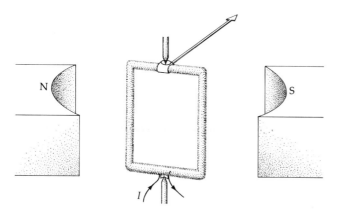

7.14 As the coil turns, the needle moves along a calibrated scale. A restoring force is furnished by a spring (not shown) attached to the coil

operation of the motor in this discussion so that the simplicity of the energy interchange mechanism will be clear. In practice, the rotating coil consists of several coils at various angles, and the current in the coil must be reversed each half revolution of the coil. You can see why this is true if you notice that, after half a turn, the two sides of the coil have exchanged places. To maintain the same turning force direction on the coil, the current must flow as shown in the diagram. However, since the two sides have been interchanged, the current direction in the coil must also be reversed if the motor is to continue running.

The magnetic force on a loop of wire is also used in the operation of meters. The internal workings of an ammeter, for example, are outlined in Figure 7.14. When the current to be measured is sent through the coil, the coil experiences a force trying to turn it because it is in the magnetic field of the two magnets. In turning, the coil distorts a spring and the coil turns only until the turning force due to the magnetic field balances the spring's restoring force. The amount the coil turns is designed to be proportional to the current flowing through the meter. A needle attached rigidly to the coil moves along a scale as the coil turns. The manufacturer calibrates the scale so it reads directly the current which flows through the ammeter.

A voltmeter is similar except that it has a high resistance. When the voltmeter is connected across a battery or other potential difference, a small current flows through it; in a good voltmeter, the current is negligibly small. However, the meter is designed so that even this small current will cause the coil to rotate. The manufacturer calibrates the scale so that it reads the potential difference across which the voltmeter is connected.

Many other devices operate on the principle that a current-carrying wire experiences a force in a magnetic field. We shall see in the next section that the force is not restricted to moving charges in a wire. It is also exerted on charges moving freely through space.

Magnetic Forces on Moving Charges

Since currents experience forces when placed in a magnetic field, we are tempted to hypothesize that the moving charges which cause the current also experience a force. Even in the mid-1800s there was speculation on this point. However, this supposition could not be tested until the late 1800s, when vacuum techniques had become advanced enough. The experiment is not easily performed in the presence of air, since the moving charges collide with the air molecules too often.

When at last it was possible to shoot charged particles through a vacuum tube in a magnetic field, it was found that the expected force on the moving charge does exist, as illustrated in Figure 7.15. Let us consider the positive charge to have an upward velocity through the magnetic field. We can use the right-hand rule we used for currents to find the direction of the force on a moving positive charge; after all, a current is simply a stream of positive charges. As we see in Figure 7.15a, the

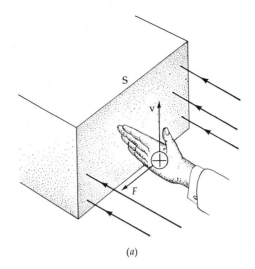

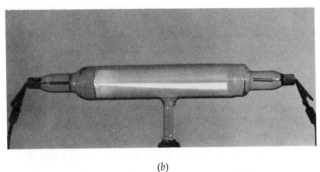

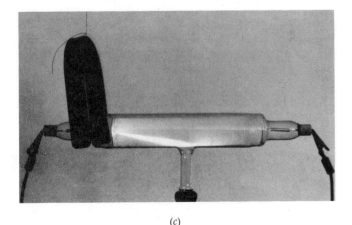

7.15 *The right-hand rule applies to charged particles moving in a magnetic field. The north pole is on the near side of the tube.*

force on the moving positive charge is given by the right-hand rule. The magnitude of the force is proportional to the strength of the field B and the speed of the particle. Furthermore, the force varies with the angle between the velocity and the field just as it does for the angle between the current in a wire and the field lines. In other words, if the particle is moving along a field line, the force is zero; if it is moving perpendicular to a field line, the force is maximum.

Although the rule we have just given is for the direction of the force on a positive charge, we can use it to find the force direction on negative charges as well. The force on a negative charge is opposite in direction to that on a positive charge. Therefore we find the force assuming the charge to be positive and then reverse the direction of the force if the charge is actually negative. To test your mastery of this point you might examine the particles shooting through the vacuum tube in Figures 7.15*b* and *c*. Using the data shown there you should be able to tell whether the charges shooting through the tube are positive or negative.

The magnetic force on a moving charge does not speed up or slow down the charge. Like the sun's gravitational pull on the earth, the magnetic force on a particle affects only the direction of motion of the charge. To see why, consider a positive charge of mass m moving perpendicular to a magnetic field of strength B as shown in Figure 7.16. The magnetic field is directed into the page; the crosses represent the tails of the receding arrows on the field lines.

At the bottom of the figure on the left, the positively charged particle is moving horizontally and toward the right. The right-hand rule tells us the magnetic field exerts a force on it which is perpendicular to both the line of motion and the field lines; the direction of the force is upward, as shown. Notice that *the force is perpendicular to the motion of the particle and hence does no work on it*. It simply deflects the particle upward.

ELECTROMAGNETISM 147

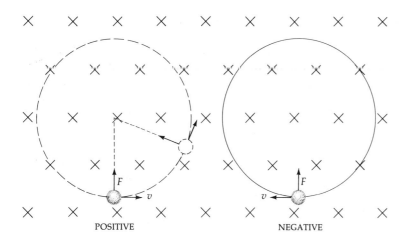

7.16 *A charged particle moving perpendicular to a magnetic field follows a circular path.*

As mentioned above, this situation is similar to the earth circling the sun or a satellite circling the earth: In all these cases a force is pulling the object out of a straight-line path. For the particle of Figure 7.16, the force is always directed toward the center of the circle (check the force direction at several points and see). As a result, the charged particle follows a circular path. The radius of this circle turns out to be proportional to the momentum of the particle, mv, divided by the product of the field strength B and the charge q. Therefore

$$r = \frac{mv}{Bq} \qquad (7.3)$$

If we know B, q, and v and measure the radius of the circle r, we can compute the mass of the particle. This is the common method for determining the mass of atomic-sized particles.

A negative particle of equal mass moving at the same speed would follow a circular path of identical size. However, since the forces on a negative charge are opposite to those on a positive charge, a negative charge would go around the circle in the reverse direction as shown on the right-hand side of the figure. Thus from the deflection of a charged particle in a magnetic field, we can determine both the mass of the particle and the sign of the charge on it.

EXAMPLE 7.1

An electron with $v = 4 \times 10^7$ m/sec is shot perpendicular to the earth's field which we can take to be 1 gauss. What is the radius of the circle in which it moves? Repeat for a proton.

Method: We need only substitute in Equation (7.3), taking care to use proper units. Since

$$B = 1/10{,}000 \text{ tesla}$$
$$m = 9.1 \times 10^{-31} \text{ kg}$$
$$q = 1.6 \times 10^{-19} \text{ coul}$$
$$v = 4 \times 10^7 \text{ m/sec}$$

the relation $r = mv/Bq$ yields

$$r = 2.3 \text{ m}$$

For a proton, m is 1,840 times larger and so r will be larger by that factor.

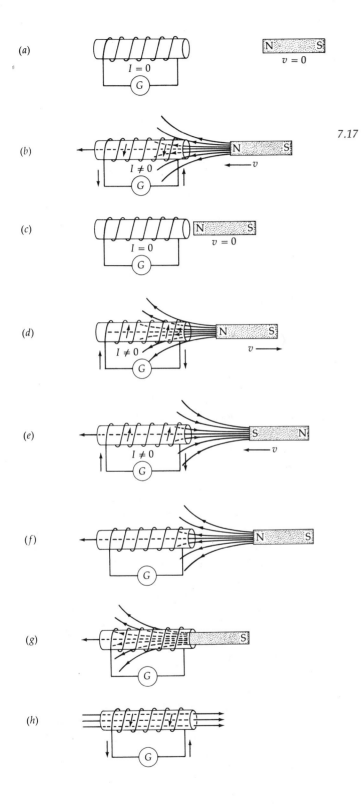

7.17 Current is induced in the coil only when the number of lines through it is changing. How can we predict which way the current will flow? (a) No change. (b) Magnet adding flux in the coil. (c) No change. (d) Magnet reducing flux in the coil. (e) If the magnet's poles are reversed, the direction of the current is also reversed; compare (e) with (b) above.

Induced EMFs

In addition to exerting forces on moving charges and current-carrying wires, magnetic fields cause another important effect: A changing magnetic field can induce a current to flow in a loop of wire. Michael Faraday (1791–1867) studied this effect extensively and published his results in 1831. They are summarized in Figure 7.17.

In the figure we see a coil of wire wound on an iron rod (since iron greatly increases magnetic effects), forming a closed circuit through a galvanometer G (a sensitive ammeter). The circuit has no battery, so as long as neither the magnet nor the coil is moving, no current flows through the galvanometer. This situation is shown in Figure 7.17a.

However, as soon as the north pole of the magnet starts to move toward the coil, a current begins to flow as indicated in (b). If the magnet stops moving, as in (c), the current also stops. When the direction of the magnet's motion is reversed, the direction of flow of the current in the coil is also reversed, as shown in (d). Moreover, if the magnet's poles are reversed as in (e), the current direction also reverses.

Faraday interpreted these experiments in terms of

MICHAEL FARADAY (1791–1867)
THE ROYAL INSTITUTION

Faraday, the son of an English blacksmith, received only a rudimentary education before finding work as a bookbinder's errand boy in 1804. Soon he became the bookbinder's apprentice, and this gave him the opportunity to read the books passing through his hands. Growing fascinated by science, he carried out such simple experiments as his meager salary allowed and attended public scientific lectures, the most inspiring of which was the series of lectures given by the great chemist, Sir Humphrey Davy. A few years later, in 1813, Faraday took the bold step of writing to Davy to request employment in his laboratory. Davy was impressed by the notes Faraday enclosed, which he had taken at the lecture series, and offered Faraday a job as a servant in his laboratory. From these humble beginnings, tutored by Davy, he advanced swiftly as a result of his obvious experimental skill and insight. In 1825, he was made director of the laboratory of The Royal Institution. His whole life was devoted to experimental investigations, particularly in electricity. He is recognized as one of the greatest experimental scientists of the 19th century.

an imaginary battery in the coil. He attributed the current induced in the coil to a hypothetical battery which we call an *induced emf*. The induced emf obviously depends upon the way the magnetic field lines are *changing* in the coil. We call the number of these lines passing through the coil the *flux*. When the flux, or number of lines through the coil, is constant as in (a) and (c), the induced emf is zero. But whenever the flux through the coil is increasing, as in (b) and (e), or decreasing, as in (d), the coil has an emf which causes current to flow in it. Faraday showed that the induced emf is directly proportional to both the rate at which the flux is changing and to the number of loops of wire on the coil. This is summarized as *Faraday's law*:

> *The induced emf in a coil is proportional to both the rate of change of flux through the coil and the number of loops of wire on the coil.*

We emphasize again that *the induced emf exists only when the number of field lines through the coil is changing.*

Notice that the direction of the current in the coil, and thus the emf, is not the same in (b) as it is in (d). We can determine the direction of the induced current and emf by the way the flux through the coil is changing. When the magnet in (f) is being moved to position (g), the induced current is in the direction shown in (h). The induced current itself produces flux through the coil to the right, whereas the motion of the magnet from (f) to (g) puts flux through the coil to the left. It is as though the coil is trying to cancel the increasing flux to the left by inducing a current in itself which causes a flux toward the right.

This is the way the coil always responds. Whenever the flux is changing, the coil induces an emf and current in itself to cancel the change. You might check your understanding of this by seeing if you can explain the directions of the induced current in (d) and (e). Part (e) may cause you trouble until you realize that the magnet is causing an increase of flux directed toward the right. The direction of Faraday's induced emf is summarized in Lenz's law:

> The induced emf is always in such a direction as to keep the flux through the coil from changing.

The Electric Generator

Induced emf's are basic to the operation of electric generators, the devices by which the power companies generate electric power. To generate a voltage, we simply change the flux through a coil. For example, we could conceivably light a light bulb with the device shown in Figure 7.18. Suppose the magnet slides back and forth, toward the coil and then away from it. The changing flux through the coil will induce an emf in the coil.

When the magnet is approaching the coil, current will flow toward the right through the bulb. When the magnet moves away from the coil, the induced emf is reversed and the current then flows toward the left through the bulb. If the magnet is strong enough and if the coil has a large number of loops and is wound on an iron rod, the induced current will be large enough to light the bulb. The bulb gives off energy in the form of light and heat.

This energy comes from the person or machine that causes the magnet to move back and forth. To see this, refer back to Figure 7.17f, g, and h. Notice in (h) that the induced current in the coil causes the coil to act like a bar magnet. When the north pole of the magnet is brought closer as in (f), the coil acts like a bar magnet with its north pole repelling the approaching magnet. Notice that when the magnet recedes from the coil as in (d), the coil tries to pull it back by inducing a south pole to attract the north pole of the magnet. Clearly, whatever moves the magnet back and forth must do work against these forces exerted upon the magnet by the

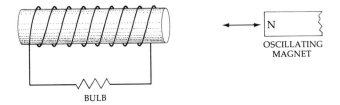

7.18 As the magnet oscillates back and forth, induced current will flow back and forth thru the bulb causing it to light.

coil. This work appears as electrical energy (i.e., battery or induced emf energy) in the coil.

The generator in Figure 7.18 would operate equally well if the magnet was motionless and the coil was oscillated back and forth. The only requirement is that the flux through the coil should change, and this will occur in either case. Since Faraday's law says the induced emf is proportional to the *rate of change* of flux through the coil, the induced emf will be large when the flux changes rapidly. Hence the induced emf will increase as the rate of vibration of the magnet is increased.

The basic operation of a commercial electric generator is shown in Figure 7.19. We see a coil of wire mounted on an axle and suspended in a magnetic field. The magnetic field can be furnished by a permanent magnet or an electromagnet. An external agent (such as a waterwheel or steam turbine) is attached to the axle and causes the coil to rotate. As we see from the diagram, the number of flux lines going through the coil changes as the coil rotates. When the angle $\theta = 0$, the field lines skim past the plane of the coil and no lines go through it; when $\theta = 90°$, a maximum number of lines go through it.

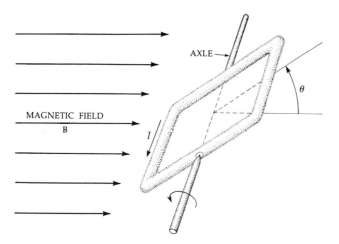

7.19 As the coil rotates on its axle, the flux changes through it; as a result, an alternating emf is generated in it.

As the flux through the coil changes, an emf is induced in the coil. For example, if θ is increasing and the coil is in the position shown, the induced emf causes current to flow in the direction indicated in the figure. (Can you show that this is the correct direction?) However, when θ becomes larger than 90°, the direction of the induced current reverses. In fact, after every half revolution the current in the coil reverses.

The emf generated in a rotating coil such as this is shown in Figure 7.20. It is called an _alternating_ emf or an _ac_ emf (ac stands for _a_lternating _c_urrent). Since most power stations generate their power this way, the voltage they furnish to your house is also ac. In effect, the emf is like a battery power source that is constantly reversing its polarity. Ordinarily the coil rotates through 60 revolutions each second and so the voltage goes through 60 cycles each second. This is why we refer to the power furnished by the power company as 60 cycles/sec ac (60 cps ac).

A commercial generator requires mechanical energy to rotate the coil. When current flows in the coil, we know that the wire (or coil) carrying a current in a magnetic field experiences a force. From Figure 7.19 and the right-hand rule, you can see that this force is directed such that it stops the coil from rotating. Mechanical energy overcomes this force and turns the coil; as it turns the coil it is converted to electrical energy. _Thus a generator converts mechanical energy to electrical energy._ Mechanical energy may come from the potential energy of falling water turning a waterwheel which rotates the coil, or from the heat energy of burning coal, or from the nuclear energy released in a nuclear reactor. We shall investigate these energy sources in later chapters.

The Transformer

Electric power generating stations are usually designed to produce electricity at high voltages. For example, it is not uncommon for a generating station to furnish electrical energy at a voltage of hundreds of thousands of volts. It is most economical to transport power at high voltages since power is VI and thus I can be kept small. However, such high voltages are not safe in homes. In addition, many of the components inside electronic devices such as radios and TV sets require widely different voltages. For these reasons a _transformer_ is necessary to change one ac voltage to another voltage.

The transformer uses induced emf's to change one voltage to another. It consists of a _primary coil_, which is connected to an ac power source, and a _secondary coil_, in which an emf is induced. They are shown in Figure 7.21. The ac voltage source in the primary circuit causes the current to constantly alternate direction in the primary coil. Hence the field lines shown will reverse their directions constantly. This causes a continuously changing flux through the secondary coil which, according to Faraday's law, will induce an emf in the coil. Since the flux reverses in a cyclic fashion, the induced emf in the secondary coil is an alternating emf or voltage.

The magnitude of the induced ac voltage in the secondary coil depends on the number of loops of

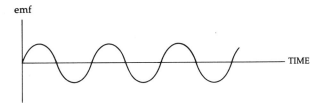

7.20 *An ac generator delivers a voltage of the form shown above.*

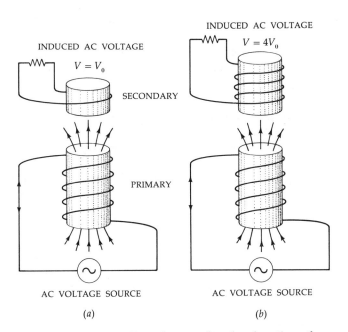

7.21 *Since the secondary has four times the number of loops in (b) as it does in (a), the induced emf in (b) will be four times larger.*

wire on the secondary coil. As shown in Figure 7.21, the induced voltage increases by a factor of 4 as the number of loops on the secondary coil in (b) is made 4 times larger than in (a). This is, of course, exactly what Faraday's law predicts. Thus we can obtain any size induced voltage we wish in the secondary coil circuit by choosing the appropriate number of loops for the secondary coil. If the secondary coil voltage is larger than the voltage in the primary, the device is called a *step-up* transformer. Typical of this type is a neon sign transformer which plugs into the ordinary 110-volt ac source and delivers about 10,000 volts to the neon sign. Every TV set has a step-up transformer to provide the high voltage needed to accelerate the electron beam in the tube.

So called *step-down* transformers have fewer loops of wire on the secondary coil than on the primary. TV sets and radios have several of these. They reduce the 110-volt power line voltage to the much lower voltages used for transistors and vacuum tubes.

Measurement of Atomic Masses

To conclude this chapter and introduce concepts important in later chapters, let us discuss how we determine the masses of electrons, protons, atoms, and molecules. When we discussed the motion of charged particles in a magnetic field, we found that if the charged particle (charge q and mass m) moves with speed v perpendicular to a magnetic field which has an intensity B, the radius of the circle on which it moves is given by

$$\text{Radius} = \frac{mv}{qB}$$

Hence, if v, q, r, and B are known, the mass of the particle can be found.

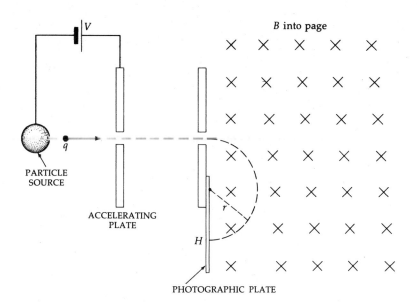

7.22 *A mass spectrograph.*

We can obtain a known magnetic field B by use of a solenoid or, alternatively, a calibrated magnet. The charged particles can be accelerated through a potential difference V to give them kinetic energy; the loss in electrical potential energy Vq will equal $\frac{1}{2}mv^2$ provided the particle's speed is not too high. Depending upon the particle, the charge q will be

$$q = 1.6 \times 10^{-19} \text{ coul} = +e$$

or perhaps twice or three times this value. In any given situation, the proper charge is known. For example, for an electron, $q = -e$, whereas for a sodium ion (i.e., a charged atom), $q = +e$. Therefore q, B, and v will be known and so m can be computed from the radius of the circular path followed by the particle.

A device to measure the radius r is shown schematically in Figure 7.22. The particle moves through a vacuum chamber to avoid collisions with air molecules. An ion (i.e., a charged atom) or other charged particle escapes through a small hole in the particle source. Since there is a potential difference V between the source and the accelerating plate, the charged particle acquires a kinetic energy equal to the loss in electrical potential energy, Vq. (You should be able to tell whether q is positive or negative in Figure 7.22 from the polarity of the battery.)

The high-speed particle passes through a slit in the accelerating plate and coasts on to the next plate. After passing through a small hole in this second plate, the particle enters the magnetic field region. As discussed before, the field causes the particle to follow the circular path shown. (Is this direction of deflection correct for the type charge you think q to be?) After completing a half circle, the particle strikes a photographic plate. A stream of such particles makes a mark on the plate where they strike it. Measuring the location of this spot on the plate allows us to determine the radius of the circle r. Knowing r, v, q, and B, we can calculate the mass of the particle.

This device, called a *mass spectrograph*, gives extremely precise values for the masses of ions and atoms. This will be important to us when we study the structure of atomic nuclei. To find the mass of electrons we use a modified apparatus, since the

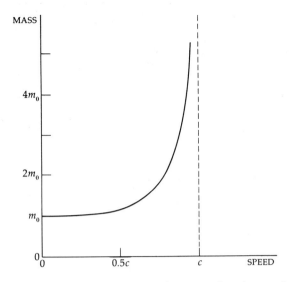

7.23   As the particle approaches the speed of light c, its mass becomes larger than its rest mass m_0.

electron mass is only about 1/1,840 as large as the smallest average atomic mass.

Measurement of the electron mass presents a puzzle. If we follow the procedure we have described, we obtain different values for the electron mass unless the potential difference V is kept small. In fact, when we measure the speeds of the electrons we find that at large values of the potential V, the measured speeds do not agree with those we compute from $Vq = \tfrac{1}{2}mv^2$.

Even when we measure the speed of the electrons directly and use the measured speed of the electron to compute its mass from the radius of the circle which it follows, we find a peculiar result. A graph of the experimental results for the mass of the electron as a function of its measured speed is shown in Figure 7.23. Notice that at all ordinary low speeds, the mass of the electron is constant. It's value is $m = 9.1 \times 10^{-31}$ kg. However, as the electron's speed comes close to the speed of light, $c = 3 \times 10^8$ m/sec, the measured mass of the electron increases. (The symbol c is used universally for the speed of light.) In fact, it appears that the electron would have infinite mass if it could go as fast as light. If a potential difference of 1 million volts accelerates the electron, the electron mass increases by three times the value found at low speeds.

This puzzling behavior, an increase of particle mass at high speeds, is also observed for protons and atoms. It is more difficult to show this experimentally for these more massive particles, since we have to accelerate them through higher voltages to make their speeds approach the speed of light. Nevertheless, detailed experiments on all sorts of small particles show that for all of them, their masses at low speeds (called <u>rest masses</u>, m_0) are smaller than their masses at very high speeds. The graph of Figure 7.23 appears to be true for any particle. As the particle's speed approaches the speed of light, the mass of the particle seems to become infinite. Einstein predicted this even before the actual measurements had been made. We will discuss Einstein's prediction and his theory of relativity in Chapter 10.

BIBLIOGRAPHY

1   BUECHE, F.: *Principles of Physics*, 2d ed., McGraw-Hill, New York, 1972. See chapter 19 for a more detailed and quantitative treatment of the topics of this chapter.

2   MAGIE, W. F.: *Source Book in Physics*, Harvard, University Press, Cambridge, Mass, 1965. Contains excerpts from the writings of Faraday and his fellow workers in magnetism. In particular see pp. 387–480.

3   DART, F.: *Electricity and Electromagnetic Fields* (paperback), Merrill, Columbus, Ohio, 1966. Contains a more detailed treatment of the phenomena discussed in this chapter.

4   BITTER, F.: *Magnets* (paperback), Anchor Science, Doubleday, Garden City, N.Y., 1959. A readable discussion of magnets written by one of the leading scientists in this field.

5 HESSE, M.: *Forces and Fields* (paperback), Littlefield, Adams, Totowa, N.J., 1965. A historical treatment of the field concept as it applies to gravitation, electricity, and magnetism. The development of this powerful concept is traced from the earliest times to its impact on present-day theories of nuclear structure.

6 MACDONALD, D.: *Faraday, Maxwell and Kelvin* (paperback), Anchor Science, Doubleday, Garden City, N.Y., 1964. An interesting, storylike presentation of the lives and works of these three famous men.

SUMMARY

Every magnet has two poles, a north and a south pole. When suspended as a compass needle, the north pole of a bar magnet points north. If two adjacent magnets are brought close, unlike poles attract and like poles repel each other.

When a compass needle, a tiny bar magnet, is placed in a magnetic field, it orients itself so that the north pole of the needle points in the direction of the field. We can represent magnetic fields by drawing lines much as we did for electric fields. These lines, called flux lines, come out of the north pole of a magnet and enter the south pole. We represent the strength of a magnetic field by the symbol B; its unit is the tesla (10,000 gauss).

Magnetic fields can be produced by current flowing in a wire. In the case of a current in a long straight wire, the field lines circle the wire. The direction in which the field circles the wire is given by the right-hand rule. Inside a long solenoid, the field of a solenoid is like the field of a bar magnet of the same size.

Ampère attributed the large magnetic effect of iron to atomic loop currents circulating within the iron. When an iron bar is magnetized, these current loops are lined up so that their individual magnetic fields are added together. The result is a strong magnetic field.

When a wire carrying a current is placed in a magnetic field, the wire experiences a force. The direction of the force is given by a modification of the right-hand rule. These forces make the coil in a motor rotate. They are also basic to the operation of ammeters and voltmeters.

When a moving charged particle enters a region of a magnetic field, the particle will experience a deflecting force unless it is moving parallel to the field lines. If its motion is perpendicular to the field lines, the particle will move in a circular path provided the field is uniform. The force on the particle is always perpendicular to the direction of motion and so it does no work. By measuring the radius of the circle, we can determine the momentum of the particle.

If the magnetic field in the vicinity of a coil of wire is changed so that the number of flux lines going through the coil changes, an emf will be induced within the coil while the flux is changing. According to Faraday's law, the magnitude of the emf is proportional to the rate of change of the flux and to the number of loops on the coil. An electric generator rotates a coil in a magnetic field and thereby generates an emf or potential difference. The voltage difference so generated is an alternating voltage.

In a transformer, a primary coil containing an ac current changes the flux through a secondary coil. The magnitude of the induced emf in the secondary coil can be varied by choosing different ratios for the number of loops on the primary coil to the number of loops on the secondary. A transformer can change ordinary 120-volt ac alternating voltage to voltages ranging from a fraction of a volt to many thousands of volts.

When the mass of a particle at rest is measured, it has a mass m_0, the rest mass. However, at a high speed, its mass is larger than m_0, and the mass becomes very large as the particle speed approaches the speed of light, $c = 3 \times 10^8$ m/sec. It appears that the particle mass would become infinite if the speed were equal to c.

TOPICAL CHECKLIST

1. How can you determine which pole of a magnet is the south pole? How are magnets and compass needles related?
2. What can be said about the forces magnet poles exert upon each other?
3. Which materials are influenced profoundly by magnets?
4. Explain how a magnet picks up a nail and how that nail can pick up another.
5. What is meant by a magnetic field? How can the direction of the field be determined?
6. Sketch the magnetic field of a bar magnet and of a horsehoe magnet. Sketch the earth's magnetic field, being careful to show the proper direction of the field lines.
7. Draw a diagram showing the magnetic field close to a straight, current-carrying wire. What is the dependence of the field strength on r and I?
8. In what unit is B measured? What is a gauss? About how large is the earth's field? About how large is B if it is 10 cm from a wire carrying a current of 5 amp?
9. Illustrate the use of the right-hand rule to determine the direction of B near a current.
10. Sketch the magnetic field of a current loop. Of a flat coil. Of a solenoid.
11. How did Ampère explain the magnetic field of a bar magnet?
12. Why does the magnetic field due to a solenoid increase a hundredfold when a soft iron rod is placed in its center? Will an aluminum rod give a similar effect?
13. What is an electromagnet? List a few devices which contain one.
14. A current-carrying wire experiences a force in a magnetic field. How do the following affect the force: current, field strength, orientation of the wire? What is the right-hand rule for this situation?
15. Sketch a coil of wire in a magnetic field and explain how the magnetic field causes it to rotate. How does a motor use this effect?
16. Explain how voltmeters and ammeters use the effect you discussed in 15.
17. Why does a charged particle move in a circular path when shot perpendicular to a magnetic field? What happens if the particle is shot parallel to the field lines?
18. Does the magnetic field change the energy of a particle shot through it? Explain.
19. In 17, the radius of the circle depends on three quantities: momentum, B, and q. What is the dependence in each case?
20. What is meant by an induced emf and when is it observed?
21. Explain the interrelation between induced emf, flux, and change.
22. What is Faraday's law for induced emf's? Give several illustrations of it.
23. What is Lenz's law? Use it in a few situations.
24. Explain how an electric generator uses induced emf. Why is the voltage produced of the ac type?
25. Show the mechanism within a generator that makes it necessary to do work to operate the generator when current is being drawn from it.
26. What is a transformer? Explain its principle of operation, and tell what it is used for.
27. How is a magnetic field used to measure atomic masses in the mass spectrograph?
28. What happens to the mass of an object as it is accelerated to speeds close to the speed of light?

QUESTIONS

1. What is meant by the statement: "The North Pole of the earth is a south pole while the South Pole of the earth is a north pole"?

2. Suppose the ceiling of your classroom was a huge south pole of a magnet and the floor was a north pole. What would the magnetic field in the room look like? How could you tell it existed?

3. Discuss the direction and magnitude of the force a current-carrying wire would experience when oriented different ways in the room of Question 2. Repeat for a current-carrying coil of wire and for a solenoid.

4. Two parallel long straight wires carrying current in the same direction attract each other. Explain why. What happens if the currents are oppositely directed? (*Hint:* One wire causes a magnetic field to exist at the position of the other wire.)

5. The earth's magnetic field is nearly parallel to the earth at the equator. Find the direction of the force on an electric wire carrying current (*a*) eastward or (*b*) northward at the equator.

6. A proton is shot into a region where a uniform magnetic field exists. Describe the proton's motion if the velocity is (*a*) parallel to the field; (*b*) perpendicular to the field; (*c*) at an angle of 45° to the field lines; (*d*) zero. Repeat for an electron.

7. A coil of wire is attached to a galvanometer as shown in Figure 7.24. Describe what the meter does if the coil is (*a*) thrust into the gap between the poles of a horseshoe magnet; (*b*) held stationary between the poles; (*c*) rotated between the poles; (*d*) suddenly pulled from the magnet region.

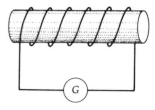

7.24

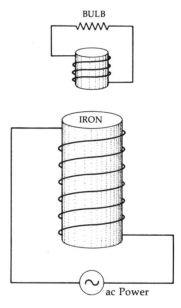

7.25

8   A coil of wire lies flat on a table. From above, a bar magnet is dropped, north pole first, toward the center of the coil. Will the induced emf cause a current in the coil to flow clockwise or counterclockwise?

9   Suppose in the situation of Question 8 the magnet is suspended north pole down from a spring so that it oscillates up and down above the table along the coil's axis. Describe the current which will flow in the coil.

10 Suppose instead of a coil of wire in Question 9, a flat circular copper plate lies on the table. It is found that if the magnet is very strong and oscillates rapidly, the copper plate becomes hot. This is the result of so called *eddy currents* flowing in the plate. Describe how these currents originate. (*Hint:* Assume that the plate consists of concentric loops of copper wire.)

11 Two circular loops of different size lie on a table with their centers coinciding. A current flows clockwise in the outer loop and a current flows counterclockwise in the inner loop. Describe the force on the inner loop because of the current in the outer loop.

12 Suppose the inner loop of Question 11 has no emf source, so that no current flows in it. If the current in the outer loop is suddenly turned off, what will happen to the inner loop?

13 Consider the situation shown in Figure 7.25. It is found that the light bulb lights even though there is no battery in its circuit. Why does it light? What is a combination of two coils such as this called? What will happen if the ac power is replaced by direct current?

14 If the upper coil in Figure 7.25 is replaced by a metal ring, the ring becomes hot if held in place. Why? If the ring is sitting on the top of the lower coil when the power is turned on, the ring flies off into the air. Why? Will these effects occur if the ring has a cut through it so that it does not make a complete circle?

15 The power company does not charge us for used energy unless we draw a current from their generators even though voltage from the generators is furnished to us constantly. Explain why they can afford to do this, using your knowledge of the way a generator changes energy from one form to another.

PROBLEMS

1   Two long straight wires are stretched parallel to each other on a tabletop. They are 10 cm apart and carry equal currents of 30 amp each. Find the magnitude of B midway between them if (*a*) the currents are in the same direction and (*b*) if they are in different directions.

2   Repeat Problem 1 for a point on the tabletop 20 cm from one wire and 30 cm from the other.

3   Find the magnetic field B inside a hollow solenoid which has 1,000 loops of wire wound on its 40-cm length. The current in the wire is 0.5 amp. Would the value for B still be correct if the coil were wound on a wooden rod? On an iron rod?

4   Find the momentum of a particle which follows a circular path with radius 2 cm when it is shot perpendicularly to a field of 1,000 gauss. If this is a proton ($m = 1.67 \times 10^{-27}$ kg), how fast is it moving? If it is an electron? ($m_e = 9.1 \times 10^{-31}$ kg.)

5   A proton ($m = 1.67 \times 10^{-27}$ kg) is accelerated through a potential difference of 2,000 volts. (*a*) What is its final speed? (*b*) How large a circle does it follow when moving perpendicular to a field of 2,000 gauss?

6   Repeat Problem 5 for the case of an electron.

Chapter 8   WAVES

Every material object acts like a wave; that is, it has wave properties. This fundamental fact of nature was discovered early in this century. We are familiar with waves on a violin string and the electromagnetic waves by which radio signals and light are transmitted. Further, wave characteristics become extremely important when we try to describe the behavior of atomic and smaller particles. All waves have similar properties, which we will discuss in this chapter. Later on, we will apply this knowledge to electromagnetic waves and to the electron's motion in an atom.

Waves on a String

All stringed instruments—the piano, guitar, violin—generate sound by the vibration of a string. These instruments provide a convenient way for us to understand the behavior of all waves. Let us see what happens when a disturbance is sent down a stretched string.

In Figure 8.1 a man sends a disturbance (or pulse) down a portion of a very long string. Notice how he does this; he quickly lifts and then lowers his hand, which is holding the end of the string. The string in turn pulls up and then down on the portion of the string to its right. As a result, the string's own movement causes the portion of the string to its right to move and the pulse travels along the string as indicated. Clearly, since the up-and-down motion of the string gives kinetic energy to portions of the string as the pulse moves by, *the pulse carries energy* from the man's hand *down the string*. We shall designate the speed of the pulse down the string by v. Experience as well as theory shows that v^2 is proportional to the tightness of the string and inversely proportional to the mass of one meter of its length.

We can send many different shapes of pulses down a string. However, a vibrating object sends out the most fundamental type of pulse. For example, a vibrating rod or a tuning fork prong oscillates in a type of motion we call *simple harmonic* or *sinusoidal* motion. Sinusoidal motion is also characteristic of such diverse systems as a pendulum, a mass at the end of a spring, and the charge on a radio antenna, as we shall see. Since sinusoidal motion is fundamental to our discussion of waves, we must take a moment to learn some of its characteristics.

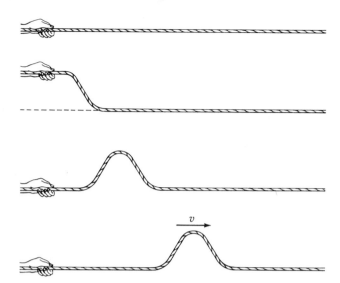

8.1 *The pulse started along the stretched string by the man's hand moves with speed v. Note how this pulse differs from the sinusoidal wave in Figure 8.2.*

We see in Figure 8.2 the trace inscribed on a moving sheet of paper by a mass vibrating at the end of a spring. The shape of this trace is called a <u>sinusoidal wave</u> or <u>sine wave</u>. All simple vibrating objects, whether a pendulum or an atom, show a curve like this when their oscillatory motion is recorded.

We designate the time taken for the vibrating object to move through one complete oscillation as the *period of vibration*. For example, in Figure 8.2, the portion of the curve between points A and C is one oscillation, and so the time taken to trace out that portion of the curve is the *period* of the oscillation. (We represent the period by τ, the Greek letter tau.) Hence when we speak of the period of a wave or oscillation, we refer to the time taken for one complete cycle of motion.

A quantity related to the period of the motion is the <u>frequency</u> of a vibration, which is the number of times the vibrating object goes through a complete cycle of the motion in unit time. For example, suppose the period of vibration is $\frac{1}{10}$ sec. Then there will be 10 cycles of the vibration every second (unit time), and the frequency is 10 cycles/sec (cps) or 10 vibrations per second (vps) or 10 hertz (Hz). These names (cps, vps, Hz) for the unit of frequency are interchangeable and are all used widely. (We shall use Hz in this text, as it is the form recommended by the American National Standards Institute.) Two interchangeable symbols are used for frequency, f and ν (Greek nu).

We have said that when the period of a vibration τ is $\frac{1}{10}$ sec, the frequency f is 10 Hz. In so doing, we used the fundamental relation

$$\tau = \frac{1}{f} = \frac{1}{\nu} \qquad (8.1)$$

which applies to all oscillating systems. For our particular example, after inverting,

$$\nu = \frac{1}{\tau} = \frac{1}{\frac{1}{10} \text{ sec}} = 10 \text{ per sec}$$

or 10 Hz.

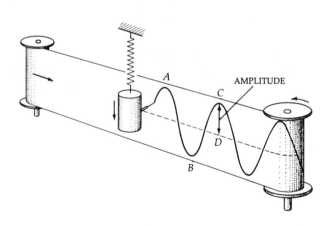

8.2 *As the mass vibrates up and down, it leaves a record of its path on the paper moving at constant speed past it. The rest position is shown by the dotted line.*

WAVES 161

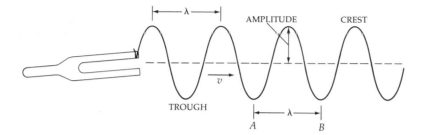

8.3 *As the tuning fork prong vibrates back and forth, it sends a wave down the string. (The wave would be of much smaller amplitude than shown.)*

Another word we shall use is *amplitude*. The amplitude of the vibration is the maximum displacement of the vibrating object from the midpoint of the vibration (or from its equilibrium position). The amplitude of the vibration and wave in Figure 8.2 is the distance CD.

Let us return to the vibration of a string. In Figure 8.3 a vibrating tuning fork prong sends a wave disturbance down a string tied to it. We can think of the *crests* of the wave (the top points as indicated) as a series of pulses similar to the one in Figure 8.1. The *troughs* of the wave (the lower portions) are simply pulses sent out by the prong during the lower part of its oscillation. The form of the string shown in Figure 8.3 (known as a sine-wave form) is what we would see in a high-speed snapshot of the string. It appears this way only for an instant. Each time the tuning fork moves down and up, it sends one pulse along the string. Thus, the wave shown in the figure is a record of the multiple pulses sent out by the prong in its previous vibrations.

We define the wavelength of the wave, λ (Greek lambda), as the distance between two corresponding points on the wave, for example, from crest to crest or trough to trough. There is a simple relation, true for all waves, between λ, ν, and v, the speed of the wave. Notice that for the trough at A to move along to the position of the trough at B, the wave must move a distance $s = \lambda$ (one wavelength). During that time, the tuning fork sends out one additional vibration, and so the time is simply $t = \tau$. Since $s = \bar{v}t$ and $\bar{v}$ in this case is equal to the constant speed v of the wave, we have

$$\lambda = v\tau = \frac{v}{\nu} \tag{8.2}$$

This relation is true for all waves, since to obtain it we assumed only that there was a wave of some sort being sent out by some kind of vibration source.

Reflection of Waves on a String

In the previous section we assumed the string to be long enough so that collision of the pulses with the object holding the string's far end was of no concern. Let us now suppose that the string is fastened

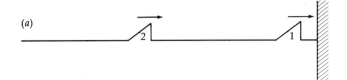

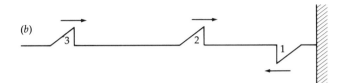

solidly to a wall and that several equally spaced pulses are sent down the string (see Figure 8.4). Experiment shows that when a pulse strikes a solidly fixed end, the pulse is reflected as indicated in Figure 8.4b. Notice that the pulse is inverted (turned upside down) as well as reversed in the process. Clearly, the energy carried down the string by the pulse is reflected along with the pulse.

8.4 *A wave reflected by a fixed end is turned upside down.*

As time goes on, reflected pulse 1 of Figure 8.4b will eventually run into incident pulse 2. Pulse 1 will try to displace the string downward, while pulse 2 will try to pull it upward. To a good approximation, the *principle of superposition* applies in such a case: *the string will displace according to the sum of the two pulses.* In computing this sum, downward displacements are negative and tend to cancel the upward positive displacements.

To see this more clearly, consider the wave in Figure 8.5a. It is a sine wave without its negative part. In (a), none of the pulses have yet been reflected. At the time the "snapshot" of the string shown in (b) is taken, pulses 1 and 2 have been reflected (and inverted). At that instant, pulse 1 is exactly between 3 and 4 while pulse 2 is between 3 and the wall. As a result, pulses 3 and 4 are undisturbed by 1 and 2 and vice versa.

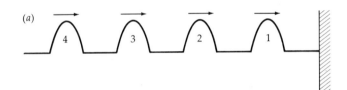

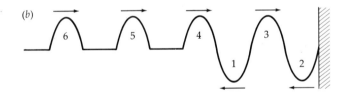

The situation still later is shown in Figure 8.5c. Now the pulses are shown by the dashed lines. Notice that the incident pulses are at precisely the same positions as the reflected pulses. Since the incident pulses pull the string up exactly as much as the reflected pulses pull the string down, the effects of the two pulses cancel at that instant; as shown by the horizontal line, the string is undisplaced. We say that the two waves, the incident and reflected, are undergoing *interference* and, at this instant, we have *completely destructive interference*. You might investigate what behavior the string at point *P* shows as time goes on, and the incident and reflected waves continue to move.

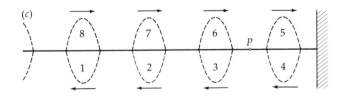

8.5 *The reflected pulses interfere with the incoming ones at certain moments.*

8.6 *The incident sinusoidal waves (dashed) are moving to the right; the reflected waves (dotted) are moving leftward. They cause the string to vibrate within the limits shown in (d).*

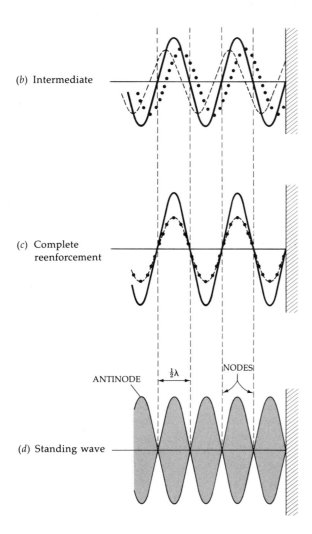

(a) Complete cancellation

(b) Intermediate

(c) Complete reenforcement

(d) Standing wave

ANTINODE $\tfrac{1}{2}\lambda$ NODES

Standing Waves

If you examine the behavior of point P in Figure 8.5c, you will see that it does not move at all as the incident and reflected pulses move past it. This is a specialized example of a general situation that arises when similar waves moving in opposite directions combine. Now let us look at a more basic situation, the interaction of sinusoidal waves like those shown in Figure 8.6a. Knowledge of how such fundamental waves behave will allow us to predict the behavior of more complex waves later.

In Figure 8.6 we see an incident sinusoidal wave (the dashed line) traveling to the right, striking the wall, and being reflected in inverted form (dotted line). In Figure 8.6a there is complete destructive interference and the incident and reflected waves cancel. However, the incident wave continues to move toward the right while the reflected wave proceeds to the left.

Looking at Figure 8.6b, we see the waves an instant after the situation in Figure 8.6a. The resultant displacement of the string is no longer zero; it now has the form of a sine wave, as shown by the heavy line. You can confirm this by adding up the reflected and incident waves at various points, in accordance with the principle of superposition.

Later still, the waves are as shown in (c); now the crests of the incident and reflected waves coincide. As a result, they displace the string by the largest amount possible, since at this moment both waves are pulling the string in the same direction.

Notice that in all three figures certain points along the string remain fixed, i.e., unmoving, at the positions indicated by the dashed vertical lines. We call these points *nodes*. Midway between the nodes, the string oscillates back and forth between the top and bottom limits indicated in Figure 8.6d. If the frequency of the vibration is above 30 Hz, we see a blur instead of the string itself, as shown in (d). Because the blur presents a steady pattern to the eye, we call it a *standing wave*. As indicated in the

8.7 The string resonates when it is a multiple of λ/2 long.

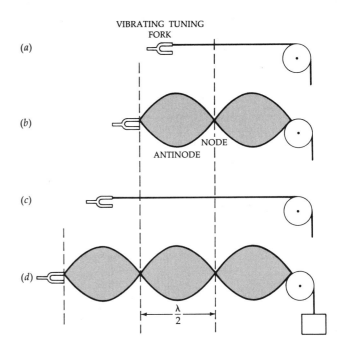

figure, the places of zero vibration along the string are nodes; the places where the string vibrates most violently are called *antinodes*. Figure 8.8 shows photographs of standing waves.

A standing wave provides a convenient way to measure wavelengths. Comparing Figure 8.6c with 8.6d and recalling that the distance between crests of a wave is the wavelength λ, we see that the distance between nodes is $\lambda/2$; therefore, if we measure the distance between nodes (or antinodes) in a standing wave, *the wavelength of the wave is simply twice this distance.*

Resonance of Waves

Let us review the previous sections briefly. When we send a wave down a string, it is reflected from the far end. As a result, two waves are moving along the string at the same time, the wave we are sending down it and the wave that was reflected back. The string undergoes a vibration which is the combined effect of these two waves. These two waves always cancel each other at certain definite places along the string. At these places, called nodes, the string remains stationary with no vibration at all. These nodal points are spaced every half wavelength along the string. The points midway between the nodes are positions of maximum vibration, and are called antinodes.

When the reflected wave returns to the end of the string from which it was sent, it is reflected once again. This twice-reflected wave goes down the string superimposed on the wave we are sending down the string for the first time. After a short time, a large number of waves will all be moving along the string at the same time. Only if these waves are in step with each other so that they all pull the string in unison will the string vibrate widely. When it does, the string shows a definite pattern called a standing wave. Let us clarify this with the experiment illustrated in Figure 8.7.

We see a tuning fork which is electrically driven and vibrates with a frequency of ν Hz. We keep the tension in the string constant by running it over a massive pulley to a weight as indicated. As a result, the speed v of the wave along the string is constant. However, if we lengthen the string slowly by pulling it over the pulley as shown, we observe an interesting behavior.

At almost all lengths, the string quivers slightly and is displaced only a millimeter or so, about as much as the tuning fork prong. However, as shown in Figure 8.7b and d, at certain special lengths, the string vibrates widely. These are the patterns we have called standing waves. When a standing wave exists on a string, we say that the string is undergoing *resonance*. Since the distance between nodes, the stationary points, is $\lambda/2$ as shown in the previous section, we see that the string will resonate when its length is $\lambda/2$, $2(\lambda/2)$, $3(\lambda/2)$, and so on. Only at these lengths do the reflected waves reinforce each other. For the string to resonate, the wave has to fit just perfectly on the resonating system. In the present case, the string will not resonate unless the string is integer multiples of $\lambda/2$ long. Only then can a standing wave exist with nodes at its two ends.

Using the phenomenon of resonance, we can measure the speed of a wave on a string. For example, suppose the tuning fork in Figure 8.7d has a frequency of $\nu = 120$ Hz, and the measured distance between nodes is 30 cm. Since the distance between nodes is $\lambda/2$, we know that $\lambda = 60$ cm. And we know from Equation 8.2 that

$$\lambda = \frac{v}{\nu} \quad \text{or} \quad v = \lambda\nu$$

Substituting values, we find

$$v = (60 \text{ cm})(120 \text{ Hz}) = 7{,}200 \text{ cm/sec}$$

Resonance and Vibration

The phenomenon of resonance is not confined to waves on a string. Anything that will vibrate will also resonate. For example, a child and his swing resonate if you push on them correctly. You know, of course, that there is a right and a wrong way to push on a swing. If you push at just any time, you will sometimes slow the swing and sometimes speed it up. As a result, the swing will vibrate a little, but not much. To make the swing vibrate widely (i.e., swing high), you must push once each oscillation and always so as to increase the motion, never to stop it.

Two important points are: (1) the swing has a certain natural frequency, and (2) you always push just as the swing starts to move away from you, that is, you push *in phase* with the natural motion of the swing. Under these conditions—pushing in phase and with the swing's natural frequency—the swing will vibrate strongly; the swing is in resonance. It is customary to call the swing's natural frequency of vibration its <u>resonance frequency</u>.

You can certainly think of many other situations where resonance occurs. The vibration of a mass at the end of a spring is an example. Or perhaps a

8.8 Can you show that the frequencies of the vibrator are in the ratio of 1:2:3:4 in the four cases?

(a)

(b)

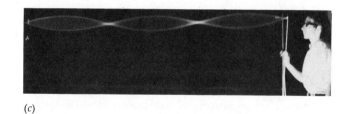

(c)

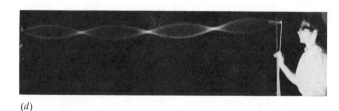

(d)

refrigerator or some other electrical appliance in your home is noisy because it vibrates at a frequency close to the resonance frequency of something nearby and thereby causes the nearby object to resonate noisily. In every case, no matter how complicated, resonance occurs when some external force adds energy to the vibrating system and causes it to vibrate more strongly.

In the case of a vibrating string, there are an infinite number of resonance frequencies for the same string. To see this, refer to Figure 8.8. The string of length L illustrated there resonates in at least four different ways. Each one of these vibration modes has its own characteristic resonance frequency. Notice that each vibration mode has a resonance pattern with nodes at the two ends of the string; because the ends are fixed, the string cannot move far at these two points. Since the distance between nodes is $\lambda/2$, where λ is the wavelength of the wave on the string, the string can resonate only if the length $L = \lambda/2$ or $2(\lambda/2)$ or $3(\lambda/2)$, etc. Stated more succinctly,

A string of length L tied at its two ends will resonate for those wavelengths λ such that $L = n(\lambda/2)$, where n is an integer.

Usually we are more interested in the frequency with which a string resonates than in the wavelength, since the frequency is the same as the sound frequency it gives off. To find the resonance frequencies ν, we need only substitute in Equation (8.2),

$$\nu = \frac{v}{\lambda}$$

where v is the speed of the wave on the string. Since at resonance $L = n(\lambda/2)$, this tells us

$$\frac{1}{\lambda} = \frac{n}{2L}$$

which gives, after substitution,

$$\nu = n\frac{v}{2L} \qquad (8.3)$$

where $n = 1, 2, 3, \ldots\ldots$

Let us apply Equation (8.3) to Figure 8.8. We see in (*a*) that the string vibrates in one segment and so $n = 1$. Hence

In (*a*): $$\nu = \frac{v}{2L}$$

But in case (*b*), there are two segments and therefore $n = 2$, giving

In (*b*): $$\nu = 2\frac{v}{2L}$$

This means the string is vibrating twice as fast in (*b*), with double the frequency, as it is in (*a*). Reasoning in a similar fashion, you should be able to show that the resonance frequencies in (*c*) and (*d*) are three and four times the frequency of the resonance shown in (*a*).

EXAMPLE 8.1

For a certain string of length 100 cm, the speed of waves along the string is 2,000 cm/sec. Find the resonance frequencies of the string.

Method: We could simply substitute in Equation (8.3) to find ν. However, to show that we need not memorize that relation, we will not use it here. At resonance, the string will vibrate in n segments, each segment being $\lambda/2$ long. Therefore, we can write at once

$$L = n\frac{\lambda}{2}$$

or

$$100 \text{ cm} = n\frac{\lambda}{2}$$

WAVES

8.9 Depending upon how a string is plucked, it will vibrate in different ways.

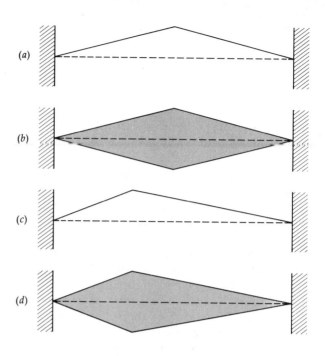

from which

$$\lambda = \frac{200}{n} \text{ cm}$$

But since we wish to find ν, we substitute in $\nu = v/\lambda$, using $v = 2{,}000$ cm/sec. Then we find

$$\nu = 10n \text{ Hz}$$

Hence the string will resonate to frequencies of 10, 20, 30, 40, 50, ... Hz. When $n = 300$, the string will be vibrating in 300 segments and its vibration frequency will be 3,000 Hz.

Vibration of Complex Systems

The question now arises: "How does a violin or piano string vibrate?" Our answer must be, "It all depends." Usually the vibration of a string fixed at both ends is complex, as you know if you have compared the sounds of a violin when played by an expert and by a beginner. The vibrating string sends a vibration into the air, and so the sounds we hear reflect the type of vibration the string undergoes. (The sound reaching a person's ear is the result of the vibration of the air near the ear.)

When a string vibrates in one of its resonance frequencies, it gives off a single, simple, pleasant tone. Examples are the tones given off in Figure 8.8a, b, c, and d. They are designated as the *fundamental tone*, the *first overtone*, the *second overtone*, and the *third overtone*, respectively. Their frequencies are in the ratio 1:2:3:4, and so the pitch of the tone goes higher from (a) to (d). If it were possible to make a violin string vibrate in exactly these ways, the violin would sound the same no matter who was playing it. However, it is difficult to make a string vibrate in these ways by bowing, plucking, or striking it.

For example, consider what happens when you pluck a string at its center, as in Figure 8.9a: the string vibrates as shown in Figure 8.9b. Or, if it is plucked or bowed near one end, as in (c), it may vibrate in a pattern somewhat like that shown in (d). Clearly, neither of these forms is a simple resonance form for the string. In these cases, the tone given off by the string is quite different from that given off when the string is vibrating simply. These complex varieties of resonance allow a skilled musician to obtain many different beautiful sounds from a stringed instrument.

Despite this complexity, the vibration of a string can be described entirely in terms of vibrations of the simple type shown in Figure 8.8. We can think of any vibrating string as performing several simple vibrations (like those shown in Figure 8.8) at the same time. To see this, let us examine the vibration shown in Figures 8.9a and b.

When a string is plucked at its center, as shown in Figure 8.9a, the result is different from the waves previously studied. A wave of this shape cannot be generated by smoothly oscillating one end of the string back and forth. Thus, unlike a sine wave, it is not considered to be a basic resonance form. To generate it as a standing wave on a string, the end

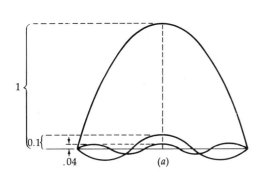 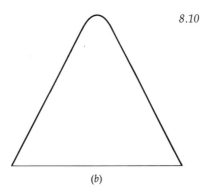

8.10 If the three sinusoidal waves shown in (a) are added together, the result shown in (b) is obtained. It is close to being a representation of the wave shown in Figure 8.9a.

must be vibrated back and forth in a complicated way. Experiments show this complex way is equivalent to several different basic resonances occurring on the string simultaneously. To show this, let us think of the string as vibrating simultaneously with three different resonance waves, the resonances shown in Figure 8.10a. If we say that the fundamental tone, second overtone, and fourth overtone have amplitudes in the ratios of 1 to 0.1 to 0.04 as indicated, then the sum of these three vibrations is as shown in Figure 8.10b. Notice how well it approximates the triangular form the string takes when it is plucked at its center. If we added waves of the fourth and higher overtones, the triangular form could be duplicated precisely. In mathematical jargon we say: "The complicated wave can be represented by a Fourier series of sinusoidal waves."

This example is typical of a general situation. No matter how complicated the string's vibration, the string is vibrating simultaneously in several of its resonance forms; by adding these resonance forms or waves together in their proper proportions, we can duplicate the actual vibration of the string. From this we conclude that the string is vibrating with several different frequencies at the same time and the sound we hear from it contains tones of all these frequencies. This mixture of numerous different frequencies at the same time gives rise to the distinctive tonal qualities of each musical instrument.

But we know that more than just the vibration of the string of a violin is important; the whole wooden structure of the violin contributes to its tonal qualities. For example, a Stradivarius violin differs from a cheap one in the wooden structure, not in the strings. The wooden panels and the open spaces within the violin resonate to certain vibrations of the violin string. As a result, some tones are further emphasized by the body of the violin. We will see why this is so after a short discussion of sound waves.

Compressional Waves; Sound Waves

In the previous sections we were mainly concerned with waves on strings. These waves are called *transverse waves*. In a transverse wave ("trans" means "across"), the particles of the string move perpendicular to, or across, the direction in which the wave is moving. A similar situation exists when

8.11 A picture of water waves on a still pond. The water waves move out along the surface of the pond while the water molecules move (mostly) up and down, perpendicular to the surface. (Courtesy of Fundamental Photographs.)

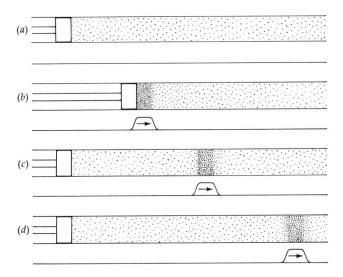

8.12 *The sound pulse sent down the tube by the piston is simply a region of highly compressed air.*

water waves move along the surface of a still pond; the water molecules move (mostly) up and down, while the wave travels out along the surface as shown in Figure 8.11. Water waves, like waves on a string, are transverse waves.

Sound waves, however, are entirely different in this respect. They are what we call *compressional* or *longitudinal* waves. To illustrate this, we show one way to generate a sound pulse in Figure 8.12. When the piston is suddenly pushed toward the right, the molecules in the air are pushed together, forming a region of high pressure, a pulse of air molecules. This pulse travels down the tube with constant speed. If you place your ear at the end of the tube, the pulse will hit your eardrum and make it vibrate. You perceive the pulse as sound. In open air, the speed with which such a pulse moves, the speed of sound, is about 1,080 ft/sec or 330 m/sec.

Notice that the motion of the air molecules is *along* the direction in which the sound wave pulse travels. This is the distinguishing characteristic of longitudinal or compressional waves. We can treat these waves just as we did transverse waves on a string, except in one respect. The difference lies in the way we plot the wave. In the case of a transverse wave on a string, the displacement of the string is perpendicular to the string itself, and so a graph of the displacement along the string looks exactly like the string itself; both have the shape of waves. But in Figure 8.12, the horizontal line of motion, which will be in the direction of the axis of our graph, is also the line along which the molecules are displaced. However, we want to plot the displacement vertically on the graph, and so the pulses plotted in Figure 8.12 do not mean that the molecules themselves are displaced vertically. The highest point on the pulse simply tells us how much the molecules have been displaced, i.e., how big the sound pulse is at that position.

Figure 8.13 shows how a loudspeaker sends out sound. As you probably know, the central, flexible diaphragm of the loudspeaker can be pulled back and forth. To do this, an appropriate electric current is sent through a coil attached to the diaphragm; this current-carrying coil then experiences a force because it is in the magnetic field of a permanent magnet mounted near the center of the speaker. As a result, if an ac current is sent through the loudspeaker coil, the diaphragm will oscillate back and forth, sending a series of sound pulses into the air. If the disturbance from the loudspeaker reaches our ears, it will cause the ear diaphragm to oscillate back and forth with the same frequency as the loudspeaker oscillation. Thus we hear a sound of the same frequency as the oscillation frequency of the loudspeaker diaphragm.

Resonance of Sound Waves

We can obtain resonance with sound waves, just as we can with waves on a string. You will recall that a string resonates if the string is just long enough so that a wave will fit properly on it. The string must be $n(\lambda/2)$ long so that the wave's nodes will fit at the two ends of the string. A similar situation exists for resonance of sound waves in tubes.

WAVES 171

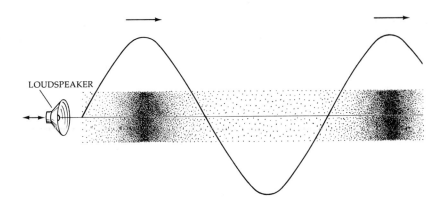

8.13 As the center section (diaphragm) of the loudspeaker vibrates back and forth, pulses are sent out into the air. In practice, the pulses are not beamed along the axis as much as indicated.

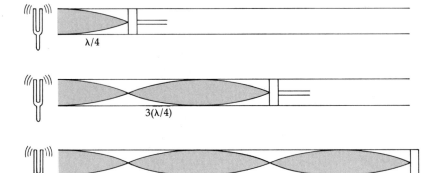

8.14 The tube resonates with a node at its closed end and an antinode at its open end.

8.15 Notice that the two pipes, though of equal length, have no resonance frequency in common.

(a)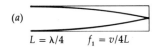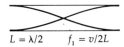

$L = \lambda/4 \quad f_1 = v/4L$ $L = \lambda/2 \quad f_1 = v/2L$

(b)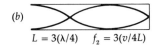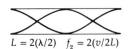

$L = 3(\lambda/4) \quad f_2 = 3(v/4L)$ $L = 2(\lambda/2) \quad f_2 = 2(v/2L)$

(c)

$L = 5(\lambda/4) \quad f_3 = 5(v/4L)$ $L = 3(\lambda/2) \quad f_3 = 3(v/2L)$

(d)

$L = 7(\lambda/4) \quad f_4 = 7(v/4L)$ $L = 4(\lambda/2) \quad f_4 = 4(v/2L)$

CLOSED   OPEN

We can demonstrate the resonance of sound waves by holding a tuning fork in front of a cylindrical tube as shown in Figure 8.14. We can adjust the length of the tube by moving the piston. We hear a sound much larger than that of the tuning fork alone when the tube is of the lengths indicated in the figure. At these lengths, the tube undergoes resonance. Notice that a node always occurs at the closed end of the tube and an antinode occurs at the open end. We might have guessed this, since the molecules at the closed end cannot move back and forth while those at the open end can move freely.

Since the distance between nodes is always $\lambda/2$, the resonance lengths are as shown in the figure. If we know the frequency of the tuning fork, we can easily find the speed of sound in the tube from measurements of these resonance lengths. For example, if the piston must be pulled out a distance of 20 cm between resonances, we know $\lambda/2 = 20$ cm, or $\lambda = 40$ cm. If the tuning fork frequency is 825 Hz, we can find the speed of the sound wave from Equation (8.2):

$$v = \lambda \nu$$

Substituting the values given, we find

$$v = (0.40 \text{ m})(825 \text{ Hz}) = 330 \text{ m/sec}$$

for the speed of sound.

A tube of fixed length, for example an organ pipe, will resonate to a large number of frequencies. The lowest resonance frequencies for an open pipe and a closed one are illustrated in Figure 8.15. Notice that the resonance frequencies of the open pipe are in the ratios 1:2:3:4 and those of the closed pipe are in ratios of 1:3:5:7. Also, the fundamental vibration of the closed pipe is at half the frequency of the open one.

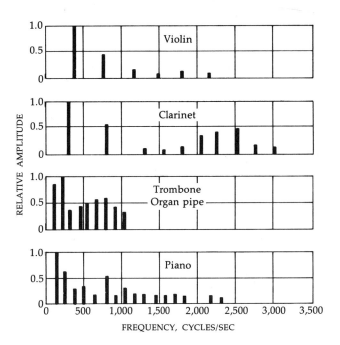

8.16 *Typical frequencies of sound given off when a single note is sounded.*

More complicated cavities will also resonate to sound. You have probably made a bottle resonate by blowing across its top. Of all the complicated air vibrations resulting at its lip, the bottle selects only a few to which it will resonate. Only sounds of these resonance frequencies are greatly amplified.

Most musical instruments have some sort of cavity to amplify certain frequencies. In a trumpet, the cavity is a certain length of pipe; the player can vary the length somewhat by use of the valves. The violin body and its cavity influence the sound given off by a violin as they resonate to certain frequencies of vibration. Usually, the sounds given off by musical instruments are not "pure," but consist of several frequencies. Examples of the sound frequencies given off and their relative amplitudes are shown in Figure 8.16. Notice that the frequencies are those we predicted for the fundamental tone and its overtones; for example, the violin tones are in the ratio 1:2:3:4:5:6. Can you see from the figure why a clarinet has a much shriller tone than an organ pipe?

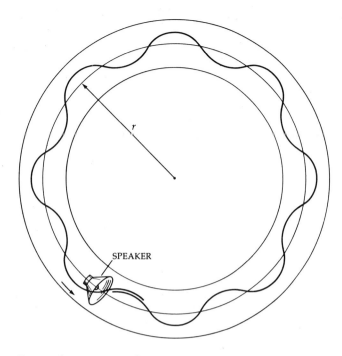

8.17 *If the circumference of the pipe, $2\pi r$, is a whole number of wave lengths long, the wave coming past the speaker will reinforce the wave from the speaker and a very intense sound will result.*

Interference of Sound Waves

As we have seen, waves can interfere with each other to produce large amplitude waves at antinodes and complete cancellation at nodes. In the next chapter we will see the effect of interference on light waves. Here we will discuss interference of sound waves.

Because sound waves can be guided through pipes, we can perform several informative experiments with them. For example, Figure 8.17 shows a tiny loudspeaker placed near the center of a hollow pipe which circles back on itself. If the speaker sends a sound wave through the pipe, the wave will eventually come back to the speaker. We assume the speaker is small enough so that the returning wave can pass by it and join the new wave it is sending out. Two limiting cases are of interest.

Suppose first that the tube is just long enough so that a crest of the wave reaches the speaker just as the speaker is sending out a crest. This situation is shown in the figure. In this case, the new wave from the speaker is in phase (in step) with the circling wave and the two waves add constructively. Each time it goes around the pipe, the wave is augmented by a new wave from the speaker. As a result, the wave going through the pipe becomes very energetic and there is an intense sound in the pipe. Clearly, for this to happen, the distance around the pipe, $2\pi r$, must be a whole number of wavelengths long.

The opposite case occurs if a trough of the returning wave reaches the speaker just as the speaker is sending out a crest. In this case the returning trough will exactly cancel the new crest. There will be no sound at all in the pipe because the waves interfere destructively with each other. Notice that for this to occur, one wave has to be a half wavelength out of phase with the other wave.

Figure 8.18 shows another interesting situation. Two speakers are driven by the same power source and so they send out wave crests at exactly the same time. Consider the waves from the two sources as they reach point P. As we see, the distance from P to B is half a wavelength shorter than from P to A. Hence the waves reaching P from the two sources are not in phase: a crest from A arrives with a trough from B and vice versa. Since the wave crest arriving from one source wants a crest at the point where the other wants a trough, the waves from the two sources exactly cancel each other at P. There will be no sound at P because of this complete destructive interference.

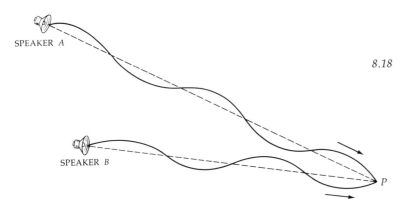

8.18 *If the sound waves are as shown, there will be no sound heard at point P.*

You can perform a similar experiment with a tuning fork. Each fork prong acts as a separate source like the two speakers in Figure 8.18. If you rotate the tuning fork on an axis about its handle while holding it at arm's length from your ear, you will find that the sound intensity varies. At certain angles, the waves from one prong will nearly cancel those from the other. Can you guess which type of fork shows the effect best, one of high or one of low frequency?

The Doppler Effect

Let us consider what happens to a wave when its source is moving. A moving source affects both sound waves and electromagnetic waves. This is the wave phenomenon that enables us to measure the recession speeds of distant galaxies in the universe; it accounts for the red shift mentioned in Chapter 1.

In Figure 8.19, we see a sound wave source sending a wave toward an observer at the left. In (*a*), the wave source is at rest relative to the observer. As a result, the source sends out wave crests equally in all directions. As the wave crests go by the observer, he hears a sound which has a frequency equal to the number of wave crests striking his ear each second. It does not matter where the observer is relative to the source, the crests travel out equally in all directions, and so he hears the same frequency sound at all points.

However, when the sound source is moving as shown in Figure 8.19*b*, the source is chasing its own crests toward the right and escaping from the crests it sends to the left. If the observer is standing at the left of the source as shown, the crests striking him are farther apart than they were in (*a*). Hence he experiences a wave of longer wavelength, and thus lower frequency, than he did in (*a*); the motion of the wave source away from the observer causes the wavelength of the wave to increase.

In the case of light waves coming from a receding source, light in the blue portion (i.e., the shorter wavelengths) of the color spectrum is shifted toward the red portion (i.e., the longer wavelengths) and so a receding light source is said to give rise to a red shift.

Suppose, however, the observer is on the other side of the source in Figure 8.19. Here the crests are bunched closer together and so they will strike the observer more frequently. As a result, the sound (or other wave type) striking him will have a higher frequency and shorter wavelength. From this we

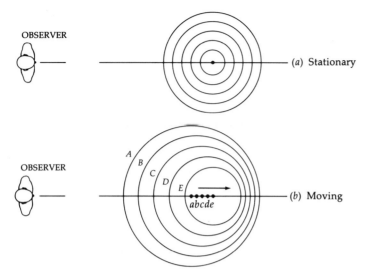

8.19 *When the moving source is at the points a, b, etc., it emits the waves shown as A, B, etc.*

see that the sound heard from a moving source depends on whether the source is approaching or receding. When it is approaching, the sound frequency is higher than normal; when it is receding, the frequency is lower than normal. This phenomenon, which occurs for all types of waves, is called the *doppler effect* after the man who first explained it in 1842.

Let us now treat this phenomenon quantitatively. Referring to Figure 8.20, we again see a wave source sending a wave toward an observer at the left. In part (a), the source sends out wave crests every τ sec, where τ is the period of the wave. During the time between the emission of successive wave crests by the source, a crest will move a distance $v\tau$ away from the source, where v is the speed of the wave. Since the distance between crests is the wavelength of the wave, we find that

$$\lambda = v\tau$$

when the wave source is stationary. This is Equation (8.2) rewritten. In Figure 8.20b, however, the source of the waves is moving toward the right with speed u. The wave is now spread out more than it was in (a). As we saw in Figure 8.19, the motion of the source has caused the crests of the wave to be further apart. To compute how large this effect is, we notice that the second crest has been moved to the right a distance $u\tau$, the distance the source moves in the time between emission of successive crests. Therefore, the wavelength of the wave is no longer $\lambda = v\tau$; it is larger by an amount $u\tau$. Let us designate it λ' (read lambda prime). We have

$$\lambda' = \lambda + u\tau$$

But since $\lambda = v\tau$ gives $\tau = \lambda/v$, we can replace τ, giving

$$\lambda' = \lambda + \frac{\lambda u}{v}$$

This can be simplified to read

$$\frac{\lambda'}{\lambda} = 1 + \frac{u}{v} \qquad (8.4)$$

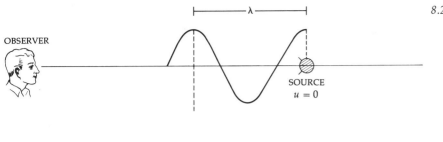
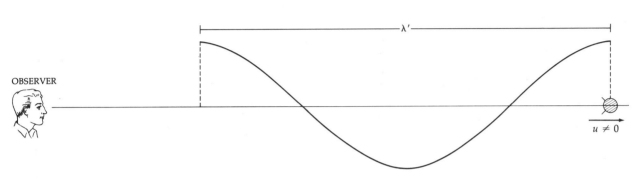

8.20 *The wave from a receding source is lengthened.*

In words, if the source is receding with speed u, the wavelength of the wave is lengthened by the factor $1 + (u/v)$, where v is the speed of the wave. When the source is standing still, $u = 0$ and so the ratio λ'/λ is unity; the wavelength is not changed.

In contrast, if the wave source is coming toward the observer, the wave is made shorter. You should be able to show that, for an approaching source, we need only replace u by $-u$ in Equation (10.4).

We can easily observe the effect of a moving source on sound waves. If we stand at the edge of a straight road and a car approaches with its horn blowing, the pitch of the horn drops dramatically as the car passes. We know that the waves reaching us from the approaching car have shortened wavelengths, and since $\nu = v/\lambda$, the pitch (or frequency) of the horn's wave is raised. As soon as the car passes, it becomes a receding sound source and the sound pitch lowers immediately. The following example will give you a hint about how the speed of a receding galaxy can be measured.

EXAMPLE 8.2

The sound of a car's horn has a frequency of 1,000 Hz when the car is stationary. How fast must the car be receding to cause the frequency to decrease to a value 900 Hz?

Method: Let us first change these to wavelengths, using $\lambda = v/\nu$ and recalling that $v = 330$ m/sec. Then

$$\lambda = 0.330 \text{ m} \quad \text{and} \quad \lambda' = 0.367 \text{ m}$$

Substituting in Equation (8.4) gives

$$\frac{0.367}{0.330} = 1 + \frac{u}{330}$$

where u will be in meters per second. Multiplying through the equation by 0.330 gives

$$0.367 = 0.330 + \frac{u}{1{,}000}$$

from which $u = 37$ m/sec. This speed is nearly 90 mi/hr.

Noise Pollution

Before leaving this discussion of sound waves, let us consider some of the deleterious effects of noise, which we can divide into two classifications. One is psychological and causes irritability, fatigue, and similar symptoms. The other, produced by very loud noise, is physical and directly damages our hearing mechanisms. Though the psychological effects of noise are equally important, they are too involved for us to pursue here. Therefore, we shall restrict our discussion to the harmful physical effects of loud noise.

To understand the physical effects of sound, we should mention a few facts about the human hearing mechanism. You probably know that the incident sound-wave pulses cause a drumheadlike membrane, the eardrum, to vibrate as they strike it. This vibration of the eardrum is magnified through a series of delicately pivoted bones and other structures which act as levers. The movements of these devices activate nerve responses which send signals to the brain; we then interpret these signals as sound. In effect, the eardrum and associated elements act as a meter to read sound intensities to our brain.

The ear is an exceptionally fine meter. Unlike most meters, which give readings over a range of a thousand or so (that is, 10^3), the ear can read sound intensities which differ by factors as large as 10^{12}. For example, a barely audible sound has an intensity only one millionth of a millionth as large as a sound large enough to cause the sensation of pain in our ear. Our ear acts as a _logarithmic meter;_ that is, when the sound intensity is made 10 times as large, our ear registers the sound to have increased by one unit. Sound intensities of 1, 10, 100, 1,000, . . . , 10^{12} are read by our ear to be 1, 2, 3, 4, . . . , 12. This is why we can hear such a large range of sounds.

However, too loud a sound reaching the ear temporarily damages the ear's ability to read weak sounds. The time taken for the ear to recover is roughly proportional to the duration of the sound and its loudness (assuming the eardrum is not broken and no other permanent damage has occurred). Hence, prolonged exposure to loud sound can seriously impair a person's hearing ability for his entire life. You may know of someone whose hearing has been permanently impaired by years of work near noisy machinery. Recent studies show that many people are suffering continuing hearing impairment from constant exposure to amplified music at too high a sound level. As our world becomes noisier, loss in hearing ability becomes an ever-increasing danger.

Another source of noise, the "sonic boom" resulting from the passage overhead of supersonic aircraft, merits specific discussion. The "boom" is the result of a very intense sound pulse built up by the aircraft moving faster than sound itself. It is a dramatic variant of the doppler effect. If we refer back to Figure 8.19b, we see a sound source moving toward the right at high speed. We assume that it sends out sound in all directions; the crests of the waves are shown as the circles around it. Since the source is chasing the crests that it sends out to the right, the crests are much closer together ahead of the source than behind it. If the source is moving with exactly the speed of sound, then the sound crests cannot get away from the front of the source. Each new crest adds to the previous ones and they build up a tremendously large sound pulse at the front of the moving source. This huge pulse of sound energy is called a _shock wave._

When an airplane moves faster than sound, the shock wave forms a cone that spreads out from the line of the plane's motion as shown in Figure 8.21. Wherever the shock wave hits the earth, and it clearly strikes all along the path of the plane, a large

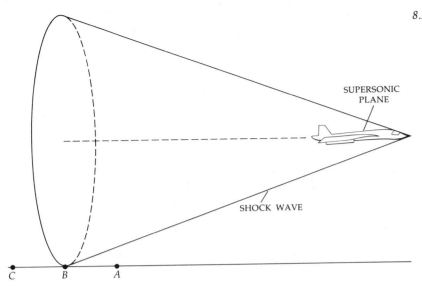

8.21 *The sonic boom has already hit point C and is moving through point B toward A.*

sound pulse is heard; it sounds like the pulse of sound from an explosion. Like the pulse from an explosion, a strong sonic boom can shatter windows and do serious harm to buildings and people. Fortunately, the speed of sound increases as the density of the air decreases. Hence a high-flying plane must be moving considerably faster than 330 m/sec (about 700 mi/hr) before a shock wave is formed.

BIBLIOGRAPHY

1. BENADE, A. H.: *Horns, Strings and Harmony* (paperback), Anchor Science, Doubleday, Garden City, N.Y., 1960. An interesting and readable account of the relation between the concepts of sound discussed here and music.

2. VAN BERGERJK, W. A.: *Waves and the Ear* (paperback), Anchor Science, Doubleday, Garden City, N.Y., 1960. Written for the layman, this book describes many of the properties of waves treated in this chapter and shows how the ear responds to them.

3. KOCH, W. E.: *Sound Waves and Light Waves* (paperback), Anchor Science, Doubleday, Garden City, N.Y., 1965. By means of vividly pictured experiments, this author shows how waves (sound and light) behave in many different situations. Nonmathematical.

4. FLETCHER, H.: *Speech and Hearing*, Van Nostrand, Princeton, N.J., 1929 and later editions. Later editions of this famous work become more mathematical but all of them provide a detailed and authoritative treatment of speech and hearing.

SUMMARY

When an object vibrates with sinusoidal (or simple harmonic) motion, it can send a sinusoidal wave down a string to which it is attached. The time taken for the object to complete one vibration is called the vibration period, τ. We designate the number of complete vibrations made in a second as the frequency of the vibration, ν. It is related to the period by $\nu = 1/\tau$.

We call the number of crests passing a given point on a string each second the frequency of the wave on the string. The distance between adjacent crests (or troughs) is the wavelength λ of the wave. If the speed of the wave on the string is v, then $\lambda = v\tau$, where $\tau = 1/\nu$. The maximum displacement of the string from its equilibrium position is called the amplitude of the wave.

When several waves exist simultaneously on the same string, the string will move in a fashion determined by the sum of the waves; thus, the string-wave system is said to obey the principle of superposition.

A string will resonate to the wave being sent down it if the wavelength of the wave is just right. A string must have stationary points, called nodes, at its two ends. Since the distance between adjacent nodes is $\lambda/2$, the resonating string must have a length of $\lambda/2$, $2(\lambda/2)$, $3(\lambda/2)$, and so on. Hence the string resonates only to certain definite wavelengths and their corresponding frequencies. The resonance pattern set up along the string is called a standing wave. The points of zero motion of the string are the nodes and the points of maximum motion are the antinodes.

Anything that has a natural frequency of vibration will resonate when pushed upon with this frequency. However, the pushing force must be in phase with the motion. A vibrating string has an infinite number of resonance frequencies whereas a pendulum only has one. The string can vibrate in several of these resonance forms or overtones at the same time. By changing the proportions of the various overtones, the quality of the sound given off by the string can be changed.

Transverse waves are like waves on strings; the vibrating object (the piece of string) moves perpendicular to the direction in which the wave travels (along the string). Compressional or longitudinal waves are typified by sound waves in the air. In a sound wave, the air molecules are moved back and forth along the direction in which the sound is moving.

Sound waves can be made to resonate in pipes and other hollow objects. For resonance to occur in a pipe, the sound wave's wavelength must be just the right size so that nodes can exist at closed ends and antinodes at open ends.

Two identical sound waves interfere destructively with each other if they are half a wavelength out of phase. In that event, the crests of one wave fall on the troughs of the other wave and the waves cancel each other completely.

When a wave source is moving relative to an observer, the waves reaching the observer are distorted. If the source is moving toward the observer, the waves are shortened and the observer notices a higher-frequency wave than normal. When the source is moving away, the waves are lengthened and the frequency is lowered. This latter effect results in the red shift. Both effects are called the doppler effect.

The ear can detect sound intensities that cover a range of about 10^{12}. Sound that is too loud impairs hearing ability; the duration of the impairment is roughly proportional to both duration and the intensity of the damagingly loud noise, provided permanent damage is not done.

A sonic boom is produced when an aircraft travels through the air at a speed near to or exceeding the speed of sound. In that case, the emitted noise (or sound wave) from the plane is unable to disperse and the sound waves pile up in front of it. This intense combined wave is called a shock wave. Like the shock wave from an explosion, a sonic boom generates a high-energy wave that can cause considerable damage.

TOPICAL CHECKLIST

1. When a pulse is sent down a string, what physical quantity moves down the string?

2. What is sinusoidal or simple harmonic motion? What is a sinusoidal wave?

3. Give the meaning of each of the following: amplitude; crest; trough; wavelength; frequency; period.

4. An object vibrates with a frequency of 10 kHz. How many cycles per second is this?

5. What is the relation between period and frequency?

6. What is the relation between λ, ν, and v for a wave?

7. Explain the meaning of the superposition principle by giving an example of its use.

8. Distinguish between constructive and destructive interference.

9. What is a standing wave? How does it originate? In relation to it, what are nodes and antinodes? How is the distance between nodes related to λ?

10. In terms of λ, how long must a string be if it is to resonate? What is meant by resonance?

11. Why does a force pushing in phase with the natural vibration of a system cause resonance while forces pushing at other times do not?

12. Define fundamental vibration; first overtone; and tenth overtone.

13. How can the complex vibration of a string be described in terms of simpler vibrations? What does this imply for the sound given off by a vibrating string?

14. What is the difference between a transverse and a longitudinal (or compressional) wave? Give examples of each.

15. Describe the mechanism by which a loudspeaker gives rise to the sensation of sound.

16. When a sound wave resonates in a tube, must the standing wave have a node or antinode at the closed end? At the open end?

17. How can two identical sound sources give rise to no sound at a particular point nearby? Why does the perceived sound intensity fluctuate as a tuning fork is slowly rotated?

18. What is meant by the doppler effect? How does it arise? Give an example of it.

19. Discuss the way in which the ear responds to sounds of widely different intensity. Why is it called a logarithmic detector?

20. What is meant by the sonic boom and how does it originate?

QUESTIONS

1. When two children hold a rope so that a third child can "jump the rope," the two children are producing both a horizontal and vertical standing wave on the rope at the same time. What is the wavelength of the wave? The frequency? Where are the nodes? Antinodes?

2. Using $F = ma$, discuss qualitatively why the speed of a wave on a string should increase with tension in the string and decrease with increasing mass per unit length of the string.

3. It is customary for a marching army to break cadence (i.e., drop out of step) when crossing a bridge. Why is this done?

4. The great singer Caruso is said to have been able to shatter a perfectly formed goblet by singing. Explain what was happening.

5. Sometimes when the load is not well distributed in a spin-type washer or dryer, the machine vibrates wildly at one particular frequency of rotation. What causes this effect?

6. Referring to the swaying bridge shown in Figure 8.22, how would you modify the structure of the bridge to save it?

8.22 *The Tacoma Narrows bridge was found after construction to have the same resonance frequencies as the wind blowing through it. Four months after its construction, it collapsed. (Courtesy of Prof. F. B. Farquharson.)*

7. Why are short, fat tuning forks high-frequency forks and long, skinny ones low-frequency?

8. Many hi-fi sets have two loudspeakers built into the loudspeaker enclosure. The larger is called the "woofer" and the smaller is called the "tweeter"; they are often in different size compartments. Explain their functions.

9. Explain how sound originates from the following objects: an exploding firecracker; hands clapping; a spoon dropped on the floor; a ticking watch.

10. Why is it impossible to hear the alarm of an alarm clock placed in a vacuum chamber?

11. Examine the construction of a child's whistle and explain why it gives off a loud sound of definite frequency. In what ways is it like a trumpet or a tuba?

12. The notes sounded by a flute, clarinet, and "sweet potato" are selected by covering and opening holes along the instrument. Explain what is going on to cause the sound to change.

13. A cylindrical tube open at one end and closed at the other resonates to several frequencies of sound. Draw the standing-wave pattern for the fourth resonance frequency counting from the lowest pitch tone. Repeat for a tube open at both ends.

14. Discuss how our lives would be changed if the ear read sound intensities in direct proportion to their loudness instead of logarithmically.

15. Suppose the human ear consisted of a cylindrical hole passing straight through the head from ear to ear with an eardrum closing the tube at its center point. How would such a hearing system differ from the one we have?

16. Most people cannot hear sounds above 20,000 Hz. Sometimes one encounters a person who cannot hear sounds with frequencies above about 8,000 Hz. What features of the ear construction are probably responsible for these limitations?

17. Ultrasonic vibrations are sound waves in air or liquid with frequencies too high to be heard by humans. Explain how ultrasonic cleaning baths can wash cloth and utensils without large-scale circulation of the water in the bath.

PROBLEMS

1. The wavelength of a certain wave is 2 ft and its frequency is 40 Hz. What is the speed of the wave?

2. The speed of sound in air is 330 m/sec. What is the wavelength of the wave sent out by a 660-Hz tuning fork?

3. When a person is listening to the tone called middle C, which is sound of frequency 264 Hz, with what frequency is his eardrum vibrating? How long does it take the eardrum to vibrate back and forth once?

4. A certain guitar string has a fundamental frequency of vibration of 198 Hz. To what other frequencies will it resonate?

5. If the player of the guitar mentioned in the previous problem divides the string in half by placing his finger at the midpoint of the string in question, what frequency sound will the string emit?

6. The speed of sound in air is about 330 m/sec. Estimate the frequency at which a 15-cm-long test tube will resonate when one blows across its lip.

7. Estimate the resonance frequency of a small Coca Cola bottle. You can check your answer experimentally by comparing the sound obtained when you blow across it with the sound from a loudspeaker connected to a variable-frequency oscillator.

8. Two loudspeakers side by side send out sounds of wavelength 50 cm long along the x axis. A listener on the axis hears alternate loud and weak sounds if one of the speakers is moved toward him while the other speaker remains fixed. Where will the moving speaker be when the weak sounds occur?

9. The speed of sound in hydrogen gas is 1,270 m/sec as compared to 330 m/sec in air. If the fundamental resonance frequency of a particular tube is 200 Hz when air-filled, what will be its resonance frequency when filled with hydrogen?

10. Two identical speakers, placed at $x = 0$ and at $x = 30$ ft, send sound toward each other. A man notices that the sound is large when he is at $x = 15$ ft but the sound nearly disappears when he moves to $x = 18$ ft. What is the wavelength of the sound coming from the speakers? With what frequency are they vibrating? ($v = 330$ m/sec.)

11. A 60-cm-long steel rod is dropped one end first onto a hard floor but is caught before it topples over. According to a bystander who claims the gift of perfect pitch, the rod emits a tone of frequency 4,000 Hz. If he is right, what is the speed of sound in steel?

12. Two identical loudspeakers side by side send out vibrations that are slightly different. One speaker vibrates with a frequency of 400 Hz while the other vibrates at 400.5 Hz. It is found that the combined sounds are alternately loud and soft; the loud sound occurs once every 2 sec. Explain why this variation in loudness occurs. (This is a phenomenon called *beats*; there are as many beats per second as the number of cycles per second by which the sources differ. Musicians use this as a very sensitive way to tune their instruments.)

Chapter 9

ELECTROMAGNETIC WAVES

Waves on a string, sound waves, and water waves are easy to visualize since they involve the motion of matter. In the mid 1800s, light was also recognized as a wave phenomenon. Scientists of the time supposed that light, like other waves then known, must be a disturbance in some material similar to air or water. They called this material the "ether." The concept of an ether was discarded as research revealed the true nature of light. Our present picture of what light is has been developed through the years as we discovered that light is basically no different from the waves associated with radio, radar, infrared, ultraviolet, x-rays, and gamma rays. All of these are forms of electromagnetic radiation.

Maxwell's Discovery

The connection between electromagnetism and light was not discovered until 1865. At that time, the noted physicist, James Clerk Maxwell (who also helped develop kinetic theory), was trying to formulate mathematically the known experimental facts about electric and magnetic fields. He discovered that all the experimental data concerning these fields could be summarized by four equations. These four equations are still accepted today as a precise formulation of the whole field of electricity and magnetism.

Maxwell's equations predicted that electric and magnetic fields should cause a wavelike disturbance moving through space. The equations predicted the speed of these waves to be 3×10^8 m/sec, which is the known speed of light waves. Maxwell therefore proposed that light actually is this predicted wave disturbance caused by electric and magnetic fields; we call these waves *electromagnetic waves.*

At that time, Maxwell and his contemporaries pictured electric and magnetic forces to be the result of strains in the ether, which they thought filled all of space. Maxwell's equations were interpreted by many (and probably even by him) to describe the motion of the ether. Since sound waves resulted from vibrations in the air, they thought that light waves must be the result of vibrations in the ether. Although such a concept can explain much of the behavior of light waves, we have discarded the ether concept, for reasons discussed in Chapter 10. We shall see that our present understanding of electric

THE BETTMANN ARCHIVE

JAMES CLERK MAXWELL (1831–1879)

Though born to wealth in Scotland, Maxwell had a somewhat spartan childhood. He was an undistinguished student until the age of 13, when his extraordinary intellectual abilities suddenly became apparent. He developed into a shy, quietly humorous man, sociable with friends but often reticent and withdrawn among strangers. After studying at Edinburgh and Cambridge, he held professorships at Aberdeen and London. When his father died, he retired to his family's estate in Scotland, where he researched and wrote his famous treatise on magnetism and electricity. In 1871, he became the first Cavendish Professor of Experimental Physics at Cambridge, and proceeded to design the famed Cavendish Laboratory. Though he is best known for his profoundly original theoretical work concerning the behavior of gases and electromagnetic waves, his experimental and mathematical talents were extraordinary as well. His contributions to science have proven as original and important as those of Newton and Einstein.

and magnetic fields explains electromagnetic radiation adequately without recourse to a hypothetical substance such as the ether.

In spite of Maxwell's conviction that light waves are a form of electromagnetic radiation, no one succeeded at that time in showing that Maxwell's waves could be produced by electrical means. It was not until 1887 that Heinrich Hertz (1857–1894) first produced electromagnetic waves in the laboratory by the use of sparks jumping back and forth between two small spheres. He was able to show that these waves could be reflected and bent much like light waves. Since then, Maxwell's identification of light as an electromagnetic phenomenon has been supported by many other experimental results. Today we make use of many types of electromagnetic waves, radio waves being one of the most prominent. Let us see how radio waves are sent out into space from a radio transmitting tower.

Radio Waves

As you know, most radio stations send out their waves from a high tower. The heart of this tower is the antenna, a long wire either running from top to bottom of the tower or suspended horizontally above the ground between two towers. To simplify our discussion, we will place two metal balls at the ends of a vertical wire and an alternating voltage source (an oscillator) in the center of the wire. The situation at one particular instant is shown in Figure 9.1.

The two ends of the antenna have been charged oppositely by the voltage source at its center. You will notice that the electric field near the antenna and along a horizontal line parallel to the earth is directed downward. (Recall that the electric field represents the force a positive charge would experience if placed at the point in question.) However, the electric field will soon change its direction because the radio station's ac voltage source will soon reverse

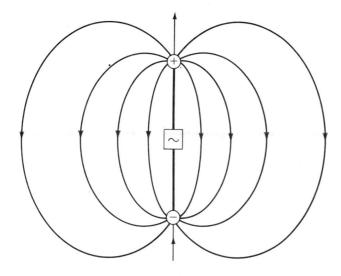

9.1 A "snapshot" of the electric field at the instant when the top of the antenna is charged positively.

the polarity of the antenna. Since the frequencies of normal radio stations lie in the range of 500 to 1,500 kHz (that is, 500,000 to 1,500,000 Hz), the charge on the antenna will reverse in a time of approximately one millionth of a second, or one microsecond. Of course, when the charge on the spheres reverses, the electric field direction near the station's antenna will also reverse.

Let us suppose (incorrectly) that the station charges the antenna in the following sequence. The top of the antenna has a steady positive charge for 10^{-6} sec. Instantaneously its charge is reversed and held that way for 10^{-6} sec Again the charge is reversed instantaneously and held for 10^{-6} sec and so on over and over again. Figure 9.2a shows the electric field as a function of the distance from the station at a certain instant. The electric field is shown by the arrows; the distance from A to B is about 300 yd (or 300 m).

At the instant the "snapshot" of the electric field in Figure 9.2a is taken, the top of the antenna is (and has been) charged positively. Therefore the electric field at A is downward, as in Figure 9.1. However, 10^{-6} sec before this instant, the top of the antenna was charged negatively; the electric field

near the antenna was then pointing upward. Since the disturbance from the antenna, the electric field in this case, travels out along the earth with the speed of light c, the upward electric field which was then at A has moved out to the region B.

Similarly, the electric field which was near the antenna when its top was last positively charged has now moved out to region C. In the same way, the fields at D, E, and F were close to the antenna at earlier times. Or if we think of men being stationed at points A, B, C, and so on, the man at A is seeing the field from the antenna as it is charged now. The man at B, however, is seeing the field from the antenna as it was 10^{-6} sec ago. (Remember, we have assumed that the oscillation takes place in such a way that the charge reverses once each 10^{-6} sec.) In the same way, the man at C sees the antenna as it was 2×10^{-6} sec ago. Like the wave along a string, the electric field out along a line into space shows the past history of the disturbance sent out by the source.

In practice, the voltage source that drives the antenna is sinusoidal. As a result, the charge on the antenna also varies sinusoidally. Thus the shape of the electric field wave sent into space by the antenna appears as shown in Figure 9.2b. The strength of the electric field is proportional to the length of the arrows.

In review, the electric field varies with distance away from the antenna at any particular instant. The sinusoidal curve drawn along the arrow tips is really a plot of the electric field strength at a certain instant as a function of distance from the antenna.

Let us consider a man standing at point F and the electric field he will observe as a function of time. Since the wave travels to the right with the speed of light c, the wave now at points E, D, C, B, and A will pass by the man at F in rapid succession. As it does, the man at F will observe the electric field point upward, then downward, then upward, and so on. In other words, he will observe an electric field

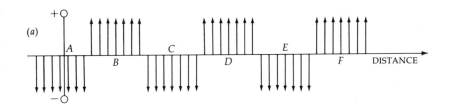

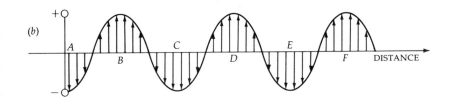

9.2 The electric field along a line perpendicular to the antenna varies with distance as shown. We do not show the decrease in field strength that results from the fact that it is spreading out into space.

which is oscillating sinusoidally up and down as time goes on. We will see later that this oscillating electric field can be used to detect the signal being sent out by the radio station; the very sensitive detector used is a radio.

The Magnetic Wave

So far we have ignored the fact that a current must flow in the antenna wire when the charge on it reverses. For the charge to vary sinusoidally, a sinusoidal current must flow in the wire. But we learned in Chapter 7 that a current flowing in a wire generates a magnetic field whose direction is given by the right-hand rule. This situation is shown in Figure 9.3.

Because the current is rapidly reversing, the magnetic disturbance being sent out from it into space also keeps reversing its direction. We find that the magnetic field, like the electric field, varies in magnitude and direction from point to point. This magnetic wave in space is also shown in Figure 9.3. Whereas the electric field from the antenna is parallel to the antenna wire, the magnetic field is perpendicular to the antenna.

The electric and magnetic fields are perpendicular to each other, as Figure 9.4 shows. The field waves move in unison to the right with the speed c. Thus, the wave sent out by the antenna is both an electric and a magnetic wave. It is for this reason that it is called an *electromagnetic* wave. Both parts of this wave carry energy that moves with the speed of light along the direction of propagation of the wave. In the next section we shall see how electromagnetic waves sent out by a radio station are detected by a radio. First, however, the following example will show us that $\lambda \cong 300$ m for a typical radio station's wave.

EXAMPLE 9.1

A particular radio station operates at a frequency of 1.2 MHz (MHz = megahertz = 10^6 Hz). Find the wavelength of its wave. Also, determine how many wave crests from the station pass a radio each second.

Method: To find the wavelength, we use Equation (8.2), that is,

$$\lambda = \frac{v}{\nu}$$

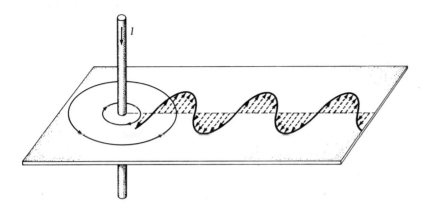

9.3 The oscillating current in the antenna wire gives rise to a magnetic field which circles the wire. As the oscillating magnetic field moves out into space, it constitutes a magnetic wave.

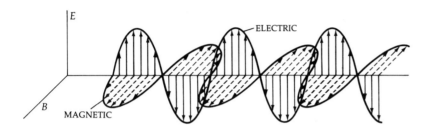

9.4 The electric and magnetic fields are in phase and perpendicular to each other in an electromagnetic wave.

In this case, $v = c = 3 \times 10^8$ m/sec and $\nu = 1.2 \times 10^6$ Hz. Substituting these values gives $\lambda = 250$ m.

The number of wave crests passing a given point in space each second is equal to the number of crests sent out by the station in a second. This is simply the frequency of the wave ν, and so 1,200,000 crests pass each second.

Radio-Wave Detection

Whether you realize it or not, radio waves are constantly passing through your body. Every station within a radius of hundreds of miles sends detectable radio waves to you. Each station is assigned a particular *operating frequency*, i.e., the frequency with which it reverses the charge on its antenna. This difference in frequencies allows us to select a particular station on our radio. The method of selecting a frequency depends upon the principle of resonance and makes use of a so-called <u>resonant circuit</u> inside each radio. For our purposes it will be sufficient to know that this circuit will respond (or amplify) strongly only if the oscillating voltage source driving the circuit has the same frequency as the resonance frequency of the circuit. The resonance frequency of this circuit in a radio is changed by turning the radio's tuning knob. Hence, when we tune a radio to a station, we are simply selecting the frequency to which it will resonate.

There are two ways of detecting radio waves. The older method uses the electric field part of the wave. As shown in Figure 9.5a, a long straight wire (the antenna) is attached to the radio and held in the region of the wave. At the particular instant shown, the upward-directed electric field charges the upper end of the wire positively and the lower end negatively. This charge will reverse with the frequency of the radio station as the electric field wave passes by the radio. Each time a charge appears on the end of the receiving antenna, an electrical "push" (i.e., energy) is given to the resonant circuit in the radio. If the natural frequency of the resonant circuit has been adjusted to the frequency of the radio station, the radio will respond strongly to the station's wave, and the radio is tuned to that station. Usually the signal from the station is weak, and so the radio is

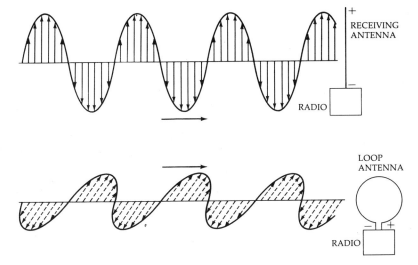

9.5 *The radio's antenna acts like a voltage source since the radio wave passing by it induces potential differences in it.*

designed to greatly amplify the response of the resonant circuit.

The second method for radio reception is now common in small radios; it makes use of the magnetic part of the electromagnetic wave. Inside the radio is a coil of wire wound on a cylinder of ironlike material. This coil is shown as the loop in Figure 9.5b. Faraday's law tells us that as the magnetic wave passes by this coil, an emf or voltage is induced in the coil by the changing flux through the coil. This induced voltage has the same frequency as the radio station. When the radio's resonant circuit is properly tuned, the radio will respond strongly to this induced emf. We should notice, however, that the induced emf in the coil will depend upon the orientation of the coil in the wave. When the coil is in the position shown in the figure, the induced emf will be maximum since the flux through the coil is largest in that orientation. However, if the radio is rotated 90° about a vertical axis, no flux will go through the coil and the radio will not respond to the wave. This explains the directional reception properties you may have noticed in some small radios.

The Electromagnetic Spectrum

We saw in Example 9.1 that the wavelength of a radio wave is of the order of 300 m. In the United States, ordinary AM radio frequencies range from 0.5 to 1.5 MHz, and the corresponding wavelengths are 600 to 200 m. However, FM radio waves and TV waves are at higher frequencies (or shorter wavelengths), frequencies close to 100 MHz.

Very high-frequency radio waves are called <u>radar</u> waves; these may have wavelengths as short as a centimeter. The frequency of a 3-cm-wavelength radar transmitter is

$$\nu = \frac{v}{\lambda} = \frac{3 \times 10^8 \text{ m/sec}}{0.03 \text{ m}} = 10^{10} \text{ Hz}$$

It is difficult to generate and receive radar waves in the 1-cm range because the waves themselves are the same size as the smallest coils of wire we can make easily. In practice, resonant circuits are replaced by resonant cavities when working in this range. The electromagnetic radar waves resonate in a metal cavity much as sound waves resonate in a tube.

ELECTROMAGNETIC WAVES   189

However, when dealing with wavelengths much shorter than a centimeter, it becomes difficult to obtain suitable resonant cavities. (Consider the difficulty we would have in obtaining resonance of sound waves of such small wavelengths.) Another problem is the movement of charges themselves. As we mentioned previously, atoms in a solid vibrate with frequencies of the order of 10^{13} Hz. This corresponds to an electromagnetic wavelength of

$$\lambda = \frac{v}{\nu} = \frac{3 \times 10^8}{10^{13}} = 3 \times 10^{-5} \text{ m} = 0.003 \text{ cm}$$

To obtain wavelengths in this range we must use oscillating atoms rather than oscillating electrical circuits as we normally think of them.

Electromagnetic radiation at these very high frequencies is called *infrared* or *heat* radiation. As implied by the latter term, heat is generated in objects when they absorb this type of radiation. Infrared radiation makes atoms and molecules vibrate when it strikes them; we recognize this atomic and molecular kinetic energy as heat energy. Clearly, much of the sun's energy is carried to us by electromagnetic waves having frequencies in the infrared region.

At still shorter wavelengths of electromagnetic radiation we come to the very narrow range of frequencies to which the human eye responds. We can actually *see* wavelengths of electromagnetic radiation in the range 0.00004 to 0.00007 cm, the range we refer to as light. The sensitivity of the eye to wavelengths in this region is shown in Figure 9.6. When dealing with such short lengths, we commonly measure them in a unit called the angstrom (Å), where

$$1 \text{ Å} = 10^{-8} \text{ cm} = 10^{-10} \text{ m}$$

and that is the length unit used in the figure. (Atoms are of the order of 1 Å in diameter.) Notice that the

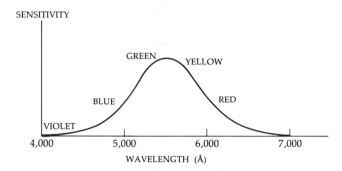

9.6 *The eye is most sensitive to greenish-yellow light.*

eye is least sensitive to the blue and red portions of the visible spectrum. We see the greenish-yellow colors best. You should know approximately how the colors are related to a wavelength; i.e., blue is short and red is long wavelength.

As we proceed to λ values shorter than 4000 Å we come to ultraviolet radiation, a radiation no longer visible to the eye. Radiations with λ less than about 100 Å are known as x-rays. Other electromagnetic radiation of very short wavelength is given off by nuclei of certain atoms as well as by nuclear reactions; it is called gamma radiation. Although gamma rays and x-rays are similar, we use the two names to describe nuclear and non-nuclear radiation respectively.

Electromagnetic radiation covers a tremendous spectrum, which is summarized in Figure 9.7. It covers the range from radio waves of nearly infinite wavelength to gamma rays of nearly zero wavelength. This wide variation in wavelengths gives rise to the great range in physical behavior which we observe. We will see later how very short wavelength radiation is related to the structure of molecules, atoms, and nuclei. But now we will consider the behavior of light waves.

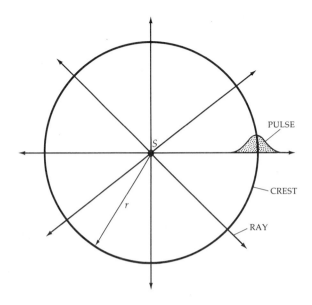

9.8 The spherical pulse spreading out from the source can be pictured by use of the crest and rays.

9.7 The electromagnetic sprectrum.

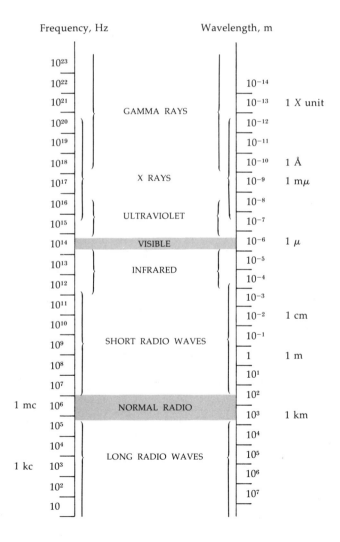

Reflection of Light

In the remainder of this chapter we shall discuss the behavior of light. However, many of the things we say will apply to all electromagnetic radiation and, indeed, to all waves. First let us learn a different way to represent a wave motion, a way based on the positions of the wave crests and the line of propagation of the wave.

Suppose a wave pulse is sent out in all directions from a central source S as shown in Figure 9.8. The position of the crest of the pulse is shown by the circle at radius r. In time, the crest will move out radially away from the source. We can show this motion by the rays indicated on the diagram. *The rays are always perpendicular to the line representing the wave crest.*

As an example of the use of such a picture, let us look at the wave disturbance moving out in all directions from a source, as shown in Figure 9.9. Consider what happens to the section of the wave between the dashed lines. The disturbance will get weaker as the wave proceeds outward, because the energy in the original wave spreads out as the wave grows in width (but not energy). This is why a light

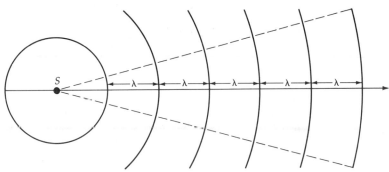

9.9 The section of the wave between the dashed lines grows wider and flatter as it moves away from its point of origin.

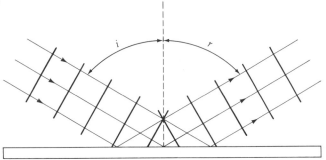

9.10 The angle of reflection is equal to the angle of incidence.

becomes dimmer as we move away from it. Notice also that the wave becomes less curved as it proceeds outward. In fact, at a point far distant from the source, the wave appears to be nearly flat. A perfectly flat wave is called a *plane wave*.

Suppose a plane wave strikes a mirror and is reflected as shown in Figure 9.10. The reflection of the wave can be represented by the wave itself or by its rays as shown. Experimentally, we find that the angle at which the light arrives (called the *angle of incidence* and designated i in the figure) is equal to the angle of reflection (designated r). This is exactly how a billiard ball moving along the incident ray would be reflected by a smooth wall. Hence it is impossible to tell from the law of reflection (that is, reflection always occurs with $i = r$) whether a beam of light consists of waves or whether it is a stream of particles. Newton and most of his contemporaries considered the particle interpretation to be correct.

As we see in Figure 9.10, a plane wave that strikes a flat mirror is reflected off the mirror as a plane wave. This is not true, however, for reflection from a concave (inward-curved) mirror. As Figure 9.11 shows, a plane wave becomes a curved wave when it reflects from such a curved mirror. The originally

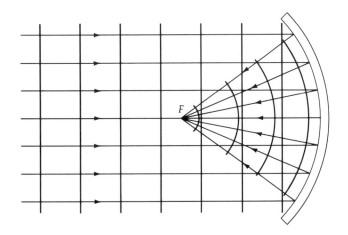

9.11 The plane waves incident from the left are converged to the focal point by reflection from the concave mirror.

parallel rays are reflected toward a single point called the *focal point* of the mirror. At the focal point F, an image is formed of the distant object from which the rays originated. For example, sunlight would form an image of the sun at F when reflected from the mirror. Similarly, if this mirror were in a large telescope, reflecting light from the sky, it would give images of the stars. The images can be recorded on a photographic plate or examined with a magnifying lens.

The reverse of this reflection process is also possible. If a small source of light is placed at F, in Figure 9.11, rays diverging from it will hit the mirror and be reflected as parallel rays. The arrows in the figure will be reversed, but since the angles of incidence and reflection at the mirror are the same we can interchange them. Thus we can obtain plane light waves from a light source placed at the focal point of a curved mirror. A flashlight provides a familiar illustration of this process.

Refraction of Light

Although mirrors can be used to form images, lenses are most frequently used for this purpose. They depend on the refractive properties of materials, and so we will now examine the phenomenon of *refraction*.

The speed of light as measured in a vacuum is $c = 2.998 \times 10^8$ m/sec. In all other media, light moves more slowly. It is common to give the relative speed of light rather than the absolute value. For this purpose we define the *index of refraction* μ (Greek mu) of a material as:

$$\mu = \frac{\text{speed of light in vacuum}}{\text{speed of light in material}}$$

Since the speed in vacuum is always the larger, the index of refraction is always greater than 1. Typical examples are given in Table 9.1. Notice, for instance, that since $\mu = 1.33 = 4/3$ for water, the speed of

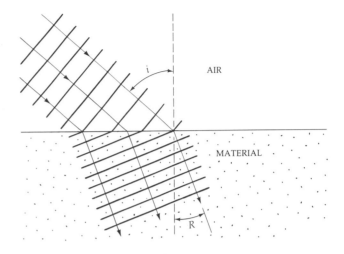

9.12 *Notice how the material slows down the waves and causes the direction of the rays to change.*

light in water is $3 \times 10^8/1.33$, or 2.26×10^8 m/sec.

Refraction results directly from the difference in the speed of light in various materials. It is the process by which the direction of the light is changed as the light ray passes from one material into another. As we see in Figure 9.12, the waves are slowed as they enter the material at the bottom. This causes the wave crests to be turned through a small angle, and so the direction of propagation of the wave is changed. Clearly, the more slowly the wave travels in the material, the larger the difference between the angle of incidence i and the angle of refraction R will be. Since a material with a high index of refraction decreases the speed of the waves most, a diamond will show this refraction effect most strongly of all the materials in Table 9.1.

Table 9.1 Indexes of Refraction

Air	1.003	Sodium Chloride	1.53
Water	1.33	Polystyrene	1.59
Ethanol	1.36	Carbon disulfide	1.63
Fused quartz	1.46	Flint glass	1.66
Benzene	1.50	Methylene iodide	1.74
Crown glass	1.52	Diamond	2.42

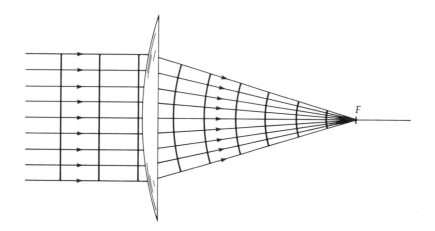

9.13 *This is called a converging lens since it converges the light. What would a diverging lens look like?*

We can arrive at a general rule from the behavior shown in Figure 9.12; when light passes from a material of low μ to a material of high μ, the angle of refraction is less than the angle of incidence. If we reverse the rays in the diagram, we see that when light goes from a material of high μ to one of low μ, the angle of incidence is smaller than the angle of refraction.

One of the most important uses of refraction is in lenses. The effect of a lens on a plane light wave is shown in Figure 9.13. Since the wave travels more slowly in glass than in air, the portion of the wave that goes through the most glass will be held back the most, as shown. The original plane waves are made curved by the lens shown here (called a converging lens), and are brought to a focus at the focal point. If the light coming to the lens is from the sun, an image of the sun is found at F. Although the exact position of the image formed by a particular lens depends upon how far away the object is from the lens, an object which is more than 10 times farther from the lens than is the focal point will have its image formed quite close to F. Keep in mind that the focal point is where the image of a distant object is formed. Images of closer objects occur some distance from this point.

A lens can also provide plane waves (or parallel rays) of light. If an extremely small light source (called a point source) is placed at the focal point F in Figure 9.13, the rays shown are reversed in direction. They come from the source F and are bent parallel to each other as they pass through the lens. In other words, the path of a light ray through a lens is retraceable, and the reverse of Figure 9.13 gives us a way for obtaining parallel rays of light.

The Prism and Color

In addition to his well-known work in the fields of mechanics and gravitation, Newton was intensely interested in the subject of light. Most of this work was published in 1704 in his *Treatise on Optics*, which you might be interested in reading. In his view, light beams consisted of a stream of tiny particles. He explained refraction, such as that shown in Figure 9.12, by assuming that the material attracted the light particles (reminiscent of gravitation). According to this corpuscular theory of light, light should move faster in glass and water than it does in air. This of course is exactly opposite to what the wave picture assumes about light's speed. In 1850, experiments to test this point finally

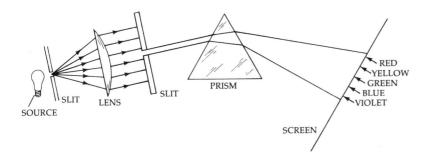

9.14 *The beam of white light is separated into its colors by the prism.*

became feasible, and they showed that the speed varies in agreement with the wave interpretation, and so Newton's corpuscle picture was discarded. We shall see in the next chapter, however, that light *does* act like a stream of particles under certain circumstances and so Newton was not completely wrong in this regard.

In spite of Newton's misconception about the nature of light, he carried out many valuable experiments in this field. It was he who discovered that white light is a combination of all the colors of the rainbow. A modification of his experiment is illustrated in Figure 9.14. The source of light could be any white-light source, such as the sun or an incandescent lamp. The light from the source emerges through a narrow slit. This slit is in turn placed at the focal point of a lens that gives plane waves, as we saw in the last section. This beam then passes through a prism, a flat triangle of glass. Upon emerging from the prism, the light has been refracted (or bent) by different amounts depending upon its color. As indicated, violet is refracted most and red least. Evidently the various wavelengths of light have different speeds in glass and are therefore refracted differently.

When slightly modified, the apparatus illustrated in Figure 9.14 can be used as a _prism spectrometer_, as shown in Figure 9.15. Notice that the second slit has been removed, and a second lens has been introduced after the prism. Since the light entering the prism consists of parallel rays, the light of any one color emerging must also have parallel rays, since each ray is refracted the same. The second lens focuses these rays to form an image at the focal point of the lens. This image is a duplicate of the object from which the parallel rays originate, the slit in this case. Hence, the second lens gives an image of the slit on the screen or photographic plate.

So far, we have assumed that only one color of light is coming from the source in Figure 9.15; but if the source is an incandescent bulb or other white-light source, it emits all colors. The prism bends the rays of each color by different amounts, and as a result, an infinite number of images appear on the screen, one for each color. The spectrum, or range of colors, seen on the screen is a blur of color ranging from red to violet as the colored images of the slit are laid out side by side along the screen. This type of spectrum, shown in band 1 of Figure 9.16, is called a _continuous spectrum_ and is characteristic of the light emitted by white-hot objects.

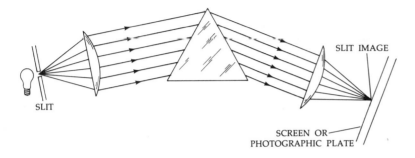

9.15 *Formation of a spectral line image in a spectrometer.*

If the same apparatus is used to examine the light from a neon sign, the spectrum is different, as the bottom band in Figure 9.16 shows. The lines you see there are the images of the slit formed by the second lens in Figure 9.15. There is one slit image for each color given off by the neon gas in the neon sign. Clearly, a luminous gas gives off an entirely different spectrum than a white-hot body. This type of spectrum is called a *bright-line spectrum* and the lines in it are called *spectral lines*. Remember that there is one line for each wavelength (or color) of light emitted by the source.

Other hot gases give off different line spectra; Figure 9.16 also shows the characteristic spectra of hydrogen and helium. Spectra are extremely important for two reasons. The first is that each element in nature gives off its own characteristic set of spectral lines. Therefore, the light emitted by the luminous gas formed when an unknown substance is vaporized in an arc can identify the substance. This method is widely used in industry and research to analyze substances. The second reason is that historically, one of the major pieces of information about atoms and their structure has been the spectra of the light they emit. As we shall see in the next chapter, the line spectra emitted by atoms provide a means for testing theories of atomic structure. But before we go on to atomic structure, we need to know a few more properties of waves.

Interference from Slits

One of the most distinctive properties of waves is their ability to interfere with each other. As we saw in Chapter 8, they can cancel each other completely at certain points. This is particularly easy to show using water waves. We need only send two waves of identical form (called *coherent* waves) into the same region so that they cross each other. In their region of overlap, they interfere. The waves in Figure 9.17a show this clearly.

To obtain the two coherent waves we can use *Huygens' principle*, a discovery made by Christian Huygens (1629–1695):

Each point on the crest of a wave acts as a new source for wave crests.

As a result, if a water wave strikes a barrier, the incident wave crests will act as new wave sources if there should happen to be a hole in the barrier.

9.16 *Bright-line spectra. The continuous spectrum shown at the top is typical of hot, incandescent objects such as the sun and a light-bulb filament. Note that the film has registered ultraviolet radiations down to the 3200 Å level, below the threshold of human vision.*

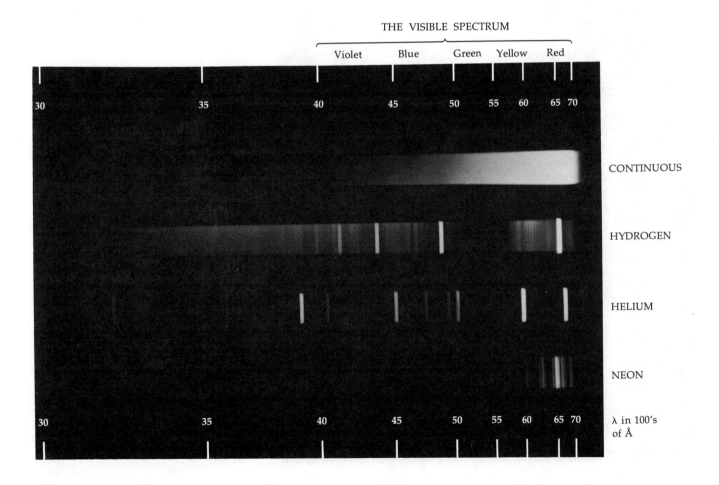

9.17 The water waves in (a) interfere with each other as they spread out from their sources. Water waves through a single slit are shown in (b) and (c). [(a) courtesy of Fundamental Photographs; (b) courtesy of the Ealing Company]

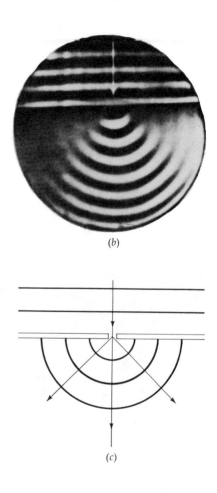

9.18 *Along the lines labeled Max, the waves from the two slits reinforce; they cancel along the lines labeled Min. [Part (a), E. Leybold's Nachfolger, courtesy of J. Klinger Company.]*

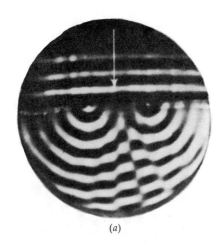

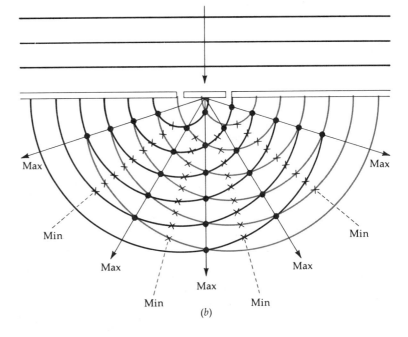

This is shown to be true in the photograph of Figure 9.17b. Notice that the hole acts as a wave source as the crests strike it. Waves move out in all directions from it in the region beyond the barrier. Part (c) of the figure is a diagrammatic representation of this situation.

Huygens' principle gives us a way to obtain two coherent wave sources. If the barrier has *two* holes in it, as shown in Figure 9.18, then two sets of identical (coherent) waves are sent into the region beyond the barrier. As we see in (a), the waves interfere with each other to provide some regions of zero disturbance and other regions of large disturbance. Figure 9.18b shows the reasons for this interference pattern.

As you examine this figure, recall that the half-circles represent the crests of the waves coming through the holes in the barrier. The troughs are midway between the crests. Hence, at the positions marked by x's in the figure, crests from one hole (or slit) coincide with troughs from the other slit. The two disturbances thus cancel each other along the dotted lines labeled "min" (for minimum). You should satisfy yourself that cancellation always occurs along these min lines as the waves travel away from the barrier. These are the regions of still water you can see in part (a) of Figure 9.18.

At the positions of the full dots along the lines labeled "max" in the figure, crest falls on crest and trough falls on trough. Along the max lines, then, the waves from the two sources reinforce each other and so the disturbance is large. You can also see these regions in part (a).

This effect should apply to all types of waves. We have already seen that the two prongs of a tuning fork or two loudspeakers can cause sound waves to interfere. The two sound sources act like the two water-wave sources, the holes in the barrier. Not surprisingly, light waves also interfere with each other under these conditions.

We need two coherent sources of light; these are provided by plane light waves striking two slits in a barrier, as shown at the top left in Figure 9.19. If a strong coherent light source sends light through slits about 0.01 cm apart, we should be able to see regions of brightness and darkness on a screen placed as shown. These dark and bright regions, called *fringes*, are the intersections of the min and max lines, with the screen. An actual photograph of the interference pattern of bright and dark fringes in such an experiment is shown at the right in Figure 9.19.

ELECTROMAGNETIC WAVES 199

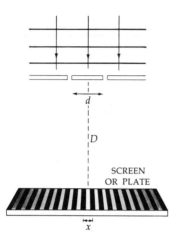

9.19 The double-slit interference pattern. (Photo courtesy of Prof. Brian Thompson, University of Rochester.)

Thomas Young first observed and measured this interference pattern in 1803. It is called *Young's double-slit experiment*, and it proved conclusively that light acts as a wave. Furthermore, Young succeeded in measuring the wavelength of light from the results of this experiment. The calculation depends on the fact that the waves from the two slits must be in phase at the bright spots in order to reinforce each other. Since the center fringe (or bright spot) in the pattern is equidistant from the two slits, the two waves reach it in phase. At the adjacent fringe on one side, the light from one slit must travel one wavelength λ farther than the light from the other slit if brightness (reinforcement) is to be observed. Hence there is a relationship between the wavelength λ, the distance between fringes x, the distance between the slits d, and the slit-to-screen distance D (see Figure 9.19). (For a more detailed explanation of this relationship, see p. 586 of reference 1, in the bibliography at the end of this chapter.) This relationship is

$$\frac{x}{D} = \frac{\lambda}{d} \qquad (9.1)$$

Since x, D, and d can be measured easily, the wavelength of light, λ, can be calculated. As stated previously, λ varies from 4000 Å for violet light to 7000 Å for very deep red.

Unfortunately, the fringes shown in Figure 9.19 are quite broad. As a result, wavelength determinations using a double slit are not very accurate. For better accuracy we use more than two slits. If identical slits are placed parallel to the original two and spaced d apart, the pattern sharpens tremendously, as shown in Figure 9.20. (Here a laser beam is the light source. It provides a bright, pencil-thin beam. Because of its small dimensions, the fringes appear as spots rather than lines.) In practice, we measure wavelengths by using a barrier with thousands of parallel slits in a region about 1 cm wide. This barrier, called a *diffraction grating*, produces extremely thin fringes and so makes possible very precise wavelength determinations.

EXAMPLE 9.2

When an unknown wavelength of light is used in a double-slit experiment, the fringes are found

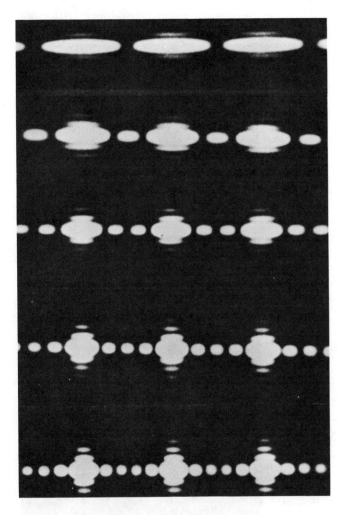

9.20 *Interference patterns obtained when a laser beam is sent through various number of slits; from top to bottom, 2, 3, 4, 5, and 6 slits were used. (Courtesy of Prof. Brian Thompson, University of Rochester.)*

to be spaced 0.10 cm apart. If $D = 100$ cm and $d = 0.050$ cm in this apparatus, what is the λ of the light used?

$$\frac{x}{D} = \frac{\lambda}{d} \quad \text{or} \quad \lambda = \frac{xd}{D}$$

Since $x = 0.10$ cm, we find that $\lambda = 5.0 \times 10^{-5}$ cm = 5000 Å. What color light is this?

Diffraction by an Aperture

In Figure 9.17 we saw that waves bend around the edge of a barrier and spread out. This phenomenon is called *diffraction*. In the case of light waves, diffraction contradicts our usual assumption that light travels in straight lines.

Ordinarily, we notice that a barrier in the path of a parallel beam of light causes a sharp, well-defined shadow. However, careful examination of the edge of the shadow with a magnifying glass shows that the shadow's edge consists of a series of interference (or diffraction) fringes, as shown in Figure 9.21. Clearly, light does not travel in *exactly* straight lines; like water waves, light waves diffract around barriers.

A more striking example of diffraction is observed when a parallel beam of light goes through a tiny slit, as shown in Figure 9.22. For a large slit width w, we expect to see a sharp shadow such that the beam width at the film is also w wide. In view of the diffraction effect shown in Figure 9.21, we also expect to see tiny fringes at the very edges of the beam. For a wide beam, this effect is negligible, so we usually assume that light travels in straight lines in situations such as this.

As the slit width w is narrowed, the beam width also decreases. However, at very narrow slit widths, the beam image on the film or screen begins to *widen*. Finally, when the value of w becomes as small as λ, the wavelength of the light being used, the image has expanded to fill the whole film. This is

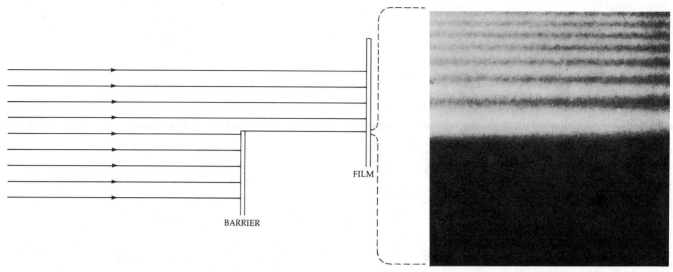

9.21 *The shadow appears sharp, but magnification shows it consists of a series of interference fringes. (Photo courtesy of Bausch and Lomb.)*

SHADOW EDGE, MAGNIFIED

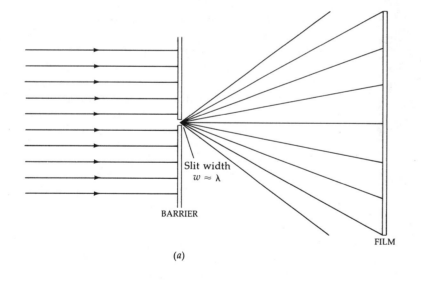

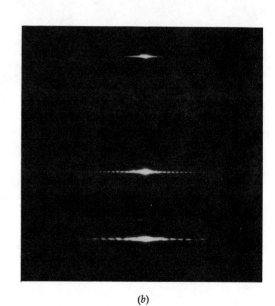

9.22 *When the slit width, w, approaches λ, the light from the slit spreads and fills the whole film. The widths of the single slits used in (b) were, from top to bottom, 0.0346, 0.0173, and 0.0087 cm. They were taken using a laser beam.*

illustrated in Figure 9.22b. If you refer again to Figure 9.17, you will see that water waves are also affected this way; they, too, diffract into the whole region beyond the slit, since $\lambda \approx w$ in that case as well. The theory for this type of behavior is developed in Reference 1.

Diffraction is a feature of all forms of waves. We can summarize it as follows: When a wave disturbance passes through an opening of size comparable to λ, the disturbance spreads out to fill the whole region beyond the opening. A similar effect exists when a small barrier, such as a rod or needle, is placed in the path of a wave disturbance. If the dimension of the barrier is comparable to λ, then the disturbance blurs the shadow behind the barrier. Figure 9.23 shows this effect photographically.

At this point, we can begin to appreciate the importance of wave diffraction. It implies that we cannot see any detail of an object that is close to or smaller in size than the wavelength of the light we are using to illuminate the object. This is true because everything we look at behaves somewhat like a slit light source. If we view an object by looking at the light it transmits, the object acts like a slit through which the light passes. Or if we view an object by the light it reflects, the object, as a source for the reflected light, acts like a slit. No matter how hard we try to get around this difficulty, we must accept the following fact.

> *It is impossible to see details that are smaller than the wavelength of the radiation used to observe them.*

This statement means that even a perfect microscope will not show details of objects smaller than 4000 Å, the wavelength of violet light. We could do better by using a gamma-ray or x-ray microscope if that were possible, since these radiations have shorter wavelengths than light. However, even if we could surmount the experimental difficulty involved in such a microscope, we would still be

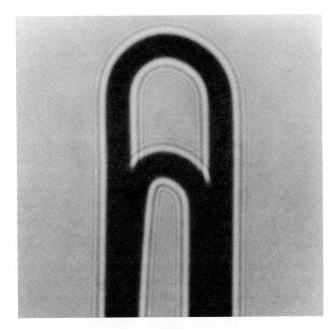

9.23 *The shadow of a paper clip (a) and a washer (b) in a beam of light. Notice how diffraction effects make the shadows indistinct. [(a) Courtesy of Fundamental Photographs; (b) courtesy of Bausch and Lomb.]*

limited by the wavelengths of these rays. What we are beginning to see here is a fact that became clear in the mid-1920s; *we can never describe any object exactly*. There is a fundamental limit to our observational abilities, and there will always be uncertainty in our knowledge of nature.

BIBLIOGRAPHY

1. BUECHE, F.: *Principles of Physics*, 2d ed., McGraw-Hill, New York, 1972. See chapters 23–25 for a more detailed discussion of the effects discussed in this chapter.

2. HOLTON, G., and ROLLER D. H.: *Foundations of Modern Physical Science*, chapters 29 and 30, Addison-Wesley, Reading, Mass., 1958. Contains some of the historical details concerning the discoveries of interference effects and the theory of light.

3. PAGE, R. M.: *Origin of Radar* (paperback), Anchor Science, Doubleday, Garden City, N.Y., 1962. A nonmathematical discussion of the principle of operation of radar together with its uses. Very readable.

4. BATTEN, L.: *Radar Observes the Weather* (paperback), Anchor Science, Doubleday, Garden City, N.Y., 1962. Explains clearly, in terms the layman can understand, how radar weather observations are made.

5. VAN HEEL, A. C. S., and VELZEL, C. H. F.: *What is Light?* (paperback), McGraw-Hill, New York, 1968. A mostly nonmathematical but detailed discussion of reflection, refraction, interference, and diffraction. Contains beautiful color photographs.

6. ANDRADE, E. N.: *Sir Isaac Newton* (paperback), Anchor Science, Doubleday, Garden City, N.Y., 1954. An interesting and readable account of the life and work of this great scientist.

7. JAFFEE, B.: *Michelson and the Speed of Light* (paperback), Anchor Science, Doubleday, Garden City, N.Y., 1960. The most accurate methods for determining the speed of light were developed by Michelson. This is a readable story of his work that demonstrates the importance of his results.

8. SHURCLIFF, W. A., and BALLARD, S. S.: *Polarized Light* (paperback), Van Nostrand, Princeton, N.J., 1964. We have not discussed this property of light but it is basic to the operation of Polaroid glasses and many other effects. This is a good introduction to the topic.

9. PIERCE, J. R.: *Waves and Messages* (paperback), Anchor Science, Doubleday, Garden City, N.Y., 1967. A clear and nonmathematical discussion of the use of waves in transmitting information over large distances.

10. WOOD, E.: *Crystals and Light* (paperback), Van Nostrand, Princeton, N.J., 1964. Discusses the interaction of light with crystals. The treatment is a little involved but not beyond the capability of the interested student.

11. FINK, D. G., and LUTYENS, D. M.: *Physics of Television* (paperback), Anchor Science, Doubleday, Garden City, N.Y., 1960. One of the best popular treatments of the subject. Tells you how TV works without getting too complicated.

SUMMARY

Maxwell established the connection between light and electromagnetic waves such as radio waves. His theory of electric behavior showed that light and electromagnetic waves generated by oscillating charges should both move through space with the same speed, 3×10^8 m/sec. Experiment later confirmed this. We now know that radio, FM, TV, radar, infrared, visible light, ultraviolet, x-ray, and gamma-ray waves are all electromagnetic radiation. They differ primarily in wavelength; the wavelengths extend from a few hundred meters for radio waves to a fraction of an angstrom (1 Å = 10^{-10} m) for x-rays and gamma rays. We designate this sequence of waves the electromagnetic spectrum.

Radio waves are generated by oscillating charges

on a wire called an antenna. The ends of the antenna are alternately charged plus and minus. This causes an oscillating electric field close to the antenna. Traveling with the speed of light c, this oscillating electric field moves out into space and subjects areas even quite distant from the antenna to an oscillating electric field. Each radio station has assigned to it a certain frequency, and it sends out a wave of this single frequency. Most radio stations in the United States operate in the frequency range 0.5 to 1.5×10^6 Hz.

Currents flow in the antenna wire as the station oscillates charges on it. These currents generate a magnetic field which oscillates with the same frequency as the electric field. As a result, the electric and magnetic fields combine to produce an electromagnetic wave moving out into space from the antenna. All electromagnetic radiation, from radio to x-rays, is a combination of electric and magnetic waves. Visible light has wavelengths between 4000 and 7000 Å.

To detect the waves from a radio station, the resonant circuit within a radio must be tuned (using the tuning dial) to the station's frequency. When a wave of that frequency strikes the radio's antenna, it causes the circuit to resonate and the radio responds to that station. The receiving antenna can be either a straight wire antenna responding to the electric field part of the wave or a loop antenna responding to the magnetic field part.

When light is reflected, the angle of incidence equals the angle of reflection. A plane mirror reflects a parallel beam of light unchanged except for its direction of travel. But a properly curved mirror reflects all the rays in a parallel beam to a particular point, the focal point. Such a curved mirror forms images of distant objects near the focal point. When a light source is placed at the focal point of a curved mirror, the reflected rays form a parallel beam of light.

The speed of light in vacuum is $c = 3 \times 10^8$ m/sec. Its speed v in any other material is always less than c. Hence the ratio c/v is always larger than unity. We call this ratio the index of refraction of the material. Typical values are 1.33 for water and 1.5 for glass.

A beam of light is deflected when it passes from air to glass. In this case, the angle of incidence is greater than the angle of refraction. This is always true when the beam enters a material of higher index of refraction. Since light rays can be retraced, the effect is reversed if light goes into a material of lower index of refraction.

When a parallel beam of light passes through a glass lens, the portion of the beam that goes through the most glass will be retarded the most. Hence a lens which is thickest at its center causes the beam to focus at the focal point. Images of distant objects are formed close to this point by the lens.

Since different colors of light are refracted by different amounts, a prism separates light into its various colors. Thus the prism spectrometer determines the color composition of a beam of light. Two major kinds of light are observed: a continuous spectrum containing all colors is emitted by white-hot objects; a line spectrum, consisting of only certain wavelengths, is emitted by glowing gases such as a neon light.

When light from a small source strikes two slits in a barrier, the slits act as sources of coherent light. This illustrates Huygens' principle, which states that each point on a wave acts as a new wave source. Beyond the slits, these two sets of waves interfere in a pattern of alternating dark and bright fringes. This is known as Young's double-slit experiment; it can be used to measure the wavelength of light.

Although we usually consider light to travel in straight lines and cast sharp shadows, this is not really true. When it goes through a small slit or past a barrier, the light spreads into the shadow. This is called diffraction. When light passes through a narrow slit, the shadow becomes indistinct. If the slit width equals the wavelength of the light, the shadow is completely obscured.

Diffraction makes it impossible to see detail smaller than the wavelength of the radiation being used for the observation. This probably means that we will never be able to describe the physical nature of any object with absolute precision.

TOPICAL CHECKLIST

1. What did Maxwell's investigations into the theory of electricity lead him to discover about light?
2. How does the electric field portion of a radio wave originate? The magnetic field portion?
3. What is an electromagnetic wave? How fast does it move in space?
4. At a given instant, describe the radio wave from a nearby station.
5. About how large is λ for a broadcast radio wave? An FM or TV wave? Radar?
6. How does a radio detect the electric field part of a radio or TV wave? The magnetic field part?
7. What is the function of the resonant circuit in a radio or TV set?
8. Why does a table radio respond better to a station when oriented one way than another?
9. List the following in order of decreasing λ: x-rays, infrared, ultraviolet, radio, light.
10. What electromagnetic radiations have been omitted from the list in 9?
11. List the colors of the visible spectrum from short to long λ and give λ values for the two extremes.
12. What is an angstrom unit?
13. How can a wave be represented diagramatically by crests? By rays? What is a plane wave?
14. What is the focal point of a focusing mirror? A lens? How can each be used to produce parallel light?
15. What is the meaning of refraction? Refractive index?
16. What is a prism? When white light passes through a prism, what happens?
17. What is a spectral line? How do the spectra of an incandescent bulb and a luminous gas differ? Define continuous and line spectra.
18. If two sets of waves are coherent, what does this tell us about the waves?
19. What evidence do we have that light is a wave phenomenon?
20. What is Huygens' principle?
21. Describe Young's double-slit experiment and discuss the origin of the interference fringes.
22. Why is a diffraction grating preferable to a double slit when determining wavelength?
23. Does light travel in straight lines? Explain.
24. What is meant by the word diffraction?
25. Discuss the diffraction patterns produced at the straight edge of a barrier and also by a single slit. How does the slit pattern depend upon λ and slit width?
26. How does diffraction limit our observational abilities?

QUESTIONS

1. Since a radio wave gives rise to an oscillating electric field, shouldn't a small positively charged ball be made to oscillate by a radio wave? Why can't we notice such an effect?
2. Radio waves pass easily around metal poles, etc., casting no definite shadow, while light casts a well-defined shadow. Why the difference?
3. Radar waves are strongly reflected by metal objects and rain clouds. Explain how this fact is used to monitor the locations of airplanes and storms.

4. At the top of the atmosphere is a layer of ions (called the ionosphere) produced by the radiation from the sun. Radio waves in the AM range are reflected by this layer, whereas the much shorter TV and FM waves are not. Explain why AM radio stations can transmit over much larger distances than TV stations.

5. Explain how a satellite hovering above the Atlantic in stationary orbit can be used to transmit electromagnetic waves from Europe to the United States.

6. When a red cloth is illuminated by white light, why does it appear red rather than white? What color would it appear if illuminated with blue light?

7. When white light shines through a glass of beer, the transmitted light is yellow. Why isn't it white? What happens when red light or blue light shines through the beer?

8. Using a small portable radio which can be opened to show the receiving coil, how could you determine whether a nearby radio station has a vertical or horizontal antenna?

9. Microwave cooking devices heat moist substances such as meat by inducing eddy currents to flow as a result of the changing magnetic field in the wave. Explain why radar waves are effective in this regard while radio waves are not.

10. It has been claimed that near an extremely powerful radio transmitting tower, sparks can be observed jumping on a metal farm fence. Can this be true?

11. How could you determine the focal length of a concave mirror. Of a converging lens? (The focal length is the distance from the surface to the focal point.)

12. A skin diver has a searchlight which he shines up toward the water's surface. Draw a diagram showing at what angles the light emerges into the air as he shines it up at various angles. Notice that when the angle between the light beam and the vertical is greater than a certain amount, the beam cannot get out of the surface. We say the beam is then *totally internally reflected.*

13. A fish below the surface of a still pond can see a fisherman on the bank by looking up at the surface at an angle of about 50° to the vertical. Why doesn't he have to come to the surface and look along the surface to see the bank?

14. When a beam of light enters a diamond, it has difficulty getting out. Explain why this is true. Why, in view of this, do diamonds sparkle? (*Hint:* See Question 12 and notice that diamond has a very high index of refraction.)

15. If you examine the image of the sun or a light bulb formed on a sheet of paper by a thick glass lens, you will notice the image has colored edges. How does this separation of the colors occur?

16. The brilliant colors seen in oil films on water as well as in soap bubbles are the result of interference. A beam of light entering the film is partly reflected at the top surface of the film and partly at the lower surface, producing two coherent waves which have traveled different distances. Explain how the red light is reflected strongly at certain places while the blue light is strongly reflected at others.

17. Antireflection coatings are widely used on the lenses of better cameras. A thin layer of transparent material is placed on top of the lens. (You can tell if a lens has this layer by looking at the light reflected from it. Since it is nonreflecting for only one color, the light reflected is colored.) Explain why such a layer can cause one color to give zero reflected intensity. (*Hint:* See Question 16.)

18 There is a limit to the width of a very fine fiber we can measure by use of a light microscope. Explain why. Can we tell that a fiber much smaller than the wavelength of light exists in the field of view of the microscope?

PROBLEMS

1 With what frequency must a station oscillate charges on its antenna if it is to send out 3-cm radar waves?

2 When listening to a radio station which operates at 1,200 kHz, to what frequency is the resonant circuit within the radio set? What is the wavelength of the waves from this station?

3 From the fact that a ray of light can be retraced in the reverse direction, show that a ray of light that enters a thick flat plate of glass at an angle of 30° will also leave at 30°. What does a flat plate of glass do to a light beam?

4 How long does it take a radar signal bounced off the moon to make a round trip from the earth (Moon distance = 3.8×10^8 m.)

5 Galileo tried to measure the speed of light by stationing a helper about a kilometer away and timing how long it took a pulse of light to travel the distance. Explain why he was unsuccessful.

6 From the fact that the area of a sphere is $4\pi r^2$, where r is the radius of the sphere, show that the light intensity (or energy striking the eye) from a light bulb should decrease approximately as $1/r^2$.

7 To design a Young's double-slit experiment for 3-Å x-rays, how far apart must the slits be so that the fringes on a screen 100 cm away are 5.0 cm apart? Since atoms are about an angstrom in diameter, how practical is such an idea?

8 Suppose a Young's double-slit experiment is carried out with coherent sound waves coming through two holes in a wall. The holes are 1.0 m apart on a horizontal line and the wavelength of the sound is 0.3 m. A man walking parallel to the wall 40 m away and opposite the holes notices the sound is alternately strong and weak as he walks along. How far apart are the points where the sound is loud?

Chapter 10

THE TWENTY-FIVE GOLDEN YEARS

The first quarter of this century was an unprecedented era of important discoveries in the fields of physics and chemistry. Among the secrets of nature revealed by the work of Einstein, Planck, de Broglie, and others were the following: mass can be created and destroyed; particles have wavelike properties; electromagnetic waves have particle-like properties. In this chapter we shall learn about the theories proposed by these men to explain such diverse behavior as the flow of energy from the sun and the radiation of light by atoms. We shall see that the insights they have given us altered not only the course of science, but that of civilization as well.

Relative Motion

Shortly before 1900, it became possible to measure the speed of light precisely. Using light-wave interference, scientists were able to compare speeds of light in various directions on the earth with extreme precision. It had been predicted (and widely accepted) that the speed of light that was travelling parallel to the earth's motion would be slightly different from the speed of light travelling perpendicular to the earth's motion. This belief was based on the mistaken notion that light was a wave motion of the ether, a nebulous material assumed to fill all space. (See reference 1 at the end of this chapter for a discussion of the history of these ideas.)

Experiments early in this century contradicted these predictions. Many experiments have since explored whether the speed of light (in vacuum) depends upon the motion of the light source and the motion of the person who is measuring the light speed. All these experiments have confirmed the following statement:

The speed of light (and of all electromagnetic radiation) through vacuum is always the same, c, and is independent of the state of motion of the radiation source or the observer.

This seemingly innocuous fact has fundamental importance.

Before the constancy of the speed of light was discovered, it was believed that light would provide a means for detecting the absolute speed of an object. To see why this was so, we must first recognize that all the speeds we normally measure are *relative speeds*. That is, we can say how fast a car is moving along the ground; but this speed is only relative to the roadway. The earth itself is moving around the sun and the whole solar system is evidently flying through space. Clearly, we can measure the speed of the car relative to the earth or the sun or even the distant stars. But these are all relative speeds, speeds measured with respect to some other object; they do not tell us the speed of the car in any absolute sense.

However, if ether filled all the universe and were at rest, then we would only need to measure the speed of the earth through the ether in order to specify the speeds of other objects in relation to it. These would be absolute speeds. Therefore, scientists around 1900 measured speeds of light to determine the speed of the earth through the ether. When they discovered that the speed of light is not affected by the movement of the light source or the observer, they realized that their hypothetical ether was not needed to explain the propagation of light waves, and they soon abandoned the concept of a stationary ether altogether.

By 1905, experimental results provided convincing evidence for the following hypothesis:

Only relative velocities can be measured.

This deceptively simple statement suggests a fundamental fact of nature: Any careful experiment performed in an isolated laboratory should give the same result, independent of the velocity of the laboratory. For example, suppose the laboratory is a closed boxcar of a train riding at constant speed along straight tracks. No matter what we do *within* the car, the result will be the same as it would be if the car were at rest on a siding. Without looking outside of the car (which we call our *frame of reference*), no experiment we can do will tell us what the speed of the car is because we can only determine speeds relative to other things (other reference frames). We cannot find absolute speeds.

Therefore, we must conclude that all reference frames moving with constant velocity are comparable. Any careful experiment done in a laboratory moving with constant velocity will yield the same result in any other constant-velocity laboratory, even though their velocities are not the same. Hence, any law of nature discovered in such a laboratory (or reference frame) will be found true in any other constant-velocity reference frame. Since one of the fundamental laws of nature is Newton's law of inertia (the first law of motion), we often call such a reference frame an *inertial reference frame*. All nonaccelerating reference frames are inertial reference frames. *The laws of nature are the same in all inertial reference frames.*

Even though we cannot measure absolute velocities, we can measure absolute acceleration. Consider two trains moving side by side down parallel tracks at the same speed. A passenger in one of the trains suddenly notices that his train seems to be accelerating slowly past the other one—or is the other train decelerating while his continues at a constant velocity? Many a passenger has experienced such a moment of confusion. If he has a suitable instrument for measuring acceleration, he can resolve his confusion without another glance at the other train. If his train is undergoing acceleration, it is being subjected to an unbalanced force, which the passenger can measure without reference to anything outside his car. (If his car is not experiencing an unbalanced force, the other train must be undergoing the acceleration he observes; but in that case, he can only measure the relative velocities of the two cars.) If the passenger's car is given a large deceleration, the forces involved are correspondingly large. Hence the passenger knows at once that it is his train which is decelerating if the train's safety braking system is suddenly engaged. Remember, unlike constant-velocity motion, accelerations require forces; these forces allow us to measure absolute accelerations.

AMERICAN INSTITUTE OF PHYSICS

ALBERT EINSTEIN (1879–1955)

As a student, Einstein was unhappy in the rigid and militaristic school system of his native Germany. His exceptional abilities and interest in physics flourished after he moved to Zurich to complete his undergraduate studies. Despite excellent references, he could not get a teaching job and found work in the patent office instead, continuing his studies in theoretical physics in his spare time. At the age of 26, he published papers on Brownian motion, the photoelectric effect, and relativity. Any one of these would have merited a Nobel Prize (he won it for his paper on the photoelectric effect). These famous studies were but the beginning of a long and brilliant career. He fled Hitler's Germany, and in 1940 he became an American citizen. Partly through his efforts, the United States assembled many of the world's best scientists in order to develop atomic bombs before Nazi Germany could do so. In addition to being the most famous modern scientist, Einstein was noted for his warm and compassionate personality and his deep interest in peace and social justice.

Einstein and Relativity

The considerations of the previous section may seem of little consequence to you. However, Einstein was able to deduce enormously important conclusions from them. They are embodied in his *theory of relativity*, a theory based entirely upon the following two postulates:

1 The laws of physical science are the same in all inertial reference frames.

2 The speed of light in vacuum is the same for all observers, and is equal to c.

All experimental evidence shows that these postulates are correct.

Although we will not attempt to explain Einstein's deduction process, it is possible to show, using a simple thought experiment, how his two postulates lead to the unsuspected results he obtained. (Simple quantitative treatments can be found in the references at the end of this chapter.)

Picture a rocket ship moving parallel to the earth's surface with a high speed, as shown in Figure 10.1. Assume its speed is v relative to the earth's surface.

From postulate 1 we conclude that the spaceman cannot tell that he is moving if we restrict his observations to the spaceship's reference frame. We shall impose this restriction upon him. Since we don't have to worry about expense, let us suppose the spaceship has a rod several miles long extending in front of it. Call the length of the rod D_s. Further, we assume the spaceman can send a pulse of light out from the nose of the ship. When the light pulse reaches the far end of the rod, it is detected by a device at the rod's end, as shown. (For example, the light pulse could be sent from a flashbulb and the detector could be a photocell.)

Now suppose the spaceman times how long it takes for a light pulse to travel from the nose of the spaceship to the far end of the rod, a total distance D_s. Call this time t_s. Since speed = distance divided by time, the speed of the light pulse can be com-

TWENTY-FIVE GOLDEN YEARS   211

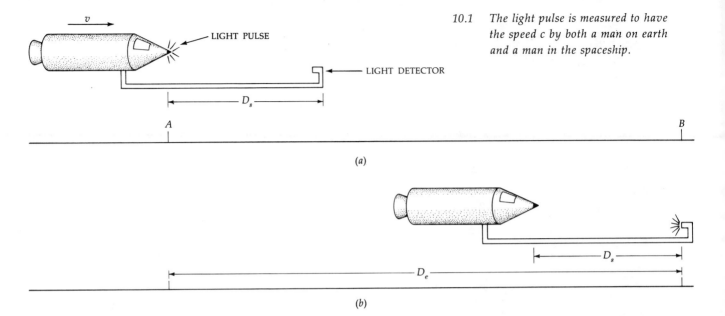

10.1 *The light pulse is measured to have the speed c by both a man on earth and a man in the spaceship.*

puted; it will be D_s/t_s. But according to postulate 2, the answer must be the known speed of light c. Hence the spaceman concludes

$$\text{Speed of light} = c = \frac{D_s}{t_s}$$

Let us now consider how an observer on the earth below the ship would view this experiment. He sees the pulse of light emitted when the nose of the spaceship is above point A on the ground. When the pulse reaches the detector at the end of the rod, as shown in (b), the pulse is above the point B on the ground. (Remember, the ship is moving swiftly toward the right and so it moved to the position shown in (b) during the time it took the light pulse to travel to the detector.) Using the symbol D_e to represent the distance from A to B, the man on earth concludes the pulse moved a distance D_e. His timing clock shows that it took a time t_e for this to happen. Hence he uses the equation speed = distance divided by time to compute the speed of light and finds

$$\text{Speed of light} = c = \frac{D_e}{t_e}$$

If we compare these two results, we see that they appear inconsistent. According to our usual way of thinking, t_s and t_e should be the same since both observers measured the same event, the time taken for the light pulse to reach the detector. But D_e is definitely larger than D_s and so D_s/t_s cannot be the same as D_e/t_e. Yet we have just been told that they *are* the same and *are* equal to the speed of light —providing we agree with Einstein's two postulates. Something must be wrong with our usual way of looking at things.

If we are willing to accept new ideas, there is a way out of this dilemma. Perhaps the ticking of the spaceship's clock is slowed down because of its motion. In that case it would read t_s to be smaller than t_e. Then, even though D_s is smaller than D_e the ratio D_s/t_s can still equal the ratio D_e/t_e, since it is also true that t_s is smaller than t_e. Hence one possible explanation is that moving clocks tick out time more slowly as a result of some previously unknown law of nature.

But there is another way out of the dilemma. Suppose the spaceman measures the rod on the ground before the flight, and then measures it again during the flight, and finds it is exactly the same length each time. He calls this length D_s and uses it in his formula. When he returns to the ground, he compares his measurement D_s with the ground observer's measurement D_e and finds they are identical. In other words, D_s appears shorter than D_e only to the ground observer watching the spaceship streak past

Table 10.1 The Relativistic Factor

v/c	0	0.01	0.10	0.20	0.50	0.70	0.90	0.99	0.999
$\sqrt{1-(v/c)^2}$	1.000	0.99995	0.990	0.98	0.87	0.71	0.44	0.14	0.045

him with a high speed v. If this were true, the spaceman would still see everything as normal, but lengths of objects moving past would appear to the stationary observer on the ground to have shrunk along their line of motion.

Admittedly, this seems like a farfetched explanation, and you can be forgiven if you do not readily accept either of these interpretations. Einstein's discussion and analysis of the problem was not accepted immediately even by his fellow scientists. Einstein showed quantitatively that both these interpretations were partly correct. Using a much more general approach, he arrived at the following universally applicable conclusions:

> When an object or timing mechanism moves with speed v past an observer, the stationary observer notices the following:
>
> 1 The moving timing mechanism is ticking out time too slowly, and is slower by a factor of $\sqrt{1-(v/c)^2}$.
>
> 2 All moving objects appear to shrink along their line of motion; lengths in the direction of motion decrease by a factor of $\sqrt{1-(v/c)^2}$, where c is the speed of light in vacuum, 2.998×10^8 m/sec.

In other words, Einstein confirms our suspicions. The spaceship clock is ticking out time too slowly; the time read on it will not be t_e but $t_e\sqrt{1-(v/c)^2}$. Similarly, the spaceship's rulers will shrink when they are laid down along the path of motion and so will measure the distance D_s to be longer than stationary rulers would measure it. Because of the combination of these two effects, D_s/t_s will still give the speed of light to be c.

Einstein's algebraic quantity $\sqrt{1-(v/c)^2}$ is called the *relativistic factor*. It is the shrinkage factor for lengths and the slowing factor for clocks. This quantity varies with v, the speed of the moving object, as shown in Table 10.1 and Figure 10.2, where the speed is given in relation to the speed of light.

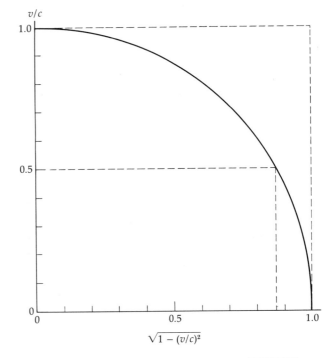

10.2 The relativistic factor, $\sqrt{1-(v/c)^2}$.

To interpret the table and graph, we should recall that the speed of light is about 186,000 mi/sec. This is about 100,000 times faster than the speed of supersonic aircraft. Hence, for the fastest airplane, v/c is less than 0.00001, and the relativistic factor is still almost exactly unity. Clearly, it will not be easy to find evidence for relativistic length contractions and time dilation (the stretching out of time by relativistic slowing of timing mechanisms). Only for atomic-size particles have we thus far been able to achieve speeds close enough to the speed of light to make the relativistic factor become much less than unity.

It is perhaps fortunate that an object has to be moving extremely fast before relativistic effects can be noticed. Life would be quite different if v/c became unity for a speed of 40 mi/hr, for example. [The physicist George Gamow gives an entertaining account of life as it would be under those conditions in his book, *Mr. Tompkins in Wonderland* (reference 2 at the end of this chapter).]

A Limiting Speed Exists

The theory of relativity makes two other important predictions. For the first, refer again to Figure 10.1. In that figure, we implicitly assume that the spaceship is moving slower than the light pulse.

However, if the ship were moving faster than the speed of light c, the light pulse, which moves out ahead of the ship, would also be moving with a speed larger than c. But this contradicts Einstein's postulate 2. We must conclude that either the postulate is wrong or the ship's speed cannot exceed c. Since a great many experiments have confirmed predictions made from the postulate, we are inclined to accept the latter alternative. Indeed, there have been many attempts to accelerate particles to speeds greater than the speed of light and all have failed. We therefore accept the following as a correct prediction of Einstein's theory of relativity:

No material object can be accelerated to speeds greater than c, the speed of light in vacuum.

This prediction is supported by the result we showed in Figure 7.23, which is redrawn here in Figure 10.3. It shows that the measured mass of a particle increases as the speed of the particle increases, and the mass of the particle becomes extremely large as the particle's speed approaches c. Since mass measures the inertia of an object, the larger the mass, the harder it is to accelerate the object. If the mass of an object approaches infinity as the speed of the object approaches the speed of light, no amount of force will be able to make it go still faster. Its speed can never exceed the speed of light.

The Mass-Energy Interconversion

Long before it was measured, Einstein predicted this increase in mass of an object as its speed increases. From the two postulates of his relativity theory, he showed that

An object that has a mass m_0 when at rest (its rest *mass) will appear to have a mass*

$$m = \frac{m_0}{\sqrt{1 - (v/c)^2}}$$

when moving with speed v.

Later, when the necessary experiments were performed, they showed that mass changes with speed exactly as this relation predicts. Notice that when $v = c$, the denominator becomes zero. Since $m_0/0 \to \infty$ (that is, the rest mass divided by zero approaches infinity), the mass of an object becomes infinite when it nears the speed of light. This is why no object can be accelerated beyond the speed of light.

Einstein analyzed this situation further, and showed that the mass increase of an object is directly related to the work done in accelerating it.

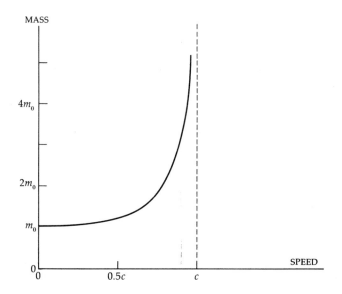

10.3 *As the particle approaches the speed of light c, its mass becomes larger than its rest mass m_0.*

He also found that the phenomenon is even more general than the relation between work and mass:

When the energy of an object is changed by an amount ΔE,* its mass will change by an amount Δm such that

$$\Delta E = \Delta m c^2$$

where c is the speed of light.

This relation is frequently written $E = mc^2$, with the understanding that a mass m is equivalent to an energy E given by the relation. This equivalence of mass and energy is basic to the production of nuclear power.

No longer can we believe that "matter can neither be created nor destroyed." Each time we give an object energy, whether it be potential, kinetic, heat, or any other type of energy, the mass of the object appears to increase slightly. When we decrease the energy of an object, its mass also decreases. Although this change in mass is too small to measure except in rare instances, the existence of nuclear power and the hydrogen bomb are convincing evidence of it, as we shall see in Chapter 12.

* Remember that Δ, delta, simply means "change in."

Time Dilation and the Twin Paradox

Although the Einstein mass-energy relation is easy to verify, the length-contraction and time-dilation predictions are less readily demonstrated. In this section we shall discuss an experiment which shows that moving clocks do, indeed, appear to tick out time too slowly. To understand the experiment, we must keep in mind that time can be measured in many different ways. For example, Galileo often used his pulse beat as a clock. In effect, his whole body was the timing mechanism. Frequently we measure time in generations, roughly the age of the parents when their children are born. This latter measure of time is similar to the measure used in the following experiment.

It is difficult to test the moving clock prediction because the clock must be moving nearly as fast as the speed of light before the effect becomes appreciable. At present, such high speeds can only be achieved for atomic- and nuclear-size objects. We therefore use an unstable nuclear-size particle as the clock for conducting the test. The particle first used for this experiment was the mu meson (or *muon*), a particle which is like the electron but which is unstable and lives only one millionth of a second. The muon in effect contains a clock which ticks out its life. After one millionth of a second, the muon changes into a new particle. We can use its average lifetime as our clock.

Muons with relatively low speed can be obtained in the laboratory as reaction products in nuclear reactions, and their average lifetimes can be measured accurately. We can also obtain muons in the laboratory moving with speeds close to the speed of light. *These muons are found to live longer than the slowly moving ones.* The clock within the fast-moving particles ticks out the lifetime of the particles more slowly. Our laboratory clocks, not being slowed in this way, tell us that the fast-moving particle has a greatly increased lifetime. The measurements show that the particle's clock has been slowed exactly as Einstein had predicted, by a factor $\sqrt{1 - (v/c)^2}$.

This experiment has led to an amazing prediction about human beings, since we age in a way governed by complex biological clocks within our own bodies. Suppose one of two identical twins boards a spaceship and flies at high speed into space. The other twin remains on earth as his brother shoots through space at speeds close to the speed of light (relative to the earth). Let us say his speed is large enough so that the relativistic factor is 0.010. Eventually, the ship returns to the earth—say after 40 years as measured by earth clocks. Of course, the twin who remained on earth has aged 40 years.

However, the clocks in the spaceship, including the biological clock in the spaceship twin's body, have read the time passed as only 0.010×40 or 0.4 years, i.e., about 21 weeks. As a result, the spaceship twin will be only 21 weeks older when he steps out of the ship to greet his twin—who has aged 40 years during the time the ship was gone. This prediction, called the _twin paradox_, seems unbelievable. Indeed, many years passed before scientists finally agreed that the reasoning is correct.* The test of the prediction remains for the future. If confirmed, as we believe from the muon experiment that it will be, the effect of time dilation may enable a man to carry out in a single lifetime a space journey spanning centuries of earth time. However, the practical difficulties are not minor. The energy necessary to accelerate a man to a speed of $c/2$ is of the order of the energy consumed by a million people in a year.

Planck's Discovery

Even before 1905, when Einstein first proposed his relativity theory, there were indications that our understanding of the physical laws of nature was far from complete. The first hint that a new scientific era was dawning came in a most unexpected and a seemingly unimportant way. In 1900, Max Planck proposed an explanation of how light is emitted from a hot, glowing oven—a proposal that eventually demonstrated that our previous ideas about vibrating systems were incorrect.

As we know, a red-hot object emits light of all colors, together with infrared and ultraviolet radiation. We can easily measure the amount of radiation emitted at each wavelength; the results for an object at two different temperatures are shown in Figure 10.4. Notice that most of the radiation coming from both objects is outside the visible range; it is in the infrared or heat radiation wavelength region of the spectrum. The wavelength of maximum radiation intensity is in the infrared at moderate temperatures but shifts into the visible region at extremely high temperatures. This agrees with our usual assumption that a white-hot object is hotter than a red-hot object.*

When we seek to understand the origin of this electromagnetic radiation, we recall that radio and radar waves are emitted by charges oscillating on an antenna. Thus it seems reasonable to assume that the much shorter wavelength radiation we are now discussing is emitted by charges oscillating in the atoms and molecules of the red-hot object. When this idea is embodied in a quantitative theory, it predicts the radiation curves of Figure 10.4 accurately, but only in the long wavelength region. In fact, all attempts at theoretical explanations of this phenomenon lead to a curve such as the dashed curve in Figure 10.4.

In other words, the theory predicts larger and

---

* New objections to it have been voiced recently. See M. Sachs, _Physics Today_, September, 1971, p. 23.

* As pointed out in Chapter 1, the average temperature of the universe about us can be related to the big-bang theory of its origin. We can determine the temperature by noting the radiation from the far reaches of space. We obtain a curve similar to those shown in Figure 10.4 and a temperature close to 3°K for space.

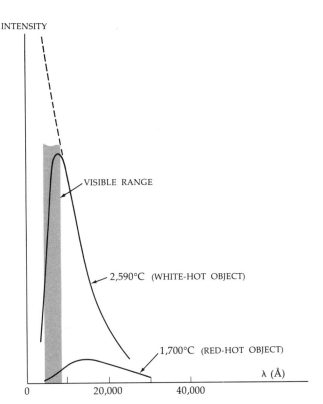

10.4   *Radiation emitted by a hot, glowing object.*

larger amounts of radiation as we proceed to shorter and shorter wavelengths, which means that a red-hot stove or oven should give off tremendous quantities of ultraviolet and x-radiation. We know that this prediction is completely false, that in fact radiation intensities decrease at short wavelengths. Because the failure of the theory to predict the proper radiation intensity in the ultraviolet was so important and puzzling, it came to be known as the *ultraviolet catastrophe*. In 1900, Max Planck developed a theory which successfully explained the observed radiation. However, he could do so only if he postulated that atoms behave much differently than had been assumed prior to that time. His discovery of this hitherto unsuspected property of atoms opened a new era in physical science.

If we assume the hot, vibrating atoms in the glowing oven act like tiny antennae, the radiation they should emit does not correspond to reality. Planck tried to find the way in which these atoms must vibrate if they are to emit the electromagnetic radiation we actually observe coming from a glowing oven. He concluded that the atoms, as well as all other objects and systems which vibrate, have a previously unknown feature controlling the vibration: the energy of the vibration is <u>quantized</u>. By this we mean the following.

Any vibrating object—a pendulum, a mass at the end of a spring, or whatever—has a certain natural frequency of vibration which we shall call ν_0 (read nu zero). Even though the pendulum (or any other such system) vibrates back and forth, making ν_0 vibrations each second, we can increase or decrease the amplitude of the vibration, thereby influencing the vibration's energy. For example, when the pendulum is swinging widely, its energy is large. As its swing dies down, its vibrational energy is lost. We ordinarily think of this as a continuous process: as the pendulum slows, it continuously loses energy. Its energy drifts through all the values from the energy of its widest swing to zero. Planck discovered that this is not correct.

He showed that the energy of the vibrator (a pendulum, atom, etc.) could have only certain discrete (distinct) values. If ν_0 is the natural vibration frequency of a particular vibrator, its energy of oscillation can only have the values $h\nu_0$, $2h\nu_0$, $3h\nu_0$, $4h\nu_0$, and so on. The quantity h is a universal constant of nature, called <u>Planck's constant</u>; its value is 6.626×10^{-34} joule-sec. Planck found that the energy of a vibrating system can have certain discrete energies but cannot take on any energy in between the allowed values, $nh\nu_0$, where n is any integer. This is what we mean when we say a vibrating system is quantized.

For example, a pendulum cannot vibrate with all amplitudes. It can vibrate widely enough so that its energy is $h\nu_0$ where ν_0 is its natural frequency of vibration; or it can vibrate twice as energetically so that its energy is $2h\nu_0$; or it can vibrate with energy $3h\nu_0$; and so on. But it cannot vibrate with any intermediate amplitude. At first glance, Planck's

AMERICAN INSTITUTE OF PHYSICS

MAX PLANCK (1858–1947)

As a young German student, Planck came close to choosing music rather than physics as his major interest; his long scientific career was devoted almost entirely to the fields of heat and thermodynamics. He discovered the quantized nature of energy while a professor in Berlin, publishing his findings in a series of papers between 1897 and 1901. He developed a formula that agreed with his experimental data on heat radiation, and then had to introduce the quantum concept in order to explain his results. At first, Planck thought his quantum concept might be nothing more than a happy mathematical accident. As others made further important progress based upon the concept, his doubts diminished, but he continued to look in vain for ways to reconcile quantum physics with the classical Newtonian view. By 1918, when he received the Nobel Prize, his preeminence as a scientist and teacher were beyond question. In 1930, he became head of the Kaiser Wilhelm Society in Berlin, but was forced to resign in 1937 because of his opposition to Hitler's persecutions. His son was executed in 1944 after an attempt on Hitler's life. After the war, when he was nearly ninety, he was reappointed head of the society, which was renamed the Max Planck Society in his honor.

discovery appears to say that a pendulum can swing back and forth through 1, or 2, or 3 cm but not through 1.5 cm or any other value intermediate to these quantized values. This contradicts our everyday experience. Only when we compute the exact quantized energy values for a pendulum do we see why Planck's discovery cannot be dismissed as unrealistic. The distances between allowed amplitudes of swing are not 1 cm as in the example just mentioned. Instead, they are of the order of 10^{-31} cm.

> Suppose a pendulum oscillates with a natural frequency $\nu_0 = 1$ Hz. If the pendulum bob (of mass m) swings to a height H, the pendulum energy is mgH. According to Planck, the pendulum energy can be only $h\nu_0$, or $2h\nu_0$, etc., and so mgH can only take on these quantized values. Therefore the pendulum can only swing to certain discrete heights given by
>
> $$mgH_1 = h\nu_0$$
> $$mgH_2 = 2h\nu_0$$
>
> and so on. Let us compute H_1, H_2, and H_3 for a pendulum with $m = 20$ gm. Since we can compute H_2, for example, from $mgH_2 = 2h\nu_0$ to be
>
> $$H_2 = \frac{2h\nu_0}{mg}$$
>
> we find
>
> $$H_1 = 3.3 \times 10^{-33} \text{ m}$$
> $$H_2 = 6.6 \times 10^{-33} \text{ m}$$
> $$H_3 = 9.9 \times 10^{-33} \text{ m}$$
>
> Notice that the allowed heights of swing differ only by 3.3×10^{-33} m.

This distance is fantastically small, many orders of magnitude smaller than the diameter of an atom. Measurement of such small changes in distance is far beyond our capability. Hence we shall never be able to prove or disprove Planck's ideas with a vibrating pendulum or any other vibrating system of conventional size. It is clear, then, why Planck's discovery of quantized oscillation energies had not been made long ago. Only in vibrating atoms, molecules, and similar tiny systems do the effects of energy quantization become measurable.

Let us summarize Planck's important contribution. His original purpose was to develop a theory which could explain the observed intensities of light and other electromagnetic radiations given off by a very hot object. He pictured each vibrating atom within the object as acting like a miniature radio antenna which sends out short wavelength electromagnetic radiation. But he could make his theory agree with the experimental results only if he assumed that the energy of each vibrator was quantized. Accordingly, he assumed that a vibrator can only have certain discrete energies, namely $nh\nu_0$, where n is any integer, ν_0 is the natural frequency of vibration of the vibrator, and h is a universal constant now called Planck's constant.

The Photoelectric Effect

Planck's supposition that the energy of an oscillator is quantized was viewed with considerable skepticism by his fellow scientists. Planck could justify his assumption only by showing that it explained the radiation emitted by a hot object. Planck himself was skeptical about his quantum concept. It might have led to the correct result in this one case because of some lucky coincidence. The idea of quantized energies for oscillators demanded much more supporting evidence before it could be considered valid. The first additional supporting evidence was presented in 1905 when Albert Einstein showed that Planck's ideas could lead to a simple interpretation of a completely different puzzle of nature, the _photoelectric effect_.

The apparatus in Figure 10.5a illustrates the principle of the photoelectric effect. Two metal electrodes are sealed in an evacuated glass tube. We make a circuit between these two plates by connecting them to a battery and a very sensitive current meter. Ordinarily no current will flow in this circuit, since there is no wire connecting the two plates in the vacuum tube.

However, if we shine a light on plate A, we find that electrons are thrown out of the metal plate by the light. These electrons are attracted across the tube to the positive plate B, and a current flows through the circuit. When the light is turned off, the electrons stop coming out of plate A and the current stops. In addition we find that the current depends upon the wavelength of the light used, as Figure 10.5b shows.

At first sight this seems reasonable, since we know that a beam of light carries energy. It would give this energy to the electrons in plate A and some of the electrons apparently acquire enough energy to escape from the plate. However, a detailed interpretation raises questions. For example, why don't wavelengths of light longer than λ_c (the critical wavelength shown in Figure 10.5b) cause electrons to escape from plate A? No matter how strong the light (as long as it doesn't melt the plate), if λ is greater than λ_c, no electrons are emitted from the plate. But no matter how weak the light, as long as λ is smaller than λ_c, electrons are emitted as soon as the light strikes the plate. It appears impossible to interpret this with Maxwell's electromagnetic picture of light.

Einstein correctly interpreted the photoelectric effect in 1905, the same year that he presented his theory of relativity. He assumed that light is emitted by oscillating atoms or molecules. If the natural vibration frequency of the atom or molecule is ν_0, then Planck's theory says that the vibrator can have only the energies $h\nu_0$, $2h\nu_0$, etc. Hence, if an atom is to send out light energy, the atom must shoot it out in small packets (or quanta), each energy quan-

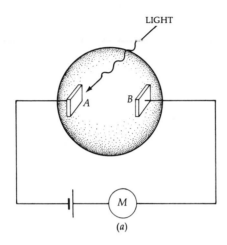
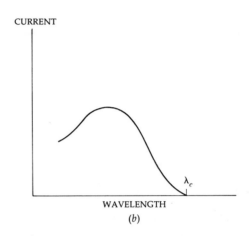

10.5 *Why does no current flow in this, the photoelectric experiment, if the light has wavelength greater than λ_c?*

tum being $h\nu_0$ in magnitude. For example, suppose an atomic oscillator has energy $5{,}217h\nu_0$, an allowed value according to Planck. If it loses energy, its energy must decrease to at least $5{,}216h\nu_0$ according to Planck so the oscillator must have lost an energy $h\nu_0$. Einstein assumed this energy was emitted as a pulse of light energy which we call a *light quantum* or *photon*. The photon energy must be $h\nu_0$.

We can find the wavelength of the emitted light by making use of the analogy between a vibrating atom and charges vibrating on a radio antenna. In the case of radio waves, the emitted electromagnetic waves have the same frequency as the frequency with which the charge oscillates on the antenna. If light waves, the electromagnetic waves emitted by vibrating atoms, behave the same way, then the frequency of the light wave ν should equal the frequency of vibration of the atom ν_0. Guided by this line of thought, Einstein assumed that photons of energy $h\nu_0$ are light pulses and that the frequency of the light waves is $\nu = \nu_0$. Since $\lambda = c/\nu$ [from Equation (8.2)] the wavelength associated with a photon of energy $h\nu$ is simply $\lambda = c/\nu$, where c is the speed of light.

Let us summarize the essentials of Einstein's assumptions:

> *A beam of light with wavelength λ (or frequency $\nu = c/\lambda$) consists of a stream of energy quanta (called photons) and the energy of each photon is $h\nu$ (that is, hc/λ).*

In other words, a beam of light consists of a stream of photons, each of which is like a tiny, condensed light pulse. If the light has a wavelength λ, then each pulse (photon) has an energy equal to hc/λ. Hence, when the beam strikes an object, it acts like a series of energetic particles with individual energies hc/λ. Notice that the photons in x-rays and ultraviolet light have much higher energies than the photons in red and infrared light, because the photon energy hc/λ is largest for the shortest wavelength radiation.

The photon concept enables us to explain the cutoff wavelength λ_c in Figure 10.5b. It takes a certain amount of energy to tear an electron loose from a metal, and, of course, current will not flow in the circuit of 10.5a unless the light can knock electrons loose from plate A. The photons in light of long wavelength have small energies and are unable to knock an electron loose upon collision. However, photons in a beam of small wavelength light have large energies and can easily knock an electron out of the plate when they collide. The dividing line between photons which have enough energy to free an electron and those which do not is the critical wavelength, the cutoff wavelength λ_c in Figure 10.5b.

There is a simple quantitative test which can be made of this general idea concerning photons and the photoelectric effect. Since the energy of a photon is $h\nu = hc/\lambda$, the shorter the wavelength of the light, the more energy the photon will have. If we shine onto a metal a beam of light with a wavelength λ, shorter than λ_c, the photons will give the electrons an energy hc/λ which is larger than the required energy hc/λ_c. The difference in these two energies should appear as the kinetic energy of the emitted electrons. Actual measurements of the kinetic energies of the emitted electrons show the difference to be exact.

From this and many other pieces of evidence, scientists have concluded that Einstein's assumption is correct. Although light is an electromagnetic wave phenomenon, the energy in the light beam comes in small pulses called light quanta or photons. The energy of the photon is related to the frequency and wavelength of the light and is $h\nu$ or hc/λ, where h is Planck's constant. Generally, the wave properties of light are most evident when the light is moving through space. The quantum (particle) nature of light is most evident when the light interacts with matter.

The Nuclear Atom

Before we investigate what happens inside an atom when it emits light, let us see how knowledge of the structure of the atom was acquired. In 1900, the accepted model of the atom was that proposed by J. J. Thomson (1856–1940). Through the years,

CAVENDISH LABORATORY

SIR J. J. THOMSON (1856–1940)

Thomson's brilliance became evident while he was still a young English schoolboy. He entered Cambridge University at the age of 14 and began to study engineering, but his interests soon led him to the frontiers of physics. His scholastic record was outstanding and, at 28, he was appointed to a professorship at Cambridge, a position he held for the rest of his life.

Although he is best known for his research into the nature of electrons and other charged particles, he was also an extraordinarily talented teacher. Not only was he awarded the Nobel Prize and a knighthood for his discoveries, but seven of his research assistants eventually received Nobel Prizes as well. Under Thomson's directorship, the Cavendish Laboratory at Cambridge became the world center of atomic and nuclear research. His son, G. P. Thomson, was one of the first to show that the electron has wave properties; he, too, won the Nobel Prize.

TWENTY-FIVE GOLDEN YEARS

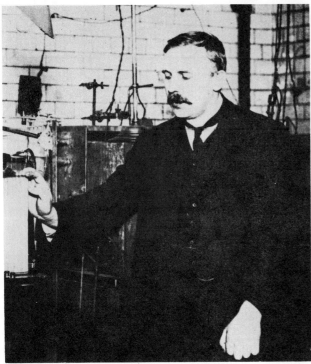

UNIVERSITY OF MANCHESTER

LORD RUTHERFORD (1871–1937)

Ernest Rutherford was the son of a New Zealand farmer. After attending college in New Zealand, he entered Cambridge and studied under J. J. Thomson. Thomson's discovery of the electron in those years gave impetus to Rutherford's long and outstanding career in atomic and nuclear physics. He was awarded the Nobel Prize in 1908 for his studies of radioactivity and atomic structure. This was still relatively early in his career, and the fact that the award was in chemistry rather than physics gave him frequent opportunities to exhibit his engaging wit. Following appointments at McGill (Canada) and Manchester Universities, he returned to Cambridge in 1919 to succeed Thomson as director of the Cavendish Laboratory. His discoveries were fundamental to our understanding of the atom and its nucleus. In addition, he is renowned for the guidance and leadership he gave to those who gravitated to his laboratories. Their many successes are in no small way a tribute to him.

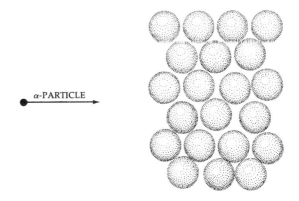

10.6 *Thomson's atom model pictured the mass and positive charge of the atom as a jellylike sphere with tiny electrons suspended in it.*

chemists had concluded that atoms contain equal amounts of negative and positive charge. Calling the charge on the electron $-e$ and that on the proton $+e$ (where e is the charge quantum discussed in Chapter 6), they knew that the quantities of positive and negative charge in the atom are Ze and $-Ze$, respectively. The number Z is the <u>atomic number</u> of the atom as listed by chemists. Using this numbering scheme, $Z = 1$ for hydrogen, $Z = 2$ for helium, $Z = 3$ for lithium, and so on. You will find atomic numbers for all atoms in the table of elements in Appendix IV on page 398.

Thomson pictured the atom as a uniform, jellylike sphere of positive charge with negative electrons imbedded in it. The electron mass was known to be about 1/1,840 of the mass of the least massive atom, hydrogen, and so the electrons contain all the negative charge of the atom but only a small fraction of its mass. This model leads to the picture shown in Figure 10.6 for an extremely thin film (a sheet of gold foil, for example).

Ernest Rutherford (1871–1937) investigated this model by shooting alpha particles through a gold film such as that shown in Figure 10.6. (Alpha particles shoot out from radioactive radium with extremely high energy; Rutherford had earlier estab-

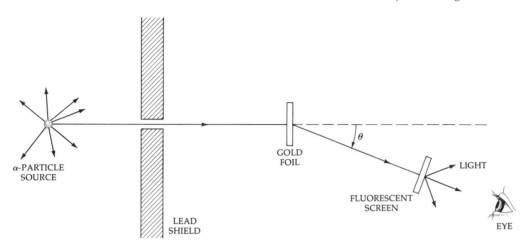

10.7 *Using this general method, Geiger and Marsden found that most of the alpha-particles went through the gold foil just as though it were not there.*

lished that these particles are simply helium atoms which have lost their electrons.) The approximate dimensions and proximities of the atoms involved are shown in proper ratio in Figure 10.6; however, Rutherford's films were about 50 times thicker than that shown. Clearly, the alpha particle must pass through numerous gold atoms before it can get through the gold film.

The apparatus used by Geiger and Marsden (then Rutherford's graduate students) to carry out this experiment in 1911 is diagrammed in Figure 10.7. After the alpha particles go through the film (if they actually do go through), they strike a fluorescent screen much like that at the end of a TV tube. When an alpha particle strikes the screen, it registers as a small flash of light at the point of impact. By moving the screen from place to place, the number of particles which are deflected through various angles can be determined.

The results of this experiment were surprising. Contrary to what Thomson's model predicted, most of the alpha particles went through the film as though it were not there. The few that were affected by the foil acted as though they had struck something massive. They were often thrown aside to large angles, some of which came close to 180°. Therefore, Rutherford postulated a new model for the atom, with which he explained Geiger and Marsden's results quantitatively.

Rutherford pictured all the Ze positive charge and nearly all the mass of the atom to be concentrated in a tiny central ball called the nucleus. We now know that the gold nucleus has a radius of about 7×10^{-15} m (0.00007 Å). Since the atom as a whole has a radius of the order of 2 Å, the atomic radius is about 30,000 times larger than the radius of the nucleus. Rutherford pictured this relatively huge volume outside the nucleus to be mostly empty space with Z tiny electrons orbiting in it. According to this picture, an alpha particle shooting toward an atom would appear as shown in Figure 10.8. Even here the sizes of the particles are grossly exaggerated. For a properly drawn atom this large, the specks representing electrons, nucleus, and alpha particle would be too small to be seen without a magnifying glass.

According to this picture, when an alpha particle shoots through an atom as shown in Figure 10.8, it has little chance of striking anything, since the atom is mostly empty space. Even a foil 100 atoms

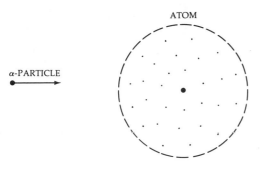

10.8 *To obtain a realistic model of an alpha particle shooting at an atom, one should decrease the size of the dots until they can only be seen with the aid of a microscope, while retaining the size shown for the atom.*

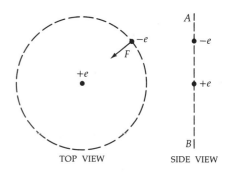

10.9 *The hydrogen atom consists of one electron and one proton.*

thick will not present much of a barrier. If the alpha particle strikes an electron, the effect will be slight, since the electron has a mass only about 1/7,000 of that of the alpha particle. Only when it encounters a gold nucleus will the alpha particle be seriously deflected. In that case, because the alpha particle's mass is only about 1/50 that of the gold nucleus, the deflection is apt to be very large. Today we have much more evidence on this point, and the basic idea of Rutherford's *nuclear model* is generally accepted. In the next two sections we will see how Bohr used it to formulate the first reasonable model of the way atoms emit light.

The Hydrogen Atom

The simplest of all atoms is the hydrogen atom, the first element in the table of elements. Its nucleus is simply a proton, and so the nuclear mass is 1.67×10^{-27} kg, the mass of a proton. We customarily designate nuclear masses in a unit called the *atomic mass unit* (u), and in this unit system the proton has a mass of $1u$. In addition to its mass of $1u$, the hydrogen nucleus has a charge $+e$, where $e = 1.6 \times 10^{-19}$ coul, the charge quantum. To compensate for the proton's charge, the hydrogen atom has one electron which contributes $0.00055u$ to the atom, a nearly negligible mass. We can therefore picture the hydrogen atom as shown in Figure 10.9.

As we see from the figure, the negative electron is attracted to the positive nucleus by the coulomb force. If nothing else were taking place in the atom, the electron would be pulled into the nucleus and the atom would collapse. This does not happen, of course. We can guess why by comparing the proton and the nucleus with the earth and sun (or the moon and earth, or a satellite and the earth). The gravitational attractive force of the sun for the earth is just sufficient to bend its normal straight-line motion into a circle; the earth keeps falling from its straight-line motion but stays at the same distance from the sun.

Therefore, we can tentatively postulate that in the hydrogen atom, the electron is circling the nucleus much as the earth circles the sun. But if we look at the atom in side view as in Figure 10.9, the electron appears to oscillate back and forth along the dotted line through the nucleus as it circles the nucleus. Since this is a situation similar to an oscillating antenna, the atom should radiate electromagnetic waves with a frequency equal to the

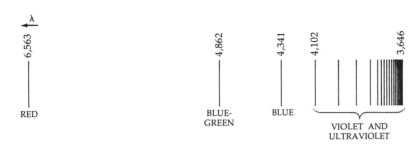

10.10 *The Balmer series in the hydrogen spectrum. (λ is in angstroms.)*

frequency with which the electron circles the nucleus. This presents two difficulties.

First, ordinary hydrogen atoms do not send out radiation. They only radiate when they are subjected to some exciting mechanism such as we shall discuss later. Second, if the atom did radiate energy all the time, it would need to obtain this energy from the potential energy of the electron. The electron would lose this energy by falling into the nucleus and so the atom would become unstable and collapse.

Hydrogen atoms will emit light, however, if the atoms are <u>excited</u> (given more energy). One way to excite the atoms is by sending sparks through a tube filled with hydrogen gas. The same principle applies to neon signs: when a high voltage is applied to the ends of a tube of neon gas, the gas glows brightly and the neon sign gives off light. The light emitted by a similarly glowing tube of hydrogen gas provides us with a tool for investigating the structure of the atom.

When the light emitted from excited hydrogen gas is separated into its constituent colors by a spectrometer, it gives a line spectrum. The visible and near ultraviolet portion of the spectrum is shown in Figure 10.10. You will recall from Chapter 9 that each line is an image of the spectrometer slit; there is one image or line for each wavelength of light. Notice how the lines are ordered. At long wavelengths they are widely spaced, but they become progressively closer together as the wavelength decreases. In the region near 3700 Å they become so numerous and close together that they are indistinguishable. Finally, at a wavelength of 3646 Å, the series ends; we call this wavelength the <u>series limit</u>.

Since the eye is only sensitive to wavelengths between about 4000 and 7000 Å, we cannot see the shortest wavelength lines of the series. However, we can photograph them, since photographic film does not lose its sensitivity at short wavelengths. When the emitted light from hydrogen is examined with instruments sensitive to other wavelengths, several series of spectral lines are found. Only one series exists at wavelengths shorter than those shown in Figure 10.10; it lies in the ultraviolet and long x-ray region. However, several other series are found in the infrared. One of these is shown in Figure 10.11 together with the visible series and ultraviolet series.

The spectral series in the visible wavelength region is called the <u>Balmer series</u>. It can be seen by the eye and was the first to be discovered. The

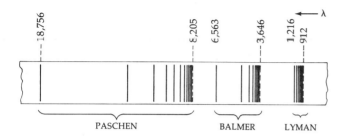

10.11 *The three spectral series of shortest wavelength given off by hydrogen atoms. (λ is in angstroms.)*

single series at shorter wavelengths is called the Lyman series, while the first series in the infrared is named the Paschen series. Because the lines are so systematically ordered, there should be a simple way to write their wavelengths.

In 1885, J. J. Balmer found that the wavelengths in the hydrogen spectral series can be represented by the following equation, the <u>*Balmer formula*</u>:

$$\frac{1}{\lambda} = 109{,}678 \left(\frac{1}{n^2} - \frac{1}{p^2} \right)$$

where n and p must be integers, that is, 1, 2, 3, etc., and λ is in centimeters.

For example, if we set $n = 2$ and $p = 3$, then λ calculated from the formula is 6563 Å; this is the red line in Figure 10.10, the longest wavelength line in the Balmer series. When $n = 2$ and $p = 4$, we find λ to be 4,862, the second line in the Balmer series. Similarly, all the lines of the Balmer series are given by successive integer values of p while n remains equal to 2. The series limit, $\lambda = 3646$ Å, is found for $n = 2$ and $p \to \infty$.

The Lyman series of wavelengths can be calculated from this formula by setting $n = 1$ and $p = 2, 3, 4,$ and so on. When $p \to \infty$, the series limit 912 Å is obtained. As you might guess by now, the Paschen series in the infrared is obtained when we set $n = 3$ and $p = 4, 5, 6,$ and so on. The series still further in the infrared are obtained for still larger values of n. This remarkably simple formula can summarize the entire spectrum given off by luminous hydrogen gas. In the next section we shall see how a particular model of the hydrogen atom can be used to explain the spectral lines which it emits.

Bohr's Model for the Hydrogen Atom

Shortly after Rutherford proposed his nuclear model for the atom, Niels Bohr, a 26-year-old Danish post-doctoral student working for him, discovered a way to obtain the Balmer formula. You will recall that a hydrogen-atom model which pictured the electron circling the nucleus as the earth circles the sun was considered untenable, because the oscillating electron should radiate energy and this should lead to the collapse of all hydrogen atoms.

Bohr circumvented this difficulty by simply ignoring it. He started with the assumption that, within the atom at least, this difficulty did not exist. As is necessary in formulating any theory, he made an educated guess and then proceeded to see if it led to anything reasonable. Bohr's guess seemed so preposterous that apparently no one else was willing even to use it as a starting point for a theory. In any event, Bohr assumed, tentatively, the following:

> *The atom has certain special orbits in which the electron can revolve without radiating energy. (We call these the Bohr orbits.)*

From this assumption he went on to say that the electron can circle the nucleus in certain stable orbits of radii r_1, r_2, r_3 and so on in increasing order of size. Since the electron will have the least energy when it is in the smallest orbit (i.e., it has fallen as close to the attracting nucleus as it can), he concluded that the electron is normally circling the nucleus in an orbit of radius r_1. We know that, given time, objects in nature tend to adjust until their potential energy is at a minimum; stones roll downhill, for example, not uphill. In the hydrogen atom, the electron is attracted to the nucleus and so the "downhill" direction is toward the nucleus. Since the atom does not collapse, there must be some minimum radius to which the electron can fall; we call this orbit's radius r_1.

Since an ordinary hydrogen atom in hydrogen gas has its electron in its innermost orbit, the electron has fallen as close to the nucleus as it can, and the atom will remain in that form for long periods of time. However, if we shoot another particle at the hydrogen atom, i.e., excite the atom, the collision may be energetic enough to knock the hydrogen atom's electron out to a larger orbit. We can excite hydrogen atoms in several ways. If hydrogen gas is heated enough, the gas atoms will move with sufficient energy to knock electrons to larger orbits upon collision. Or if we place a high voltage across a tube of gas, we can shoot charged particles through the gas with enough energy to excite the hydrogen atoms when collisions occur. In whatever way the atom is excited, its electron rotates about the nucleus in an orbit larger than the atom's innermost orbit.

Once an electron has been thrown out to one of these larger orbits, it can lose energy by falling back down to one of the smaller orbits. Bohr's second assumption was that the electron loses its energy as it falls to a smaller orbit by throwing out a pulse of light energy. Using Einstein's photon concept, Bohr postulated that if the energy lost by the electron in falling from orbit p to orbit n was $E_p - E_n$, then the energy would be radiated as a photon of energy $h\nu$ given by

$$h\nu = E_p - E_n$$

AMERICAN INSTITUTE OF PHYSICS

NIELS BOHR (1885–1962)

Bohr received his Ph.D. from the University of Copenhagen in 1911 and went to Manchester the following spring to work in Rutherford's laboratory. He brought with him a good understanding of the quantum ideas of Planck and Einstein, and proceeded to apply them to Rutherford's nuclear atom. His daring assumption that electron orbits are quantized revolutionized atomic theory, though most physicists accepted it slowly. In 1913, he returned to Copenhagen to become the first director of the Institute for Theoretical Physics. This "Copenhagen school" soon became a meeting-place for theoretical physicists from all over the world; it was particularly influential in forming our present interpretation of quantum mechanics. During World War II, he escaped from German-occupied Denmark and took part in the United States effort to develop the fission reactor and the nuclear bomb. He was greatly concerned about the need to control the enormous military and political power of the atomic bomb. Following the war, he devoted himself wholeheartedly to the development of peaceful uses of atomic power, and to attempts to control the threat posed by atomic and hydrogen weapons.

In summary,

> When an electron falls from orbit p to a smaller orbit n, the atom emits energy in the form of a photon and the photon's energy is given by $h\nu = E_p - E_n$.

We can recast this relation for the photon's energy by using the relation between frequency and wavelength, $\nu = c/\lambda$. Then,

$$\frac{hc}{\lambda} = E_p - E_n \quad \text{or} \quad \frac{1}{\lambda} = \frac{1}{hc}(E_p - E_n)$$

Notice that the latter equation has the same general form as the Balmer formula. If we choose the orbit sizes properly, the orbit energies E_p and E_n can be given values so that this relation coincides with the Balmer formula.

Bohr computed the energy of the electron in its orbit and found that he could duplicate the Balmer formula for $1/\lambda$ if he represented the orbit radii r_n by the following equation:

$$mvr_n = \frac{nh}{2\pi} \quad \text{with } n = 1, 2, 3, \ldots$$

where m is the mass of the electron, h is Planck's constant, and mv is the momentum of the electron in its orbit. If we multiply by the radius of the orbit concerned, r_n, the resulting quantity mvr_n is called the <u>angular momentum</u> of the electron. Bohr had to assume, then, that the angular momentum was quantized. In orbit 1, the angular momentum of the electron is $h/2\pi$, where h is Planck's constant. For orbit 2, the angular momentum is $2h/2\pi$; it is $3h/2\pi$ in orbit 3, and so on. In summary, then, Bohr's third assumption was:

> The stable electron orbits are those orbits for which the angular momentum of the electron is an integer multiple of $h/2\pi$.

Notice that here too the fundamental constant of nature, Planck's constant, plays an important role.

Let us review briefly. Bohr assumed that the electron in the hydrogen atom could circle the nucleus in a series of special orbits. The innermost orbit has a radius r_1, and, for unexcited hydrogen atoms, the electron stays in it. However, if the electron is given energy in a collision, it can be thrown to larger orbits. These orbits have radii $r_2, r_3, \ldots, r_n$. When an electron in one of these excited atoms falls to a smaller orbit, it loses energy in the form of a light photon; the photon's energy $h\nu$ is equal to the energy lost by the electron. For the emitted light photons to duplicate the spectrum emitted by hydrogen atoms, the radii of the orbits must be chosen so that the angular momentum of the electron in the nth orbit is $nh/2\pi$. In other words, the radius of the nth orbit, r_n, must be equal to $nh/2\pi mv$, where v is the speed of the electron in that orbit.

We see, then, that Bohr was able to provide a simple model for the hydrogen atom which predicts the wavelengths of the light emitted by the atom. It also predicts what wavelengths the atom will absorb. Furthermore, Bohr's value for r_1 is about 0.53 Å, which turns out to be correct for the size of the hydrogen atom. Only later did minor discrepancies between Bohr's predictions and experiment become evident. Many of these were explained by assuming elliptical orbits rather than circular orbits for the electron. However, the theory is seriously deficient in one respect. Bohr could give no physical justification for his stable orbits and the sizes he had to assign to them. He could only say that his rule for finding the radii gave the correct answer.

De Broglie Waves

The solution to the puzzle of the stable orbits was found in 1923 by Louis de Broglie; this work constituted his doctoral thesis. It occurred to him that Bohr's orbits seemed very much like resonance conditions. For example, we saw in Chapter 8 that sound waves going around a circular path inside a tube reinforce each other only if the circle is a whole number of wavelengths long (see Figure 8.17). Could the Bohr orbits occur at positions where some sort of wave resonance takes place? Could the electron actually behave like a wave in this circumstance?

Although this possibility seems far-fetched, it is not without precedent. We already know that light has both wave and particle properties. If a wave phenomenon such as light can behave like a stream of photons, perhaps particles, too, have a dual personality. Maybe electrons, under certain circumstances, can show wave properties.

De Broglie tried to find an analogy between the photon-wave concept of light and a possible electron-wave model for electrons. In particular, he wanted to find out what wavelength should be assigned to an electron. We know that the wavelength of a wave λ is related to its frequency ν and speed v by $\lambda = v/\nu$. Since $v = c$ for light, this becomes $\lambda = c/\nu$. If we multiply numerator and denominator of the right-hand side of this equation by h, Planck's constant, we have

$$\lambda = \frac{hc}{h\nu}$$

But $h\nu$ is the energy of the photon, and so we have a particle concept entering in. According to Einstein, the total energy of a particle of mass m is mc^2. Hence we can replace $h\nu$, the energy of the photon, by mc^2. The wavelength of the photon is then given by

Photon: $\quad \lambda = \dfrac{h}{mc}$

De Broglie suggested that this same relation applies to electrons, provided that c, the photon speed, is replaced by the electron's speed v and m is the electron mass. He therefore postulated that an electron sometimes behaves like a wave and that its wavelength is given by

$$\lambda = \frac{h}{mv} \quad (10.1)$$

10.12 *If the orbit length $2\pi r$ is a whole number of wavelengths, the wave will reinforce itself when it returns to the starting point A. In the case shown, $2\pi r = 4\lambda$.*

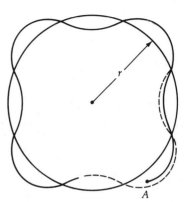

where h is Planck's constant and mv is the momentum of the electron. This wavelength is called the *de Broglie wavelength* of a particle. Let us see what it implies for Bohr's orbits in the hydrogen atom.

As we saw in Figure 8.17, a wave disturbance going around a circular orbit will reinforce itself if the orbit is an integer number of wavelengths long. For example, in Figure 10.12 we see an orbit which is 4λ long. As a result, the wave is in phase with itself when it gets back to point A, and reinforces itself strongly. We therefore suspect that the circumference of the nth Bohr orbit, namely $2\pi r_n$, should be n whole wavelengths long. It seems reasonable to suggest that

$$n\lambda = 2\pi r_n$$

If we replace λ by its value as given by de Broglie for a particle, this relation becomes

$$n\frac{h}{mv} = 2\pi r_n$$

Or, after clearing fractions and dividing by 2π,

$$mvr_n = \frac{nh}{2\pi}$$

AMERICAN INSTITUTE OF PHYSICS

LOUIS DE BROGLIE (1892–   )

Louis, the seventh duke de Broglie, was born in Dieppe, France. He did not turn to the study of science until after he had received a degree in history at the Sorbonne in Paris. World War I interrupted his studies, and he served from 1914 to 1919 as a specialist in the new and rapidly developing field of radio communications. Upon his release from the army, he began his work in theoretical physics. In his doctoral thesis, published in 1924, he presented his astonishing theory that particles have wave properties. This revolutionary theory, which won de Broglie a Nobel Prize in Physics in 1929, is fully accepted today. However, the fundamental implications of this wave nature are still a subject of much conjecture. De Broglie, who has served as Permanent Secretary of the French Academy of Science since 1942, still continues to explore the many basic questions his own theory has created.

This is exactly Bohr's condition for a stable orbit! Therefore, if we assume that the electron has associated with it a wavelength as postulated by de Broglie, the de Broglie waves reinforce on the positions of the Bohr orbits. Hence we have found a reason for the Bohr orbits: *they are the orbits in which reinforcement of the electron's de Broglie waves occur.*

Particle-Interference Experiments

If de Broglie was correct and electrons do act like waves, we should be able to do an interference experiment with electrons. To design such an experiment, we should know the approximate wavelength of the waves, since, as we saw in Young's double-slit experiment, the slit separation should be about the same size as the wavelength. An electron which has been accelerated through a potential difference of 100 volts has a speed of about 5×10^6 m/sec; its de Broglie wavelength is

$$\lambda = \frac{h}{mv} = \frac{6.6 \times 10^{-34}}{(9 \times 10^{-31})(5 \times 10^6)} \cong 1.5 \times 10^{-10} \text{ m}$$

The electron will have a very short wavelength indeed. Since atoms are about this size, it will be impossible to get slits small enough to do a Young's double-slit experiment. These lengths are more like x-ray wavelengths of electromagnetic waves. However, de Broglie waves are as different from electromagnetic waves as sound waves are. Nevertheless, we can do an interference experiment with x-rays, and the same experiment could conceivably work for de Broglie waves.

If a beam of x-rays is shot at a crystal of sodium chloride or a similar crystal as shown in Figure 10.13, the atom layers within the crystal (called crystal planes) will partly reflect the x-rays. This situation is not as simple as reflection from a mirror, since only a small fraction of the beam is reflected by each

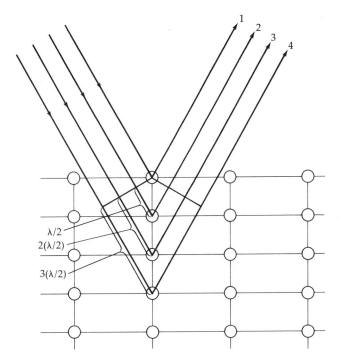

10.13 *The path length differences between rays 1, 2, 3, and 4 are whole numbers of wavelengths. As a result, the rays will reinforce.*

layer. Usually the reflected beams from the many parallel crystal planes interfere with each other so that they essentially cancel. However, in the particular situation shown in Figure 10.13, the reflected beams differ in path length by λ, 2λ, 3λ, and so on so that they will all reinforce each other. Hence, at certain special angles, the x-ray beam will be strongly reflected; at all others the reflection will be minimal. In fact if we know the separation of the crystal planes, we can use the angles at which strong reflection occurs to measure the wavelengths of the x-rays. (The reverse procedure is also important. If we know λ for the x-rays, we can use the angles of strong reflection to ascertain the crystal's architecture.)

Since the de Broglie waves associated with electrons have wavelengths about the same length as the distance between the crystal planes (called crystal lattice spacings), they too should interfere this way when a beam of electrons is reflected from a crystal. Experiments carried out to test this have agreed perfectly with the prediction. Hence we have convincing proof of de Broglie's assumption that electrons have wave as well as particle properties.

At this stage scientists wondered if the electron is the only particle that has the properties of de Broglie waves. Perhaps protons, neutrons, atoms, molecules, and even baseballs have de Broglie waves. We now know this is true. The experiment illustrated in Figure 10.13 (called an <u>electron-diffraction</u> experiment) has now been carried out with many kinds of atomic particles. In all cases the interference effect is just what we would expect for de Broglie waves. However, it is impossible to do the experiment with marbles or baseballs.

Suppose we compute the de Broglie wavelength for a marble rolling along a table at a speed of 0.50 m/sec. If the mass of the marble is 10 g (0.010 kg), the de Broglie wavelength of the marble is

$$\lambda = \frac{h}{mv} = \frac{6.6 \times 10^{-34}}{(0.010)(0.50)} = 1.32 \times 10^{-31} \text{ m}$$

But this length is billions of billions of billions times smaller than an atom. We cannot obtain slits this close together, obviously. And crystals have far too large a lattice spacing to allow the interference effect to show up. So there is no experiment we can devise which will allow us to see the interference effect between marbles which de Broglie postulates. Clearly, this is why the wave character of particles was not noticed previously. All ordinary-size particles have de Broglie wavelengths so short that no interference effects can be seen. Only in the case of atomic particles is m small enough and the de Broglie wavelength $\lambda = h/mv$ large enough for us to observe interference and wave properties.

The Heisenberg Uncertainty Principle

We firmly believe that all particles have the wave properties one would predict for their de Broglie waves. Newton's laws of motion are actually correct only for particles so large that their wave properties

AMERICAN INSTITUTE OF PHYSICS

ERWIN SCHRÖDINGER (1887–1961)

The story is told (to the author by Peter Debye) that after de Broglie presented his ideas to members of the physics faculty at Zurich, they asked their colleague Schrödinger to spend the next week trying to figure out what de Broglie had been saying. They chose wisely, for when Schrödinger reported to the faculty, he had already worked out the equation that de Broglie's waves must satisfy; it is now known as the Schrödinger equation. He and his associates carried quantum theory ahead rapidly in the following years. Schrödinger was born and educated in Vienna, where he graduated from the university in time to become an artillery officer in World War I. His important contributions to quantum physics were made before 1926, when he published a series of papers expounding the basics of wave mechanics. In 1927, he succeeded Planck as professor of theoretical physics at the University of Berlin. Leaving Hitler's Germany a few years later, he spent many years at the University of Dublin; the group he formed there became noted for its work in developing the subject of wave mechanics.

cannot be seen. Of course, his laws are also correct for small particles in cases where interference and diffraction are not important. In many cases, for example, even light can be considered as a stream of particles that obeys Newton's laws: it moves in nearly straight lines and bounces off barriers. Atomic particles will not show their de Broglie wave properties under such circumstances, so we still use Newton's laws to describe the behavior of these particles in such cases.

Unfortunately, the motion of waves is much harder to treat than the motion of particles. Even the basic equations for wave motion are complicated differential equations. One, called *Schrödinger's equation*, was formulated in 1926 by Erwin Schrödinger (1887–1961) to describe de Broglie waves; properly, it should be used rather than $F = ma$ to describe particle motion. This equation is easily solved only in certain simple cases. When applied to motions of ordinary particles, marbles, and baseballs, it gives the same result we would find from $F = ma$. Obviously we will continue to use $F = ma$ for such situations because it is infinitely easier. However, Schrödinger's equation is more precise since it embodies both $F = ma$ and the de Broglie wave behavior. Newton's laws are only an approximation to the exact way nature behaves.

When we accept the fact that particles behave as de Broglie waves, we face a previously unsuspected fact of nature: our knowledge of particle motions will always be uncertain. To see why this is so, let us explore how accurately we can locate the position of a particle. We will consider the most ideal situation imaginable, even though practical considerations may not allow us to build the apparatus to do the experiment.

We already know from Chapter 9 that even using electromagnetic radiation we cannot see detail smaller than a wavelength λ in size. As a result, there will always be an uncertainty at least λ large in the position of an object. However, it is conceivable that we could use a moving particle to locate

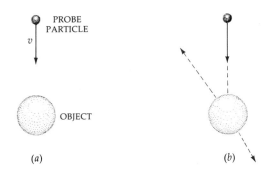

10.14 The collision of the probe particle with the object causes the momentum of the object to be uncertain.

AMERICAN INSTITUTE OF PHYSICS

WERNER HEISENBERG (1901–   )

Heisenberg was born in Wurzburg, Germany, and received his Ph.D. from Munich University in 1927. Starting from an entirely different approach than Schrödinger, he developed quantum mechanics in an alternate mathematical framework, which is widely used today. Although he is best known for the uncertainty principle, it was his development of the field of quantum mechanics that won him the Nobel Prize in 1932. During World War II, Heisenberg was appointed Director of the Kaiser Wilhelm Institute in Berlin, a post that put him in charge of German research on atomic weapons. He believed the war would end long before any nation could develop an atomic bomb and felt that Germany did not have the resources necessary to produce the bomb in any case. As a result, the German atomic research team spent the war years developing atomic power plants rather than bombs. Shortly after the war, Heisenberg became Director of the Max Planck Institute for Physics at Göttingen; he now lives in Munich, where he is currently working on a unified field theory.

the position of an object to higher accuracy. For example, if we shoot a probe particle at an object as shown in Figure 10.14a, we could locate the object by noticing whether or not the particle hits it. But the particle also has wave properties and so it will show interference and diffraction effects. Therefore, the particle will be unable to locate the object any closer than a small distance about λ large. Let us call this small distance Δs, since it is the uncertainty in the distance to the position of the object. Then under the best conditions $\Delta s \cong \lambda$, where the symbol $\cong$ is read "approximately equal to."

Perhaps we should use particles going very fast; then, since $\lambda = h/mv$, the wavelength can be made very small so Δs will be smaller. However, in order to "see" the object, the probe particle must hit it. The collision will change the momentum of the object, as shown in Figure 10.14b. When the probe particle bounces off the object, the object's momentum can change by about the amount of momentum the probe particle had, namely mv. The object's momentum will be uncertain by an amount $mv = \Delta p$ (you may recall that p often designates momentum).

Let us now form the product of the smallest possible uncertainty in position of the object, $\Delta s = h/mv$, and the uncertainty in momentum, $\Delta p = mv$. We then have

$$\Delta s \, \Delta p = \frac{h}{mv} mv = h$$

The uncertainties will actually be larger than this, since we ignored all practical measuring difficulties. Therefore, in general,

$$\Delta s \, \Delta p \geq h$$

where $\geq$ means "greater than or equal to."

A similar argument shows that it is impossible to measure time and energy simultaneously to infinite accuracy. For example, if the uncertainty within which we can measure the energy of an object is ΔE and the time at which it had this energy is uncertain within a time Δt, then

$$\Delta E \, \Delta t \geq h$$

In other words, the energy of an object will always be uncertain.

These two statements of uncertainty constitute the famous *Heisenberg uncertainty principle*, proposed by Werner Heisenberg in 1927. There appears to be no way to circumvent them. Nature is so constituted that we can never know an object's position or energy precisely. The philosophic argument concerning the possibility of man's being able to predict exactly the future of the universe is now settled. According to the uncertainty principle, we cannot even predict the exact behavior of a single object.

The uncertainty principle is another example of the importance of Planck's constant. It determines the intrinsic uncertainty in the universe. Fortunately Planck's constant is a very small number. If it had a value of 100 joule-sec instead of 6×10^{-34} joule-sec, the effects on our lives would be most disconcerting. To see why, you might compute the uncertainty in position of a 100-kg man running across a football field at a speed of 5 m/sec. Make a reasonable assumption about the uncertainty in his speed and consider his mass to be nearly exactly known. (In his book *Mr. Tompkins in Wonderland*, George Gamow uses this idea as one of the strange features of Wonderland.)

BIBLIOGRAPHY

1. BUECHE, F.: *Principles of Physics*, 2d ed., McGraw-Hill, New York, 1972. A more complete treatment of relativity is found in chapter 7; the other material of this chapter is found in chapters 26 and 27.

2. GAMOW, G.: *Mr. Tompkins in Wonderland, or Stories of c, G, and h.* Macmillan, New York, 1940. A fantasy. As its title implies, this is a popular treatment of relativity. An amusing account of what life would be like if the speed of light were only several miles per hour.

3. GAMOW, G.: *Thirty Years that Shook Physics* (paperback), Anchor Science, Doubleday, Garden City, N.Y., 1966. An interesting account of the discoveries mentioned in this chapter written by a famous physicist who was a friend of Bohr, et al. Non-mathematical.

4. EINSTEIN, A.: *The Meaning of Relativity*, Princeton University Press, Princeton, N.J., 1950. Written for the layman by the man who proposed it.

5. ANDRADE, E.: *Rutherford and the Nature of the Atom*, (paperback), Anchor Science, Doubleday, Garden City, N.Y., 1964. An interesting account of the life and feelings of those working in Rutherford's laboratories during these momentous days.

6. OLDENBERG, O.: *Atomic and Nuclear Physics*, 3d ed., McGraw-Hill, New York, 1961. More detailed treatments of the photoelectric affect (chapter 5), the Bohr atom (chapters 6 and 7), and matter waves (chapter 12).

7 ANDRADE e SILVA, J., and LOCHAK, G.: *Quanta* (paperback), McGraw-Hill, New York, 1969. A beautifully illustrated, nonmathematical introduction to the concept of quanta of all kinds.

SUMMARY

Absolute motion cannot be determined. We can only know motion relative to some frame of reference. Any precise experiment done in a laboratory moving with constant velocity will yield the same result in any other constant-velocity laboratory, even though their velocities are not the same. Since the law of inertia applies to all constant-velocity reference frames, we call them inertial reference frames.

A precise determination of the speed of light yields the value c independent of the state of motion of the light source or the observer. This experimental result proved the concept of an all-pervading ether to be useless.

1 The laws of physical science are the same in all inertial reference frames.

2 The speed of light in vacuum is the same for all observers.

From these two postulates, Einstein was able to prove that

1 Nothing can move faster than c, the speed of light in vacuum.

2 Moving clocks appear to a stationary observer to run slow by a factor $\sqrt{1 - (v/c)^2}$, the relativistic factor. This effect is called time dilation.

3 Moving objects appear to a stationary observer to shrink along the line of motion by the factor $\sqrt{1 - (v/c)^2}$.

4 A mass with rest mass m_0 appears to have a mass $m_0/\sqrt{1 - (v/c)^2}$ when moving with speed v.

5 When the energy of a mass is changed by ΔE, the mass will change according to $\Delta E = \Delta m c^2$.

These effects are most easily noticed for atomic-size particles, since these are readily accelerated to speeds v approaching c.

From his attempts to explain the wavelength dependence of the radiation given off by a very hot object, Max Planck discovered that the energy of all oscillators is quantized. If the natural frequency of vibration of the oscillator is v_0, then the oscillator can only take on energies in steps of magnitude hv_0 where h is a universal constant called Planck's constant. For oscillators of ordinary size, these steps are far too small for us to see that the energy is quantized. Only atomic systems show the effect clearly.

When light of sufficiently short wavelength falls on a surface, electrons are emitted from the surface in a phenomenon called the photoelectric effect. To explain this effect, especially the existence of a cutoff wavelength, Einstein proposed the photon concept of light. We now accept his proposal that a beam of light consists of a stream of photons or light quanta. Each photon carries an energy hv or hc/λ, where λ is the wavelength of the light. Short wavelength radiation consists of high-energy photons.

The nuclear model of the atom pictures a tiny nucleus at the center of the atom. This nucleus contains nearly all the mass and a positive charge Ze, where Z is the atomic number of the atom concerned. Floating outside the nucleus are the Z electrons of the atom. Most of the atom is empty space.

When hydrogen atoms are made to give off light, they emit a line spectrum. The wavelengths of these lines fall into several series, one of which lies partly in the visible wavelength range and is called the Balmer series. It is possible to represent all these series of lines by a simple formula, the Balmer formula.

Niels Bohr presented the first moderately successful theory for the behavior of the hydrogen atom's electron. In particular, he was able to derive the Balmer formula provided he made the following assumptions:

1 The electron revolves about the nucleus in certain orbits and does not radiate energy while it is in these orbits.

2 When an electron falls to a lower orbit, the energy it loses is emitted as a light quantum.

3 The radii of the stable orbits are those for which the angular momentum of the electron is an integer multiple of $h/2\pi$.

The justification for Bohr's orbits was provided by de Broglie, who found that all material particles have a wavelength associated with them. The de Broglie wavelength of a particle of mass m and speed v is given by h/mv. It is possible to do interference experiments with electrons, thereby demonstrating their wave nature.

All waves show interference and diffraction effects, and therefore all particles, because of their wave nature, show these effects also. This introduces an inherent uncertainty into nature expressed by the Heisenberg uncertainty principle. According to it, we can never know everything precisely about even the simplest particle. As a result, it is impossible, even in principle, to forecast the behavior of any particle exactly.

TOPICAL CHECKLIST

1 What do precise measurements of the speed of light in vacuum tell us about the effect of motion of the source or observer?

2 Can absolute velocities be measured? What kind of velocity do we ordinarily measure?

3 What is meant by a reference frame? What is an inertial frame? Why are inertial frames of interest to us?

4 Can absolute accelerations be measured? Explain.

5 What are the two basic assumptions in Einstein's theory of relativity?

6 What appears to happen to a meter stick moving past us with speed v if the meter stick extends (a) along the line of motion; (b) perpendicular to the motion?

7 Define the relativistic factor and show on a graph how it depends on v. About how large must v be before the factor is reduced to 0.9? To 0.10?

8 What happens to clocks moving past us with speed v? Give an example of an experiment, already done, which proves this.

9 Define time dilation.

10 Describe the twin paradox.

11 What does relativity predict about the maximum speed of objects? What happens to the mass of an object as it nears this speed? How does this help explain why a limiting speed exists?

12 What is the meaning of the relation $E = mc^2$?

13 Can matter be created? Destroyed? Explain.

14 What did Planck have to assume to arrive at a suitable theory accounting for the radiation emitted by a hot object?

15 According to Planck, a child swinging back and forth on a swing with frequency v_0 can have only certain definite energies. What are they? Why is it impossible to notice this effect in such a case?

16 What is the photoelectric effect?

17 How do we relate photon energy to wavelength of the radiation?

18 Why can ultraviolet radiation eject electrons from metals while infrared radiation cannot? What is the interpretation of λ_c?

19 What is the nuclear model of the atom and how was Rutherford led to propose it?

20 Describe the spectrum of radiation emitted by excited hydrogen atoms. What is a spectral series? The series limit?

21 What is the Balmer formula? How can it be made to represent the Balmer series? The Lyman series? The Paschen series?

22 Describe Bohr's model for the hydrogen atom, taking care to state his basic assumptions.

23 According to Bohr, a line in the Balmer series is emitted when the electron in a hydrogen atom falls from an outer orbit to the $n = 2$ orbit. What does the electron do when the first line of the Balmer series is emitted? The second line? The series limit line?

24 Repeat 23 for the Lyman series in which the electron falls to the innermost orbit.

25 What property did de Broglie attribute to electrons and other particles? What is the de Broglie wavelength of a particle?

26 How can we show that electrons and similar particles have wave properties? Why can't we do the same with large objects?

27 How can we explain Bohr's orbits in terms of the wave properties of the electron?

28 What is the Heisenberg uncertainty principle? Discuss its philosophical implications.

QUESTIONS

1 The whole theory of relativity is based upon two fundamental assumptions. What are they and why do we think they are justified?

2 A pendulum hangs at rest from the ceiling of an enclosed truck. By observing the pendulum, can we distinguish if the truck is standing still, moving at constant speed, accelerating, or decelerating?

3 Suppose the speed of light was 9 ft/sec (about 6 mi/hr) instead of 3×10^8 m/sec. Assuming that the theory of relativity applied to this new world, discuss the modifications one would notice in everyday life.

4 Two spaceships are orbiting the earth in opposite directions and collide head-on. Describe the collision in each of the following reference frames: the earth, one spaceship, the other spaceship.

5 A fly sits on a rotating phonograph record. Since he is a well-educated fly, he can do many of the experiments a man can do. As he does them, he obtains some rather puzzling results. What are they and how would you explain to him that they are really in conformity with the laws of nature. (He is so well educated that he speaks English, but his science background is rather weak.)

6 It would seem that a spaceship in free space could go faster than the speed of light by the following reasoning. Periodically a rocket on the ship would accelerate it and, since the men in the spaceship could measure accelerations, they would know their increase in speed. If they repeated this process over and over again, finally the sum of all of the increases in speed would be greater than the speed of light. What if anything is wrong with this?

7 A freely falling body above the surface of the earth has the same mass at the start of its fall as it does near the end, when its speed is quite large. Shouldn't its mass be larger at the end since it then has more kinetic energy?

8 Suppose a hypothetical spaceship is moving parallel to the earth at a speed of $0.99999c$. When it passes man A on the earth, it sends out a pulse of light from the front of the ship. Man A measures this pulse as moving with speed c relative to the earth, only slightly faster than the spaceship; man B in the ship must measure the speed of the pulse to be c.

Devise an experiment by which B can measure the speed of the pulse, and explain why he gets the same value as man A, who claims the pulse and ship are only slowly separating.

9 A tiny particle called the neutrino (this is not the common neutron) was first detected in the 1950s; many such particles are constantly shooting through the earth and our bodies. These particles have no rest mass and yet they have energy. How fast must they be moving? (*Hint:* Zero divided by anything but zero gives zero.)

10 A man in a spaceship shoots past the earth with a speed close to c. What will he notice as he rotates a meter stick within the ship from a position perpendicular to the line of motion to a position parallel to the direction of motion?

11 Although it is impossible to tell whether or not one is moving with a constant velocity greater than zero, many experiments can be done to show whether one is accelerating (or rotating). Describe one such experiment.

12 A man sealed in a rocket ship on earth is given a sedative and put to sleep. His ship then leaves the earth and moves into distant space at a speed close to that of light. When he awakens, what effects will he notice in the ship due to this high speed? Why?

13 What happens to the color of the light given off by a flashlight when the battery is failing? Why does this happen?

14 When you burn your finger, the heat energy supplied to your flesh gives the molecules so much kinetic energy that they tear each other apart, thereby damaging the flesh. With this in mind, why is ultraviolet light much more effective in causing sunburn than infrared light? (*Hint:* Consider the energies of the photons as they strike the molecules of flesh.)

15 Not only does a photon carry an energy $h\nu = mc^2$, but it also carries a momentum $mc = h\nu/c$. How can you interpret the fact that a beam of light striking a surface exerts a force on the surface? Why is the force larger on a shiny surface than on a black surface?

16 In a photoelectric-effect experiment it is found that λ_c is 4000 Å for one substance and 6000 Å for another when the substances are used as materials for the plate. What can you say about the electrons in the two substances?

17 An invisible object hangs in an open window. How could you learn the cross-sectional area of the object by shooting thousands of peas at the window with a pea shooter? What is the similarity between this experiment and the experiment of Geiger and Marsden?

18 One of the experiments carried out in the early 1900s was the following. A beam of x-rays of known wavelength is sent through a block of carbon. Some of the x-rays are scattered off to the side and a portion of these are found to have a wavelength longer than that of the incoming beam. This effect, the *Compton effect*, can be interpreted in terms of photons bouncing off the electrons in the block. Explain why such a picture could lead to the observed result.

19 Why doesn't a bottle of hydrogen gas give off light as it sits on a shelf?

20 When an alpha particle is shot through a container filled with hydrogen gas, a very faint pulse of light emerges from the container. Explain its origin.

PROBLEMS

1 A spaceship is moving away from the earth at a speed of 0.9 the speed of light, that is, $0.9c$. After a year has gone by on earth, how much has a man in the spaceship aged?

2. At what fraction of the speed of light must the spaceship in Problem 1 be moving if the spaceship man is to age only one-tenth as fast as an earth man?

3. An object has a mass m_0 when measured at rest on the earth. It is then carried into space on a ship and acquires a speed $0.9c$ relative to the earth. When the astronaut measures the object's mass on a balance in the ship, what will he find the value of the mass to be? How large will it appear to be as viewed by the man on earth?

4. How fast, relative to c, must an electron be moving if its mass is to appear to be $10m_0$? What would its energy then be?

5. By what fraction does one change the rest mass of an object when the object is carried from sea level to a point 2,000 m above sea level?

6. A fat man decides to use Einstein's mass-energy principle to lose weight. To that end, he lifts a 220-lb (100-kg) set of weights 2 m high and then *drops* them to the ground. How many times must he lift the weights to lose 10 kg (22 lb) by virtue of mass-energy conversion? He dropped the weights rather than lowering them slowly because he was an A student in science. What was his (incorrect!) reasoning in this regard?

7. The positron is a particle created in certain nuclear processes (see Chapter 11); it has the same mass as the electron, but an opposite charge. For that reason it is sometimes called a positive electron. When it comes close to an electron, the two destroy each other and the mass is all changed to energy. How much energy is given off in the process? (electron mass = 9×10^{-31} kg)

8. Suppose Planck's constant was 0.10 joule-sec and vibrational energy was quantized in units of $h\nu$. How would a 100-gm ball at the end of a 0.5-Hz pendulum behave?

9. From the fact that relativistic mass m is related to energy by mc^2, find the relativistic mass of a 6600 Å (that is, red-light) photon.

10. Using the formula for the Lyman series, show that the first line of the series has a wavelength of 1216 Å while the series limit is 912 Å. According to Bohr how do these two wavelengths arise?

11. Knowing that the wavelength of the first line in the Balmer series is 6563 Å, show that the energy difference between the second and the third Bohr orbits is 3.4×10^{-19} joule.

12. Find the energy difference between the first and second Bohr orbits from the fact that the longest wavelength line in the Lyman series is 1216 Å.

13. If you were to build a scale model for the hydrogen atom using a ping-pong ball for the nucleus, how far from the ping-pong ball should the electron be placed?

14. An ion has to fall through a potential difference of about 20 volts before it has enough energy to tear an electron loose from a hydrogen atom. In hydrogen gas at atmospheric pressure (atm) an ion goes only about 10^{-5} cm before it collides with an atom. However, at a gas pressure of 1/1,000 atm, an ion goes about $1,000 \times 10^{-5} = 10^{-2}$ cm before collision occurs. Using these facts, explain why a tube of hydrogen subjected to a potential difference of about 10,000 volts will not glow until the pressure is reduced to about 1/1,000 atm.

Chapter 11

THE ATOMIC NUCLEUS

In the previous chapter we discussed the behavior of the electrons in the hydrogen atom. Before we explore the electronic structure of more complex atoms, we will investigate the composition of the atomic nucleus. We will find that the number of protons within it determines the element to which a given nucleus belongs. We shall also consider radioactivity and nuclear reactions.

Measurement of Nuclear Masses

We know that the atomic nucleus is a tiny entity at the center of the atom. Its radius ranges from 10^{-14} to 10^{-15} m depending upon the atom. This is far smaller (about 1/50,000) than the radius of the atom as a whole. Only the electrons in an atom are outside the nucleus. Since an electron has a mass only 1/1,840 as large as a proton, the mass of the electron is nearly negligible in comparison to the rest of the mass of the atom.

Long before the masses of individual atoms could be measured, chemists were able to determine the relative masses of atoms by comparing the ways in which the atoms combine. For example, they knew that two atoms of hydrogen combine with one atom of oxygen to form water; thus,

$$2H + O \rightarrow H_2O$$

Here H is the chemical symbol for hydrogen and O stands for oxygen. Thus, the chemical formula for water is H_2O, which means that the water molecule consists of two hydrogen atoms and one oxygen atom. (In a later chapter we shall see how this knowledge was acquired.)

By burning hydrogen gas in oxygen to form water, we can show that 16 g of oxygen are consumed for each 2 g of hydrogen burned. This indicates that two hydrogen atoms have 2/16 as much mass as one oxygen atom. Hence we conclude that the oxygen atom is 16 times more massive than a single hydrogen atom. For example, if the mass of a hydrogen atom were 1 g, the water molecule H_2O would have a mass of 18 g since 2 g of hydrogen would combine with 16 g of oxygen to form one molecule of water. This would indicate that the oxygen atom had a mass of 16. The same reasoning applies to any assumed mass for the hydrogen atom. In all cases, the oxygen atom's mass is 16 times larger than the mass of a hydrogen atom.

Similar reactions exist for all the other elements known to man, although often elements other than hydrogen are involved. The important fact to note is that the elements always combine in definite proportions in any given chemical reaction. This is known as the *law of definite proportions*. In this case, it enables us to determine the relative masses of the elements in the way we have just outlined for hydrogen and oxygen. Therefore, we can list all the atoms with their relative masses as determined by various chemical reactions.

It seems natural to assign to the hydrogen atom a mass of 1 on a relative scale and measure all the other atomic masses by comparison. However, for reasons we shall discuss later, the chemical mass of a hydrogen atom is actually assigned the value of 1.00797 on the atomic mass unit scale. In atomic mass units the chemical mass of the hydrogen atom is $1.00797u$, and oxygen has a mass of $15.9994u$. Several other common elements are listed in Table 11.1 on page 243. A complete table of the elements is given in Appendix IV.

Let us now investigate the charges in an atom. We already know that the nucleus contains positive charge which must be equal to the negative charge of the electrons outside the nucleus, since the atom is electrically neutral (i.e., has as much negative as positive charge). Since all charges are multiples of the charge quantum e, the nucleus will have a charge Ze where Z is an integer. We call Z the *atomic number* of the element (see Table 11.1). For example, the uranium nucleus has atomic number $Z = 92$. Its nuclear charge is $+Ze = 92e$; it must have 92 electrons outside the nucleus to balance the nuclear charge. The atomic numbers Z of the elements were known before Rutherford proposed the nuclear model of the atom. Calculations of Z from the Rutherford, Geiger, and Marsden scattering experiments discussed in Chapter 10 agreed perfectly with the known values.

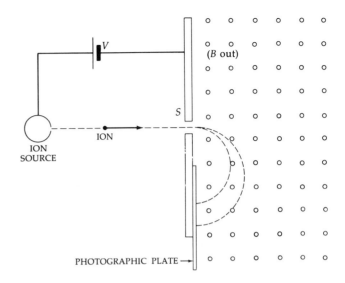

11.1 *The ion of largest mass follows the circle of largest radius. All portions of the ion path are in vacuum.*

Until about 1900, only the relative masses of the atoms were known. J. J. Thomson (1856–1940) made the first precise, direct measurement of atomic masses in 1913 (we shall not describe his method, since it has been superseded by similar but more accurate methods). Today, the masses of atoms and molecules are measured routinely in a *mass spectrograph* as shown in Figure 11.1. (Recall our discussion in Section 7.9, pages 153 to 155.) In the ion source, an electron is torn loose from one of the atoms to be measured, thereby forming an *ion*, an atom that has an unbalanced charge. (This can be done by heating the atoms or by sending a spark through the gas so that the atoms collide.) Some of the ions emerge from the small opening in the ion source, and are then accelerated through the potential difference V, so that they proceed through slit S at high speed.

After passing through the slit, the ion moves perpendicular to the magnetic field B. As a result, its path is bent into a circle, as we saw in Chapter 7. (Using the right-hand rule, can you show that the ion in Figure 11.1 is a positively charged particle?) When it completes the semicircle, the ion strikes a photographic plate or some other detector which records the position of the particle. From this position, we can determine the radius of the ion's path. More massive particles have larger radius paths than less massive ones. From the radius of the path, the strength of the magnetic field B, the voltage V through which the ion was accelerated, and the charge of the ion, we can determine the mass of the ion.

Isotopes and Nuclear Structure

When the first mass-spectrograph measurements were taken, a strange result often appeared: chemically pure elements were often found to consist of atoms with two or more different masses. For example, when pure chlorine gas is used in a spectrograph, two different spots appear on the photographic plate, as in Figure 11.1, indicating the presence of chlorine ions of two different masses. These masses are $35u$ and $37u$, and their abundance is in the ratio of about 3:1. We call atoms such as these *isotopes* of the same element.

Since chlorine atoms all behave the same chemically, and since chemical behavior depends only upon the electrons in the atom, we conclude that isotopes have identical electronic structure. As a result, the charge on the nucleus must be the same for the isotopes. Apparently isotopes differ only in the masses of their nuclei. The measured isotopic masses in Table 11.1 confirm this view.

For example, there are two common types, or isotopes, of hydrogen. (A third isotope of hydrogen exists, but it is unstable and is seldom encountered.) Both have a nuclear charge of $+e$ and have one electron. However, one has a nuclear mass of 1.0 and the other has a nuclear mass of 2.0. (Recall that the electron mass is negligible in comparison.) The mass found by the chemists is the average mass of these two isotopes. This is more easily seen in the case of chlorine, where the isotope masses are 35 and 37. Chlorine in nature consists of 75 percent of the former and 25 percent of the latter. Therefore, chemists give an atomic mass of 35.5 for chlorine, since this is the average they obtain for these two isotopes.

Looking through the elements of the periodic table in Table 11.1, we see that the isotopic masses are almost integers. The nucleus of hydrogen has a mass of 1 and consists of a single proton. You will recall that the proton has a charge of $+e$; its mass is $1.0078u$, which is equivalent to 1.672×10^{-27} kg. This mass is slightly less than that of the hydrogen *atom* because the proton mass does not include the mass of the hydrogen atom's electron.

We see from Table 11.1 that the helium atom has a mass of about $4u$ and its nucleus has a charge of $Ze = 2e$. These two units of charge and two of the units of mass can be attributed to two protons in the nucleus, but the extra two units of mass are still unaccounted for.

For many years, scientists thought that the remainder of this nucleus consisted of two more protons (which would furnish two more units of mass) and two electrons (which would cancel the extra two charges from the protons). The electron mass could be neglected. However, certain facts eventually brought this theory into question. For example, if the electron is inside the nucleus, its de Broglie wavelength must be smaller than the nuclear diameter. For this to be true, the electron would have to have a momentum some thousands of times larger than that of any electron known at the time. Later it was found that the electron has magnetic properties that cannot be accommodated by this picture of the nucleus. Hence the existence of a third particle, in addition to the proton and electron, was suspected. Like the proton, its mass must be $1u$, but it can have no charge. This particle, called the *neutron*, was finally isolated and investigated in 1932.

We now believe that nuclei are built from protons and neutrons. At present, we still do not know

*Table 11.1 Abbreviated Table of Elements**

Element	Symbol	Atomic Number Z	Atomic Mass A (Chemical)	Isotope Mass u	% Abundance of isotope as found in nature
Neutron	n	0	1.00867	1.00867	
Hydrogen	H	1	1.00797	1.008	99.985
				2.014	0.015
Helium	He	2	4.0026	4.0026	100−
Lithium	Li	3	6.939	6.015	7.5
				7.016	92.5
Carbon	C	6	12.01115	12.0 exactly	98.9
				13.003	1.1
Nitrogen	N	7	14.0067	14.003	99.6
Oxygen	O	8	15.994	15.995	99.8
Fluorine	F	9	18.9984	18.998	100
Neon	Ne	10	20.183	19.992	90.9
Sodium	Na	11	22.9898	22.99	100−
Chlorine	Cl	17	35.453	34.97	75
				36.97	25
Silver	Ag	47	107.870	106.90	51
Gold	Au	79	196.967	196.97	100
Lead	Pb	82	207.19	207.98	52
Radium	Ra	88	226	226	
Uranium	U	92	238.03	235	0.7
				238	99.2

* Notice that one isotope of carbon is assigned the mass of exactly 12.0u and all other masses are listed by comparison. Carbon is assigned the standard value since its mass is easier to compare with that of the other elements than is that of hydrogen.

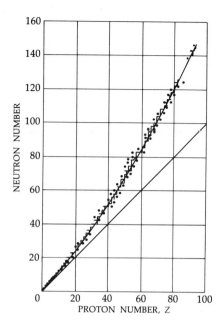

11.2 *The straight line shows what the proportion of neutrons to protons would be if the number of neutrons and protons were the same. The other line shows the actual ratio, because the heavier elements have more neutrons than protons.*

exactly how these particles are arranged and how they behave in the larger nuclei. Nevertheless, we can quickly find the makeup of any nucleus with the proton-neutron model. For example, the gold nucleus (see Table 11.1) has $Z = 79$ and so it has 79 protons. The mass of this nucleus is $197u$ of which $79u$ is furnished by the protons. It therefore has a mass $197 - 79 = 118u$ due to the neutrons in the nucleus, and so the nucleus must have 118 neutrons. The gold nucleus therefore consists of 118 neutrons and 79 protons; outside the nucleus there must be 79 electrons to make the atom electrically neutral.

As we see from Table 11.1, the number of protons is nearly equal to the number of neutrons in the nuclei of the lighter elements. However, in the more massive elements, the number of neutrons exceeds the number of protons. A major reason for this is that the positively charged protons repel each other with a force that becomes greater as the number of protons increases. In a sense, the extra neutrons dilute the protons so that they are less densely packed and thus reduce this repulsion. In Figure 11.2, we show the proportions of neutrons to protons for the stable isotopes found in nature. It is possible to synthesize nuclei other than those shown, but these are unstable and eventually throw out particles from the nucleus, thereby changing to more stable forms. Presumably, at the time the earth was formed, a large variety of unstable nuclei existed. However, most have decayed to form stable nuclei during the billions of years since. Only those nuclei which are held together strongly enough to exist unchanged over this long period of time are now found in nature.

Binding Energy of the Nucleus

None of the forces we have studied so far can explain how neutrons and protons are bound to each other. The electrostatic coulomb force actually causes the protons within the nucleus to repel each other. The gravitational attraction between the neutrons and protons is far, far smaller than the electrical repulsion force, as we pointed out in Chapter 6. In spite of this, the nucleus does hold together. Some other kind of force must operate here. We call it the *nuclear force*.

The attraction of the nuclear force is essentially the same between neutron and neutron as it is between proton and proton. (Between protons, the coulomb repulsion force is also present.) The particles of the nucleus (protons and neutrons) are called *nucleons*; the nuclear force is essentially the same for all nucleons.

The nuclear force is an extremely short-range force. Unlike the gravitational and electrostatic forces, it is not proportional to $1/r^2$; that is, it is not an inverse square law force. It decreases much more quickly with distance than these other forces. For all practical purposes, the nuclear force between two nucleons is zero if the particles are more than 10^{-14} m apart. Thus the force is restricted almost entirely to the small region occupied by the nucleus.

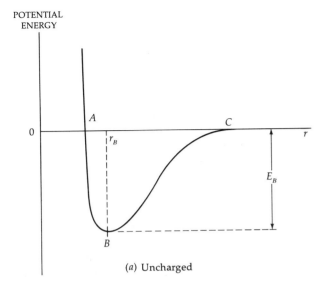
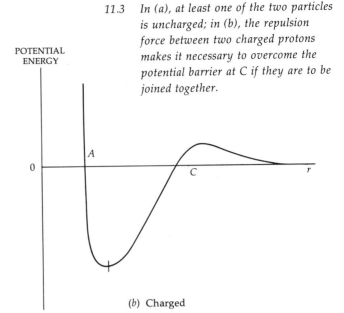

11.3 In (a), at least one of the two particles is uncharged; in (b), the repulsion force between two charged protons makes it necessary to overcome the potential barrier at C if they are to be joined together.

We can represent the effects of the nuclear force by means of an energy diagram, as in Figure 11.3. Consider the potential energy of one nucleon as it is brought close to another. As is customary, we define the zero level so that the potential energy of the particles is zero when they are infinitely far apart. Hence the potential energy curve is at zero when r, the separation of the particles, is large. What happens when one of the nucleons is fixed while r, the distance between the particles, is decreased by bringing the second nucleon closer? Initially, let us assume that the two nucleons have no charge. The behavior of their potential energy is shown in Figure 11.3a.

When the nucleon is brought close enough so that r has decreased to about 10^{-14} m, point C in the figure, the attractive force of one nucleon for the other begins to be felt, and it is pulled toward the stationary nucleon. Hence, from C to smaller r values is downhill. The nucleon loses potential energy as it moves from C toward B.

However, at a certain separation, r_B, the particles are so close that they almost touch. At this distance, they begin to repel each other. To move them closer together, we must do work to push the moving nucleon closer than the distance r_B, so the potential energy curve rises as we go to still smaller particle separations. If we recall from Chapter 5 that the system represented by such a curve is analogous to a bead sliding on a wire, we see that the equilibrium separation of the two nucleons is r_B. Work must be done upon them if they are to be either separated or pushed closer together.

In part (a), at least one of the two particles is not charged. If the particles carry like charges, the coulomb repulsion force must also be considered. The potential energy curve for this case is shown in Figure 11.3b. Here, work must be done against the repulsion force until the nucleons are pushed close enough together so that the nuclear attraction force becomes predominant. Hence the potential energy curve has a slight uphill hump in it near C, if you think of it again in terms of the bead-on-a-wire analogy. Clearly, the two nucleons must be shot at each other at high speeds if they are to surmount the hill at C.

For a practical example consider the isotope of hydrogen with a mass of $2u$ (it is frequently called deuterium). Its nucleus consists of a proton and a neutron which originally were far apart. This is exactly the situation shown in Figure 11.3a. The neutron is attracted by the proton, and it falls downhill in energy as it is brought closer. If left to itself, the neutron would fall to the bottom of the potential energy hill, point B in Figure 11.3a. Therefore, the proton and neutron of the deuterium, or *heavy-hydrogen*, nucleus are separated by a distance r_B.

Clearly, this nucleus is stable; it will not fly apart by itself. To tear these two particles apart, some outside agent must pull the neutron back up the potential hill and out to large r values. This requires the energy E_B shown in the figure. The energy needed to tear a nucleus apart is called the *binding energy*.

Thus far we have considered a nucleus composed of only two particles. However, the diagrams shown in Figure 11.3 can apply to larger nuclei as well. For example, suppose a new particle is brought into the heavy-hydrogen nucleus. The diagram would change a little at very small r values because the new particle would have two impenetrable particles to contend with, but generally it would be the same. For larger nuclei we define *the binding energy of the nucleus* to be the *energy needed* (or work done) *to tear all the nucleons apart*. In the next section we shall see how large this energy is.

The Mass Defect and Binding Energy

When we measure the masses of the nuclei with high accuracy, we find a discrepancy. For example, the measured mass of deuterium, the heavy isotope of hydrogen, is $2.01410u$. We would expect its mass to equal the sum of the masses of its one neutron and one proton. However, that sum is $2.01650u$, rather than the measured mass, $2.01410u$. Apparently, $0.0024u$ of mass is *gained* as the nucleons are pulled apart. In other words, *mass must be created when a nucleus is torn apart and its protons and neutrons are separated*.

We can interpret this phenomenon in terms of Einstein's mass-energy relation, $\Delta E = \Delta mc^2$. When the nucleus is dismantled, work equal to the binding energy must be done to tear its constituent particles apart. This work appears as additional potential energy resident in the particles themselves. As we learned in Chapter 10, when we give energy to an object, we increase its mass in conformity with Einstein's mass-energy relation. Hence, mass is created as the nucleus is torn apart.

Let us now digress for a moment to become familiar with the energy unit most widely used when dealing with atoms and nuclei. It is called the electron volt (eV) and it replaces the joule or foot-pound. The electron volt of energy is the amount of energy a charge quantum e acquires when it falls through a one-volt potential difference. Recall from Chapter 6 that when a charge q falls through a potential difference V, it acquires an energy qV joules. If $q = e$, the energy is eV joules. Therefore

$$\text{energy in electron volts} = \frac{\text{energy in joules}}{e}$$

If the energy of the charge is eV joules, then in terms of electron volts, this energy is eV/e, or V electron volts. As a particle of charge e falls through a potential difference V, then, it acquires an energy of V electron volts. In this unit of energy, the energy equivalent of one atomic unit of mass is

$$1u \rightarrow 931 \times 10^6 \text{ eV} = 931 \text{ MeV}$$

where MeV stands for million electron volts. In other words, when one atomic unit of mass is converted into energy, 931 million electron volts are released.

In the case of heavy hydrogen, the mass needed to provide the binding energy and tear the nucleus apart is $0.00240u$. Since $1u$ is equivalent to 931 MeV of energy, this amount of mass is equivalent to

$$0.00240 \times 931 = 2.23 \text{ MeV}$$

We could provide this energy to the nucleus by hitting it with another particle of very high energy—another proton, for example. This proton must

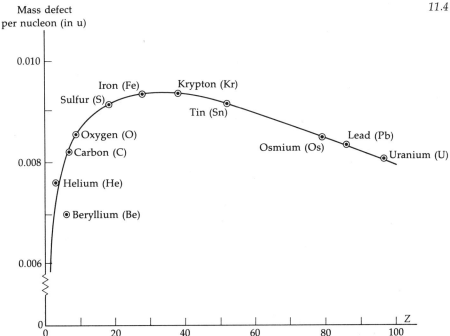

11.4 *The mass defect per nucleon is highest for elements in the mid-range of atomic masses, and so they are the most stable.*

give the heavy-hydrogen nucleus at least an energy of 2.23 MeV to tear it apart. To do so, it will have to be accelerated through a potential difference of at least 2.23 million volts. (In actuality, its energy would have to be much larger than this since some energy will remain as kinetic energy after the collision.) Particle energies this high became available in the early 1930s when the first nuclear accelerators were built.

All nuclei, since they are stable, have a binding energy. Because it takes this amount of energy to tear the nucleus apart, the individual protons and neutrons have a larger mass (and more energy) than the nucleus does. This difference in mass between the free particles and the nucleus is called the <u>mass defect</u>. We usually divide the mass defect by the number of nucleons in the nucleus to measure the average mass lost by a particle as it is placed in the nucleus. A graph showing the mass defect per nucleon for various elements is shown in Figure 11.4. We see from Figure 11.4 that the elements in the center of the curve are the most stable. (In terms of atomic mass, they fall midway between the lightest and heaviest elements, as we shall see when we examine the table of the elements in Chapter 12.)

Since the mass defect per nucleon for elements near iron (Fe) and krypton (Kr) is the largest, these elements require the most energy to tear the nuclear particles apart. When the reverse process occurs and neutrons and protons are joined together to form the nuclei, large amounts of energy must be given off.

The Fusion Reaction

When neutrons and protons are joined together to form stable nuclei, energy equivalent to the mass defect must be given off. We call such a reaction, where two or more particles are fused together to form a larger nucleus, a *fusion reaction*. It is the source of energy in the hydrogen bomb; it also provides the energy of the sun and the stars.

We believe that there are at least two major fusion reactions going on in the sun, but one appears to predominate. At the extremely high temperatures and pressures inside the sun, two hydrogen nuclei (protons) can strike each other hard enough to overcome their coulomb repulsion. As a result, they fuse together in the following reaction:

$$_1H^1 + {_1H^1} \rightarrow {_1H^2} + {_1e^0} + \text{energy}$$

THE ATOMIC NUCLEUS

Here we have used the common nuclear symbolism in which the mass (in atomic mass units) and the charge (in multiples of e) are appended to the symbol for an element in the following way:

$$_{charge}X^{mass}$$

The particle $_1H^1$ is the proton, $_1H^2$ is a _deuteron_ (heavy hydrogen), and $_1e^0$ is a positive electron (called a _positron_). Notice that we represent the positron's mass by a zero since it is very small.

The deuteron then collides with another proton to react as follows:

$$_1H^2 + {_1H^1} \rightarrow {_2He^3} + \text{energy}$$

where the product is helium 3, a light isotope of helium. It in turn reacts in the following way upon collision with another helium 3 particle:

$$_2He^3 + {_2He^3} \rightarrow {_2He^4} + 2{_1H^1} + \text{energy}$$

Notice that the overall reaction combines four protons ($_1H^1$) to produce one atom of helium. In effect, hydrogen nuclei are transformed into helium nuclei and energy.

We can compute the energy released in this fusion reaction from the difference in mass between four protons and the resulting He^4 nucleus. (The positron combines with a negative electron and in the resultant annihilation process is changed into pure energy.) That is,

$$4 \text{ protons} \rightarrow 4.0291$$
$$He^4 \rightarrow 4.0026 - 2(0.00055) = 4.0015$$

where we subtract the mass of two electrons from the helium _atom_ mass given in Table 11.1 to obtain the mass of the nucleus.

From this, the mass loss is $0.0276u$, which corresponds to an energy release of

$$931(0.0276) = 26 \text{ MeV}$$

This is a sizable amount of energy; it is 10 million times larger than we can obtain by reacting this amount of material chemically or by burning it in oxygen.

Since the oceans are full of protons from water (H_2O), the fusion reaction provides a nearly infinite energy source for us. However, a serious difficulty still prevents us from putting the fusion process to practical uses. The trouble lies in the fact that protons (and all other nuclei) carry positive charge. We know that the nuclear attractive force will hold particles together only when they are about 10^{-14} m apart. But the coulomb long-range repulsive force holds the particles apart; work must be done against it to bring the protons close enough to bind together.

In the center of the sun, the temperature is so high that the kinetic energy of the protons is often sufficient to overcome the repulsion force between them. But the necessary temperature is of the order of tens of millions of degrees, a temperature at which all materials have melted and turned to gas. Clearly, no vessel can hold the reacting protons at such temperatures. Therefore, researchers currently are trying to contain the protons and electrons (in the form of a gas of charged particles called a _plasma_) by means of magnetic fields. In this way, the particles would be held in space and would touch no material object. No practical "magnetic bottle" has yet been devised to contain this reaction, which is identical to the reaction used in the hydrogen bomb. To scale down this explosive process for peaceful use as a source of power is extremely difficult, although no one can say at this stage that it is impossible.

The Fission Reaction

The future of mankind was greatly altered by a discovery made in the Berlin laboratory of Otto Hahn (1879–1968) in 1938. Hahn, a chemist, had

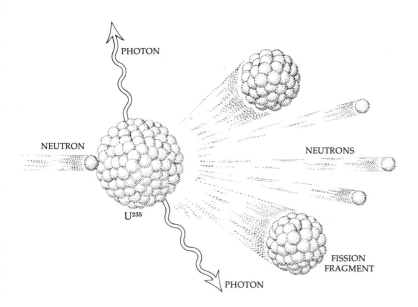

11.5   *The U^{235} nucleus splits into two or more fragments in the fission process.*

been working for several years with radioactive elements, following the series of elements into which they decayed. His associates during much of this work were Lise Meitner and Fritz Strassmann. On one December day, Hahn and Strassmann found that when a neutron came near a uranium nucleus, the nucleus assimilated the neutron and then proceeded to split into two or more parts, as shown in Figure 11.5. The process by which a heavy nucleus splits into nuclei of elements near the center of the table of elements is called *fission*.

To understand the importance of the fission process, let us refer again to Figure 11.4. There we have plotted the mass lost by each nucleon as it became part of a nucleus. Notice that the neutrons and protons in the nuclei of the intermediate elements (such as Fe and Kr) have each lost more mass than the nucleons within the very heavy nuclei such as uranium (U). Suppose now that by the fission process a uranium nucleus splits into two nuclei close to Fe in the figure. During fission, the nucleons in the original U nucleus *must lose mass* since they cannot have as much mass when they are in a nucleus of intermediate size. Hence, mass must be *destroyed* as the uranium nucleus undergoes fission.

But from Einstein's relation, $\Delta E = \Delta mc^2$, we know that a large amount of energy must be released as the uranium nucleus undergoes fission. Such a source of energy might well be of great value to mankind. Unfortunately, uranium is far less prevalent on the earth than protons, and so the fission reaction is less attractive than fusion in that respect. However, the fission reaction is relatively easy to carry out in a controlled situation.

At the time of Hahn's discovery, the world was in political turmoil. Hitler had firm control of Germany and Austria, Mussolini was dictator of Italy, and Russia was ruled by the iron hand of Stalin. Many scientists had fled from Germany, Austria, and Italy to the United States in order to escape the anti-semitism and repressive policies of their dictatorial rulers. Among these were Albert Einstein and Enrico Fermi. With many other scientists of the time, they saw that Hahn's discovery could lead to a new and potentially vast energy source.

Early in 1939, word reached Fermi, Einstein, and others that Hitler had stopped the sale of uranium from mines in Czechoslovakia, which Germany had just taken over. They guessed that Hitler planned to exploit this new energy source. (Actually, Hitler did not emphasize this project; even Hahn was allowed to continue with other work and had no real connection with the utilization of his discovery.) At the urging of a group of scientists, Einstein wrote to President Franklin Roosevelt on August 2, 1939, describing this new energy source and pointing out that "This new phenomenon would also lead to the construction of bombs, and it is conceivable—although much less certain—that extremely powerful bombs of a new type may thus be constructed. A single bomb of this type, carried by boat and exploded in a port, might very well destroy the whole port together with some of the surrounding territory."

Although the United States did not enter the war until 1941, President Roosevelt acted on Einstein's proposal at once and assembled a large fraction of the world's nuclear scientists to work on this project secretly. Finally, on December 2, 1942, Enrico Fermi and his group at the University of Chicago produced a controlled energy source using uranium; this date, the day of operation of the first _nuclear reactor_, may well be one of the most important dates in history. As you undoubtedly know, a great deal of money and effort was spent before the first uranium bomb was tested secretly in New Mexico three years later. Though Hitler had been defeated, the United States was still at war with Japan and so two of the bombs were exploded over Japan in August, 1945. This weapon was so horribly destructive that Japan capitulated days later.

The development of the atomic, nuclear-fission bomb led in turn to the development of the hydrogen, thermonuclear-fusion bomb. There are now more than enough nuclear and thermonuclear bombs to destroy everyone on earth.

The fission reaction itself can be written

$$_0n^1 + {}_{92}U^{235} \rightarrow Ba + Kr + energy + neutrons$$

This is one of many possible sets of products from this reaction. The products are usually highly

AMERICAN INSTITUTE OF PHYSICS

ENRICO FERMI (1901–1954)

Born in Rome, Fermi took his doctorate in Pisa, and was appointed Professor of Theoretical Physics at the University of Rome in 1927. He had a remarkable talent for seeing and explaining things clearly, and for developing working approximations where it was not possible to be mathematically precise. He was also known for his extraordinary charm and his infectious zest for life. His work in the field of radioactivity included the discovery that the speed of nuclear reactions might possibly be controlled; this was a crucial step in the development of nuclear power. Fermi was awarded a Nobel Prize in 1938, and went directly from the awards ceremony in Stockholm to America, rather than return to Mussolini's Italy. After three years at Columbia University, Fermi went to the University of Chicago, where he directed the construction of an atomic pile in an abandoned squash court. When this pile produced a nuclear reaction that could be sustained and controlled, the atomic age was born.

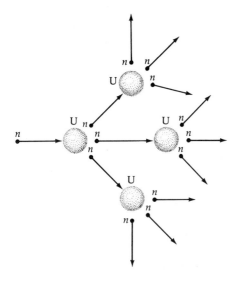

11.6 A chain reaction can be initiated by a single neutron.

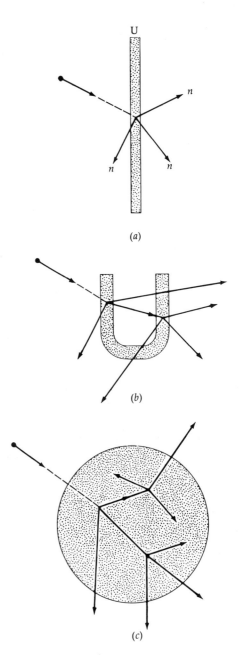

11.7 As the amount of uranium is increased, more of the neutrons from the fission process are likely themselves to cause a fission process. At a certain critical mass, fission reaction is self-sustaining (b). Beyond the critical mass, the reaction will be explosive (c).

unstable and give off other radiation and particles. Because two or three neutrons are usually given off in the fission process, a single neutron can start a <u>chain reaction</u> in uranium as shown in Figure 11.6. The first neutron produces 3 more; these produce 9 neutrons which produce 27 in the next step of the chain. If each step of the chain required a time of 0.001 sec, at the end of 0.06 sec there would be more than 10^{28} fission processes. This process would consume more than four tons of uranium. Clearly, a reaction such as this would occur with explosive violence, and that is just what happens in the fission bomb. (Fortunately, a few of the emitted neutrons are not released until about a minute after the nucleus explodes. This makes it possible to control the reaction more easily.)

To be a practical, stable energy source, however, the reaction must proceed slowly and uniformly. The factor that influences whether the reaction will die out, remain steady, or cause an explosion is illustrated in Figure 11.7. In part (a), the incident neutron causes one fission process, but the neutrons produced by the process escape without further reaction and so the reaction stops. As the size of the uranium sample is increased, more of the product

THE ATOMIC NUCLEUS

neutrons can react productively, as we see in Figure 11.7b and c. At a certain *critical mass*, one of the three neutrons from each fission process can start one other fission process. As a result, when the system "goes critical" (and each fission process produces one new fission process), the fission process continues at a steady pace. This is the condition desired in a nuclear reactor. If the reacting uranium mass exceeds critical size, the reaction will be explosive. In a fission bomb, two or more subcritical masses are shot together to form a mass greater than the critical size, and the bomb explodes.

Fortunately, it is not difficult to design a fission reactor which is essentially explosion-proof; as a result, nuclear power reactors are now coming into widespread use.* In these power reactors, the energy given off in the fission process appears as heat, which then produces steam to generate electric power. Such reactors also produce intense beams of gamma rays, high-energy neutrons, and electrons, as well as many other forms of radiation. This intense radiation is of use in bombarding other stable nuclei, thereby creating new, often extremely radioactive nuclei.

Radioactivity

As you know, some nuclei are unstable and decay to other nuclei. Only a few such nuclei still exist in measurable quantities on the earth. If there were others, they have long since decayed to unmeasurable amounts. We know that the nucleons in the nuclei of elements of intermediate mass have lost the most mass and energy. Hence it should not surprise us to find that very heavy nuclei sometimes spontaneously lose energy by throwing out part of their nucleons in a radioactive decay process. Indeed, all elements with atomic numbers Z larger than 83 are radioactive (i.e., unstable).

We define the instability of any radioactive substance by stating its *half-life*. This is the length of time taken for half the nuclei to decay in any given sample. In other words, if we start with a mass m of radioactive material, only $m/2$ of this material will be left unchanged at the end of a time equal to its half-life. The decay of a given nucleus is a random process, comparable to flipping a coin. If you flip a coin, the chance that it will turn up heads is 50-50, regardless of what has happened in previous flips. Similarly, the chance that a particular nucleus will decay in a time equal to its half-life is 50-50. The fraction of nuclei remaining unchanged after a time t is shown by the graph in Figure 11.8.

Half-lives of radioactive elements found on earth are extremely long, of course, or they would have decayed to negligible amounts in the 4.5×10^9 years that the earth has existed. For example, the half-life of radium is about 1,620 years while that for uranium 238 (an isotope of uranium) is about 4×10^9 years. All radioactive elements that occur naturally on the earth eventually decay to lead. They decay step-by-step, emitting alpha particles (helium nuclei), beta particles (electrons), and gamma rays as they decay.

For example, if we start with a sample of pure radium (which is formed by decay of uranium), we find it undergoes the following series of reactions. (Notice in these equations that the masses and the charges of the products must equal the mass and the charge of the starting material. The half-life for each element is also given.)

1   $_{88}Ra^{226} \rightarrow {}_{86}Rn^{222} + {}_2\alpha^4$ (1,620 yr)
where Rn is a gas, radon. Then

2   $_{86}Rn^{222} \rightarrow {}_{84}Po^{218} + {}_2\alpha^4$ (3.8 days)
The polonium nucleus (Po) then emits an alpha particle.

---

* Nuclear power reactors require vast amounts of water for cooling, and they create dangerously radioactive waste products which must be disposed of with extreme care. For these reasons they are not ideal power generators.

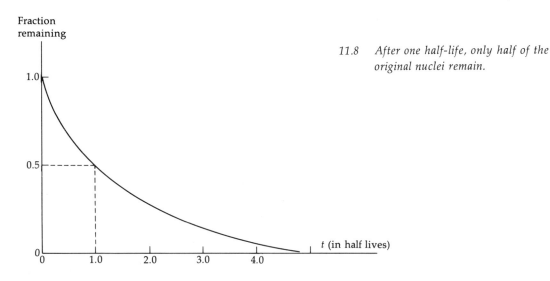

11.8 *After one half-life, only half of the original nuclei remain.*

3    $_{84}Po^{218} \to {}_{82}Pb^{214} + {}_{2}\alpha^{4}$ (3.1 min)

This isotope of lead (Pb) is not stable and emits a beta particle (an electron).

4    $_{82}Pb^{214} \to {}_{83}Bi^{214} + {}_{-1}e^{0}$ (26.8 min)

The bismuth (Bi) isotope emits an alpha particle,

5    $_{83}Bi^{214} \to {}_{81}Tl^{210} + {}_{2}\alpha^{4}$ (19.7 min)

and eventually this isotope of thallium (Tl) decays as

6    $_{81}Tl^{210} \to {}_{82}Pb^{210} + {}_{-1}e^{0}$ (1.32 min)

7    $_{82}Pb^{210} \to {}_{83}Bi^{210} + {}_{-1}e^{0}$ (19 yr)

8    $_{83}Bi^{210} \to {}_{84}Po^{210} + {}_{-1}e^{0}$ (4.85 days)

9    $_{84}Po^{210} \to {}_{82}Pb^{206} + {}_{2}\alpha^{4}$ (138 days)

This final isotope of lead (Pb) is stable, and so the series ends here. The energies of the emitted alpha particles range from 4 to 9 MeV.

In addition to the alpha and beta particles, radioactive nuclei also emit gamma rays. You will recall that these are electromagnetic radiations of very short wavelength. They are emitted when a nucleus settles to a lower energy state and throws out an energy pulse. This process is similar to a hydrogen atom throwing out radiation as the electron falls to a lower energy state. Although we do not fully understand the situation within nuclei, we know that a gamma ray is shot out when the nucleons within the nucleus settle to lower energy states. The stable nuclei found in nature are in their lowest energy state and so they do not radiate gamma rays. However, reactors are intense gamma-ray sources since the fission reaction forms many extremely unstable products.

The Tunnel Effect

As measurements of the energies became available for alpha particles emitted from radioactive substances, a puzzling fact came to light. In a nucleus such as uranium, the potential energy curve for an alpha particle near the nucleus is as shown in Figure 11.9. As usual, the alpha particle is repelled by the coulomb force when r, the distance between nucleus and alpha particle, is quite large. At r_0 the nuclear attraction force takes over and the alpha particle is pulled into the nucleus. All this seems reasonable.

11.9 *The inherent uncertainty in nature expressed by the uncertainty principle allows the alpha particle to tunnel through the barrier from A to B.*

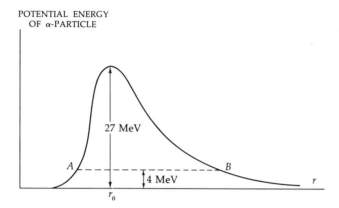

However, measurements show that the height of the potential barrier (or hill) at r_0 is about 27 MeV. Therefore, an alpha particle that succeeds in escaping the nucleus by surmounting the barrier should then fall down the potential hill as it recedes from the nucleus. Having fallen down a 27-MeV hill, the emitted alpha particle should have an energy of about 27 MeV. Instead, the emitted alpha particle is found to have about 4 MeV of energy.

George Gamow first explained this observation in 1928, using de Broglie waves and the Heisenberg uncertainty principle. Although classical physics leads us to believe the particle must go over the top of the barrier, the wave nature of the particle makes its behavior somewhat uncertain. In particular, if the particle has an energy of 4 MeV in the nucleus, its momentum is of the order of 5×10^{-20} (kg)(m)/sec. But the uncertainty relation

$$\Delta p \, \Delta s \geq h$$

tells us that its position will be uncertain. The *least possible* uncertainty in position Δs will be obtained if we substitute for Δp (momentum) the *largest* possible value. Since the uncertainty in momentum will not exceed the momentum itself, we can find the least possible value of Δs by

$$(5 \times 10^{-20}) \Delta s = h = 6.6 \times 10^{-34}$$

from which

$$\Delta s \cong 1 \times 10^{-14} \text{ meters}$$

But this is larger than the distance from A to B through the barrier at the edge of the nucleus. As a result, the alpha particle's position is so uncertain that there is a definite chance the alpha particle can be found at B, outside the nucleus, rather than at A, even though it does not have the requisite energy to surmount the barrier.

In effect, the alpha particle can tunnel through the barrier from A to B. This is called the <u>tunnel effect</u>. Although this reasoning may seem artificial, a more detailed computation in terms of the actual wave properties of particles gives the same result. We have here a process, forbidden in classical physics, which is explained by de Broglie's waves.

Nuclear Bombardment

One way of learning about the structure of nuclei is to shoot high-energy particles such as neutrons, protons, and electrons at them. This is how we find curves such as the one shown in Figure 11.9. Three basic types of machines have been constructed for this purpose; they depend on the fact that a charged particle acquires a high energy when it is accelerated through a large potential difference.

The first type, typified by the Van de Graaff generator, accelerates the charged particle between two electrodes maintained at potential differences of a few million volts. The general design of the Van de Graaff machine is shown in Figure 11.10. Because it is difficult to prevent sparks from jumping between the two electrodes, these machines are limited to potential differences of several millions of volts. Hence they produce particles with energies up to about 20 MeV, a low energy by present standards.

(a)

(b)

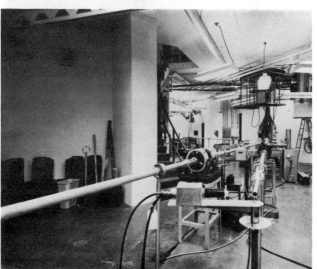

11.10   The accelerating tower of the University of Kentucky's 6MV Van de Graaff generator, shown in (a), is enclosed in a huge pressurized tank during operation. The particle beam exits into a room below the tower, where it is deflected magnetically and carried through the beam tubes shown in (b). Positive ions with accurately controlled energies up to 22 MeV emerge from the tubes. (Courtesy of Professor B. D. Kern.)

The second type of accelerator is a linear accelerator, consisting of a long line of metal tubes along which the particle beam shoots. Perfectly synchronized oscillators are placed at the gaps between the tubes so that the charged particle is accelerated as it passes from tube to tube. The linear accelerator in Figure 11.11 is a 20-billion-eV machine for accelerating electrons.

A third type of machine, used more widely for high energies, is illustrated in Figure 11.12. In this type, the particles are held in a circular path by a magnetic field perpendicular to the path. For the 70-billion-eV machine in Figure 11.12, this magnetic field is provided by C-shaped magnets around the perimeter of the circle. Each time a particle goes around the circle, oscillating voltages accelerate it through a potential difference. It achieves its high energy by falling through the same voltage difference many times as it circles around and around the circle. These devices, which are called by various names including synchro-cyclotron, bevatron, and cosmotron, are becoming increasingly expensive to build. For example, one circular accelerator now going into operation near Chicago will reach energies of 500 billion eV (500 GeV) and will cost nearly a

(a)   (b)

11.11   Two views of the Stanford University Linear Accelerator. By the time an electron has travelled 2 mi through the beam tube, hidden behind the auxiliary equipment at the left in (a), its energy is 20 GeV, i.e., 20,000 MeV. The air view (b) shows why bicycles are provided for personnel who have to travel from one end of the beam tube to the other. (Courtesy of Stanford University.)

11.12 *The 70 × 10⁹ eV Serpukhov accelerator, which is located about 60 mi southwest of Moscow, accelerates protons to speeds in excess of 0.9999c. This view, taken during construction, shows a small portion of the circular path the protons follow. (Notice the two men in the center.) The protons are held in this path by magnetic fields generated by the huge magnets along its perimeter. (Courtesy of Sovfoto.)*

billion dollars. The diameter of its circle is about a mile.

The high-energy particles produced by these machines and by reactors are used to bombard nuclei and other particles. They produce many nuclei which do not exist in nature. These nuclei are radioactive, and as they emit various radiation and particles in decaying to a stable nucleus, they give valuable information about the construction of the nucleus. In addition, the collisions produced by these high-energy particles have led to the detection of many previously unsuspected particles.

Particles accelerated to billions of electron volts of energy still do not exceed the speed of light. Since the speed of a million-eV electron is already very close to the speed of light, the additional energy must go into increasing the mass of the electron. Let us compute the mass of a 1-GeV electron.

We know that the increase in energy of a particle ΔE is related to its increase in mass Δm by $\Delta E = \Delta mc^2$. In this case, ΔE is 1×10^9 eV, which we can change to joules by multiplying by e, 1.6×10^{-19}. Therefore

$$\Delta E = \Delta m\, c^2$$

gives

$$(1 \times 10^9)(1.6 \times 10^{-19}) = \Delta m\,(3 \times 10^8)^2$$

Solving for Δm, we find

$$\Delta m \cong 1.8 \times 10^{-27} \text{ kg}$$

But the rest mass of the electron is only 0.0009×10^{-27} kg and so the mass of the electron is $m_0 + \Delta m \cong 1.8009 \times 10^{-27}$ kg. If we divide this number by the electron's rest mass, we will find how many times larger its actual mass is than its rest mass:

$$\frac{m}{m_0} = \frac{1.8 \times 10^{-27}}{0.0009 \times 10^{-27}} = 2{,}000$$

In other words, a 1-GeV electron has a mass 2,000 times as large as its rest mass. Its speed is just slightly less than the speed of light.

THE ATOMIC NUCLEUS

Particles

Until 1935, only four basic particles were known to man: the neutron, the proton, the electron, and the positive electron (or positron). If we include the photon as a particle, though it has a rest mass of zero, the total is raised to five. In an attempt to find a theory for the nuclear force, the Japanese physicist Yukawa predicted the existence of another particle in 1935. We know that the electromagnetic force field has the photon associated with it. Yukawa predicted that, in a similar way, the nuclear force field would have a particle of mass about 200 times larger than the electron associated with it. The particle is called the *pi meson* or *pion*.

After World War II, when scientists could once again devote their energies to peaceful purposes, they renewed their search for the pi meson. They expected that this particle would make its appearance when a nucleus was shattered by collision with a high-energy particle. They did find the pi meson predicted by Yukawa; they also found that many other unexpected particles were created in high-energy collisions. A partial listing of these particles is given in Table 11.2.

These particles are grouped in particle-antiparticle pairs. The electron and positron pair is an example; the positron is the antiparticle of the electron. The particles of these pairs are distinguished by the fact that they annihilate each other. Since all the stable materials we have on earth are made of regular particles, an antielectron or antiproton or antineutron can exist on earth only until it touches its partner particle; as a result, all antimatter has long since disappeared from the earth. However, we are not sure that some distant galaxy does not consist of antimatter since it is identical in all respects to ordinary matter. It would be easy to determine, however, if a voyager in space was composed of antimatter. How?

Two particles known to us have zero *rest* mass; their relativistic mass, which consists of both the

AMERICAN INSTITUTE OF PHYSICS

HIDEKI YUKAWA (1907–   )

Yukawa was born and raised in Tokyo, and studied physics at Kyoto University. He became a professor at Kyoto in 1939, and since 1950 has been director of Japan's Research Institute for Fundamental Physics. In 1935, he postulated the existence of a particle that would furnish the "glue" that holds a nuclear force field together. Postwar research led to the discovery of this pi-meson particle, and of many other particles as well. Yukawa is also noted for his prediction that a nucleus can capture one of the electrons in the first Bohr orbit of the atom. This phenomenon, called K-capture, is now known to occur. In 1949, Yukawa received the Nobel Prize for his discoveries.

rest mass and energy mass, is due entirely to their energy. They are the photon (which we have discussed often) and the neutrino (not to be confused with the neutron). W. Pauli first postulated the neutrino in 1930, and Enrico Fermi used it in a theory long before it was finally found in the mid-1950s. Fermi surmised that such a particle must exist from his attempts to explain beta-particle emission from radioactive nuclei. Unless such a particle did exist, the law of conservation of momentum would not be true for beta-decay reactions. Rather than give up this fundamental law, Fermi postulated the existence of a particle whose name, neutrino, reflects the fact that its Italian proponent considered it to be a "little neutron." Since it has zero mass and no charge, it is extremely difficult to detect. We believe that neutrinos constantly shoot through the whole earth, scarcely ever being stopped in their journey.

Many particles, including the electron, proton, and neutron, act as though they are tiny magnets. You will recall that current loops are the sources of magnetic fields. We can picture the magnetic properties of the electron, for example, as resulting from a spinning motion of the charged electron about an axis through its center. For this reason, the magnetic and rotational properties of the particles are described by a number called the *spin* of the particle. The spin of a particle turns out to be either zero or an integer multiple of $h/4\pi$. (Here again we see Planck's constant appearing in nature.) Table 11.2 lists the spins of the various particles. Notice that even the neutron has a spin and magnetic properties, a fact which leads us to suspect that the particle possesses internal structure. We shall see in Chapter 12 that the spin of a particle is a very important factor in determining how the particle behaves.

Uses of Radioactivity

Any high-energy radiation—x-rays, gamma rays, beta particles and so on—can tear apart living cells

Table 11.2   Classification and Properties of Particles

Name	Charge	Rest Mass (MeV)	Half-life (sec)	Spin*
Leptons:				
Photon	0	0	∞	1
Electron	$\pm e$	0.51	∞	$\frac{1}{2}$
Mu meson (muon)	$\pm e$	105.7	1.5×10^{-6}	$\frac{1}{2}$
Neutrino	0	0	∞	$\frac{1}{2}$
Mesons:				
Pi meson (pion)	$\pm e$	140	1.8×10^{-8}	0
	0	135	1.4×10^{-16}	0
K meson	$\pm e$	494	0.8×10^{-8}	0
	0	498	0.7×10^{-10}	0
eta meson	0	548	$< 10^{-16}$	0
Baryons:				
Proton	$\pm e$	938.2	∞	$\frac{1}{2}$
Neutron	0	938.5	646	$\frac{1}{2}$
Lambda hyperon	0	1,115	1.7×10^{-10}	$\frac{1}{2}$
Sigma hyperon	$\pm e, 0$	$\cong 1,190$	$\cong 10^{-10}$	$\frac{1}{2}$
Xi hyperon	$\pm e, 0$	$\cong 1,315$	$\cong 10^{-10}$	$\frac{1}{2}$
Omega hyperon	$\pm e$	1,676	$\cong 10^{-10}$	$\frac{3}{2}$

* In multiples of $h/2\pi$.

on impact. As a result, undue exposure to any of these radiations is hazardous. In the early days of x-rays, the people who used them did not take sufficient precautions, and many workers in this area suffered burns and contracted cancer.

We sometimes fail to realize that a single particle or photon can disrupt a cell. This is not important for the tissue of the skin, for example; but disruption of a cell that plays a part in heredity can influence the way a human embryo forms. When a baby is born with physical defects, it is likely that at least some of these defects are the result of damage to the gene cells which govern the child's development. As you sit reading this, high-energy particles shoot through your body almost every second. They come from the natural radioactivity in the materials about you and from cosmic rays, radiation striking the earth from outer space. This radiation is not strong enough to be harmful. However, man has produced x-rays and other radiation in much higher intensity than is found in nature. Especially during the child-bearing years, one should avoid unnecessary exposure to x-rays and other high-energy radiation. Of course, certain parts of the body are more susceptible to damage than others. X-rays of teeth and limbs have only minor effects, for example.

The disruptive function of radiation is widely used to kill cancer cells. High-energy x-ray machines are often used for this purpose, as is radiation from radium and radon. More recently doctors have used the radiation from radioactive isotopes made in reactors. Perhaps the most-used substance of this kind is a radioactive isotope of cobalt, Co^{60}. This isotope emits gamma rays with energy 1.2 MeV and has a half-life of 5.3 years.

In industry, cobalt 60 is used as a source of x-rays (more properly gamma rays) which detect internal flaws in large castings and similar objects. In addition, many industrial chemical reactions now use gamma rays and other radiation to initiate the reactions. Other industrial uses of radioactivity range from determining the composition of the walls of an oil-well hole to measuring the materials present in smog.

Radioactivity is also used in radioactive tracer investigations. This process employs harmlessly small amounts of radioactive materials for such purposes as following the motion of a particular constituent of food through the human body. For example, if a radioactive isotope of iodine is put in a person's food, sensitive radiation counters will soon detect radiation coming from the thyroid gland in his neck. Hence we learn that iodine tends to localize in the thyroid gland. Radioactive tracers are used frequently in studies of plants and in other biological investigations.

Radioactivity can also be used to determine the age of the earth. You will recall that the radioactive series found in nature terminate in lead. For example, the radium series on page 252 ends in lead 206 (Pb^{206}). Since this is not the most common isotope of lead, we suspect that all the lead of this isotope found in the vicinity of radium results from this decay series. The parent element for this series is uranium 238, which has a half-life of 4.5 billion years. On the earth, the amount of Pb^{206} is nearly the same as the amount of U^{238} in rocks where they are found together. We conclude from this that half the U^{238} originally present on the earth has decayed, and so the earth is one half-life, or 4.5 billion years old. Similar calculations using U^{235}, which decays to Pb^{207}, and thorium, which decays to Pb^{208}, give comparable ages for the earth. Another decay series, rubidium 87 decaying to strontium 87, also leads to the conclusion that the earth is about 4.5 billion years old.

We can also use radioactivity to determine the age of materials containing carbon, although here we are restricted to a shorter time scale. This method is based upon the fact that when nitrogen in the upper atmosphere is struck by neutrons in the cosmic rays, it undergoes the reaction

$$_7N^{14} + {_p}n^1 \rightarrow {_6}C^{14} + {_1}H^1$$

to form carbon 14. The carbon-14 isotope is unstable and has a half-life of 5,730 years. It circulates through the atmosphere, sometimes in the form of carbon dioxide, CO_2. Plants take up this isotope of carbon along with C^{12}, the normal form, and the C^{14} becomes part of the plant.

After the plant dies, the C^{14} is lost slowly as it undergoes radioactive decay; after 5,730 years, only half will be left. Before atomic-bomb testing raised the world's radiation levels, the ratio of C^{14} to C^{12}

in growing plants was 1.5×10^{-12}. Hence, if the ratio of C^{14} to C^{12} in an old piece of wood is examined and found to be only half this large, 0.75×10^{-12}, then we know that half the C^{14} has decayed. As a result, we would conclude that the wood was 5,730 years old. This technique, <u>radiocarbon dating</u>, is valuable for carbonaceous materials which are several thousand years old.

You probably know some other uses of radioactivity we have not mentioned. (For example, why is the dial of your watch luminous?) We should also recognize its dangers. Any large-scale use of nuclear weapons will contaminate most of the earth with radioactivity. It is doubtful that mankind could survive the long-term radiation from such a holocaust. Even though only a small portion of the earth would be destroyed by the explosions, those not killed by blast and fire would be forced underground to escape the deadly radioactivity carried to all parts of the world by the winds and tides. Assuming that mankind is intelligent (and lucky) enough to avoid such a disaster, other less spectacular perils must be considered. As more nations acquire nuclear weapons, extended testing of these devices could lead to seriously high levels of atmospheric radioactivity. Moreover, our seemingly insatiable quest for energy sources will necessitate much wider use of nuclear reactors. Without extensive safety precautions, accidents at these installations could seriously contaminate the surrounding areas. Moreover, the safe disposal of the radioactive waste products from reactors is also a matter for concern.

BIBLIOGRAPHY

1. MANN and GARFINKEL: *Radioactivity and its Measurement* (paperback), Momentum Book Series, Van Nostrand, Princeton, N.J., 1966. Describes the techniques scientists use to detect the ejected particles from radioactive nuclei; not particularly easy reading.

2. HUGHES, D.: *The Neutron Story* (paperback), Science Study, Doubleday, New York, 1959. An authority in the area of neutron behavior tells the story in a simple, engaging manner.

3. WILSON R. and LITTAUER, R.: *Accelerators* (paperback), Science Study, Doubleday, New York, 1960. Although this is a little out-of-date, it gives an excellent account (with pictures) of the history of these devices.

4. OLDENBERG, O.: *Introduction to Atomic and Nuclear Physics*, 3d ed., McGraw-Hill, New York, 1961. Chapters 14–18 give a detailed but nonmathematical discussion of nuclear physics.

5. CURIE, E.: *Madame Curie*, Doubleday, New York, 1937. A biography of the discoverer of radium written by her daughter; this is a classic.

6. ROMER, A.: *Discovery of Radioactivity and Transmutation*, Dover, New York, 1964. A readable account of the discovery of radioactivity in both its natural and induced forms.

SUMMARY

The nuclei of atoms have radii ranging from 10^{-14} to 10^{-15} m. They are composed of protons and neutrons. Each nucleus is characterized by two numbers, its atomic number Z and its mass number. The nucleus contains Z protons balanced by the Z electrons in the outer part of the atom. To find the number of neutrons in the nucleus, we subtract Z from the mass number.

Atomic masses are measured in atomic mass units (u). Proton and neutron masses are close to $1u$.

Since the atoms of an element are classified according to their electronic structure, all the atoms of a given element possess the same number of electrons. Thus the number of protons in the atomic nucleus of a given element must always be the same. However, the nuclei can have different numbers of neutrons and still belong to the same element. Such nuclei are said to be isotopes of the element in question. When the masses of the isotopes are determined by a mass spectrograph, it is found that all isotopes have almost integer masses in atomic mass units.

The nuclear force which binds the nucleus together is a very strong attractive force. It operates over a range restricted to about the diameter of a nucleus and is, therefore, a short-range force. It is not as well understood as the coulomb and gravitational forces.

Nuclei of the intermediate range of atomic number have their protons and neutrons more strongly bound together than do the low- and high-atomic-number nuclei. Because of this difference in binding energy, mass is converted to energy as low-Z nuclei are fused together to form intermediate-Z nuclei. This is the source of energy for the hydrogen bomb and the stars. It has not yet been possible to achieve a steady fusion reaction energy source on earth. Because energy is also released when high-Z nuclei are split in a fission reaction, we have developed practical nuclear fission reactors, which produce many radioactive products in addition to heat energy.

Radioactive nuclei are characterized by their half-life and by the particles they eject. The half-life is the time taken for half the original nuclei to decay. All the elements with Z larger than that of lead have radioactive nuclei. They eject alpha and beta particles as well as x-rays. Many other radioactive nuclei can be produced by bombarding stable nuclei with high-speed particles.

TOPICAL CHECKLIST

1. What is the law of definite proportions?
2. Hydrogen (H) and chlorine (Cl) combine to form HCl, hydrochloric acid. How can this reaction be used to determine the relative masses of the H and Cl atoms?
3. What does the atomic number Z for an element tell us about its atoms?
4. How does the mass spectrograph measure the masses of ions? Atoms?
5. What is an isotope?
6. Compare the mass and charge of the following particles: electron, proton, neutron.
7. Referring to Table 11.1, what can you say about lithium as found in nature? About silver?
8. Define nucleon. Describe the force which holds nucleons together in the nucleus. Compare it with other fundamental forces.
9. Draw and explain the energy diagram for two nucleons as a function of their distance apart. How do the charges of the particles affect the diagram?
10. What is the binding energy of a nucleus? Referring to it, why must mass be created when a nucleus is torn into its individual protons and neutrons?
11. How is the mass defect related to the binding energy?
12. What is the meaning and the use of the energy unit, the electron volt?
13. What is the fusion reaction and how can it be used to obtain an energy source?
14. Where does the energy radiated by the sun come from? How are the sun and the hydrogen bomb similar? Dissimilar?
15. Discuss the fission reaction, pointing out why it can be used as a source of energy.
17. What is the chain reaction? How is it related to the idea of a critical mass or size?
18. Describe the history of a radioactive substance such as radium.
19. Define half-life.
20. What happens to a radioactive atom when it emits an alpha particle? A beta particle? A gamma ray?

21 What is the tunnel effect? How does it show that the older classical physics is inapplicable to atomic and nuclear particles?

22 Why and how do nuclear physicists use large accelerating machines? How do such machines achieve very high-energy particles?

23 What is antimatter? How can it be distinguished from regular matter?

24 Describe some of the properties of the positron and neutrino.

25 List as many uses of radioactivity as you can.

26 How can radioactivity be used to measure the age of old wooden objects? Of the earth?

QUESTIONS

1 How do chemists decide to which element a given atom belongs? For example, how can they tell silver from sodium? What feature of the atom's structure do the chemists use when naming the atom?

2 What is the primary difference between two isotopes of the same chemical element? Is this difference easily observed chemically? How can the difference be seen?

3 Complete the following nuclear reaction:

$$_7N^{14} + {}_2He^4 \longrightarrow O^{17} +$$

[This was one of the very first nuclear reactions carried out in a laboratory (Rutherford, 1918).]

4 Consider the hypothetical element $_{21}X^{69}$. How many neutrons, protons, and electrons would each of its atoms have?

5 For the element discussed in 4, how would the two numbers associated with it change if its nucleus emits an alpha particle? A beta particle? A gamma ray? A positron?

6 One of the radioactive products formed in a fission reaction is $_{36}Kr^{90}$. The main stable isotope of krypton is Kr^{84}. Compare these two nuclei in terms of neutron and proton number. Will the radioactive Kr^{90} nucleus emit an alpha or a beta particle?

7 A friend told me he believed it soon would be possible to take water (or hydrogen gas) and submit it to a simple chemical reaction which would fuse the protons together and form deuterium. Do you concur?

8 Radium has a half-life of 1,620 years. If the age of the earth is 4.5×10^9 years, then only a fraction $10^{-840,000}$ of the original radium should be left on the earth. This is a totally negligible amount. Why, then, do we still find radium on earth?

9 Do you think Einstein, Fermi, and the other scientists who urged Roosevelt to develop nuclear energy and the bomb acted with good judgment? Should they have refused to work on such a destructive weapon? Should they have worked only on reactors and not on bombs? What about the development of the fusion bomb after World War II? Should the United States be engaged in fusion research now?

10 List as many things and processes as you can which are a result of science and which did not exist before 1940. Place the harmful quantities in one column and the beneficial ones in another. You may use a separate column for quantities you consider uncertain, but use this column sparingly.

11 Most of our knowledge about nuclear size is obtained by shooting high-energy particles at the nucleus and seeing how they bounce off. From these experiments, the nuclear size as measured by shooting-in neutrons is larger than the size found by shooting-in charged particles. What does this indicate about the distribution of mass and charge within the nucleus?

12. Slow neutrons cause fission of uranium more effectively than fast neutrons. Therefore, in a reactor, the neutrons are slowed and reflected back to the reacting core by use of what is called the *moderator*. This is usually water or hydrocarbon material. Why is this used rather than lead?

13. Much of the destruction caused by a nuclear explosion results from the direct heat, radiation, and shock waves from the explosion itself. However, this is only a portion of the total damage. What are some of the other serious effects?

14. One of the most serious effects of a nuclear explosion is the result of "fallout," the settling of radioactive dust out of the atmosphere. How is the radioactivity produced? What factors influence the severity of the effect?

15. Since a 1-MeV electron already has a speed very close to the speed of light, why build expensive machines to speed it up further?

16. In treatment of cancer deep within the body, we wish to kill the cancer cells by gamma radiation. How can this be done without seriously harming the tissue between the cancer and the skin? (For example, how can one destroy a brain cancer without damaging the rest of the brain?)

17. It is feared that seepage occurs between a sewer line and a water well some distance away. How can radioactivity be used to prove whether or not this is the case?

18. A biologist wishes to feed sugar to a plant and measure where it goes in the plant and how long it takes to get there. Outline a method he might use.

PROBLEMS

1. In the nuclear reaction

$$_7N^{14} + {}_2He^4 \rightarrow {}_8O^{17} + {}_1H^1$$

the mass of the reactants is $18.01142u$, while the mass of the products is $18.01263u$. Estimate the energy the incident alpha particle must have to cause this reaction. Rutherford carried out this reaction using alpha particles emitted from a radioactive substance.

2. Referring to Table 11.1, how much energy would be liberated in the following reaction?

$$6_1H^1 + 6_0n^1 \rightarrow {}_6C^{12}$$

Express your answer in millions of electron volts, MeV.

3. From Figure 11.4, estimate how much mass would need to be created as an iron $_{26}Fe^{56}$ nucleus is torn into neutrons and protons.

4. In Problem 3, how much energy would be needed to separate all the nucleons?

5. When an electron falls through a potential difference of 2×10^6 volts, what will its energy be? Repeat for a proton and for an alpha particle. Express your answer in millions of electron volts, MeV.

6. How fast will a proton be going if it is accelerated through a potential difference of 2×10^6 volts?

7. From Figure 11.8, how long will it take for three-fourths of a given mass of radium to decay? The half-life of radium is 1,620 years.

8. What fraction of a substance remains unchanged after one half-life time? After two? After five?

9. The average thermal energy of a gas molecule at room temperature is about 1/40 eV. What temperature would be needed to raise the average energy to 1 MeV? Why are radioactive decay phenomena insensitive to any temperature change of even a few thousand degrees?

10. Show that the tunnel effect can be interpreted in terms of the $\Delta E \, \Delta t$ portion of the uncertainty principle instead of using the $\Delta x \, \Delta p$ portion as we did in the text.

Chapter 12

ATOMS AND THE PERIODIC TABLE

In this chapter we will study the hydrogen atom further, and discuss all the atoms of the periodic table of the elements. We will see that the placement of the elements in this table follows directly from the electron arrangement within the atoms. We will learn how the electron-shell structure of the atom can be predicted by Pauli's exclusion principle. In this chapter we will study the behavior of single atoms; how atoms combine to form molecules is the subject of Chapter 13.

The Hydrogen Atom

Bohr's model for the hydrogen atom was only a crude first approximation. Although Bohr was able to predict accurately the energy levels for the atom and the light wavelengths emitted by it, he could not explain why his assumed orbits should exist. Only when de Broglie showed that the electron has wave properties did the reasons for the success of Bohr's assumptions become clear. We know now that the electron cannot be pictured as a simple ball-like particle orbiting about the nucleus. A proper picture of the electron should account for its wave nature as well as its particle nature. To do this, it is necessary to solve Schrödinger's equation, which replaces $F = ma$ in a correct description of the motions of small particles within atoms. Fortunately, we do not have to carry out the complex mathematics involved in such a correct treatment if we wish simply to understand the atom qualitatively. We shall therefore proceed with a nonmathematical description of the atom as provided by quantum (or wave) mechanics.

The wave-mechanical solution of the hydrogen atom shows that the Bohr orbits have about the same relation to reality that the max lines (or maxima) in the double-slit interference pattern have to the whole pattern. You will recall that the positions of the maxima in the double-slit pattern are the locations of complete reinforcement of the waves from the two slits. Similarly, from de Broglie's explanation, the Bohr orbits are the positions at which the electron waves interfere constructively. In the double-slit case, light photons are found in other places besides the positions of maxima in the interference pattern. This explains the fuzzy nature of the double-slit interference pattern. The electron, like the photon, is also found at locations other than the positions of the resonance maxima (the Bohr orbits). As a result, the wave nature of the electron causes the electron orbits to become diffuse.

In view of these considerations, the Bohr concept of sharp, well-defined electron orbits is untenable. Although the electron is often found in the general vicinity of the Bohr orbits, its exact location remains uncertain. We know already from the uncertainty principle that we cannot locate an electron within an atom exactly. According to the wave theory embodied in Schrödinger's equation, then, the Bohr orbits are far too simplified a picture of electron behavior.

Although Bohr's concept of electron orbits is not accurate, his theory predicts correctly the energies that the electron can assume within the atom. Consequently, his explanation of the light emission process within atoms is basically correct. You will recall that he pictured the electron throwing out an energy pulse $h\nu$ as it falls from one energy level to another. Since this concept depends only on the energy of the electron and not on Bohr's concept of orbits, it is still useful.

Many other physical properties of the hydrogen atom depend on the electron within the atom. Since Bohr's theory gives the electron's energies correctly, it often provides an easily visualized, accurate description of such properties. We will therefore make use of it for many purposes. However, when we refer to "Bohr orbits" in such discussions, we are only making a convenient choice of words. A more appropriate expression would be "... the light emitted as the electron falls from the energy predicted by Bohr for his third orbit to the energy given by Bohr for his second orbit." For convenience, we shall say "... the light emitted in a transition from the third to the second Bohr orbit."

Atomic Energy Levels

The energies of the electrons within the atoms explain many properties of atoms. The hydrogen atom is the simplest, since it has only one electron. We saw in Chapter 10 that the energies this single electron can have are given by Bohr's theory for the atom. The Bohr orbit energies are

$$E_n = -\frac{13.6}{n^2} \text{ eV} \quad (12.1)$$

Where n is an integer which represents the orbit of the electron.

You will recall that $n = 1$ when the electron is in the first (smallest) Bohr orbit. When it is infinitely far from the nucleus, the electron is in an orbit for which $n \to \infty$, and we say it has zero potential energy when it is far away from the nucleus. In the case $n \to \infty$, Equation (12.1) predicts $E_n \to 0$. As the negative electron falls "downhill" to the nucleus under the attractive force of the positive nuclear charge, it loses energy. Since its original energy was zero when it was far from the nucleus, the electron has fallen to *negative energies*. Thus the energies given by Equation (12.1) are negative. Clearly, the electron will have fallen to its lowest energy when it reaches the first Bohr orbit, for which $n = 1$. From Equation (12.1) we see that the atom energies corresponding to the various Bohr orbits are

$$E_1 = -13.6 \text{ eV}$$

$$E_2 = -\frac{13.6}{4} = -3.4 \text{ eV}$$

$$E_3 = -\frac{13.6}{9} = -1.5 \text{ eV}$$

$$E_n = -\frac{13.6}{n^2} \text{ eV}$$

$$E_\infty = -\frac{13.6}{\infty} \to 0$$

An energy of -13.6 eV means the electron has fallen downhill causing the atom to lose 13.6 electron volts of energy. In other words, the potential difference equivalent to the energy difference between an electron in the first Bohr orbit and an electron free from the atom is 13.6 volts. To put it another way, to tear the electron out of the first Bohr orbit and pull it loose from the atom, we must give it an energy of 13.6 eV. We would have to hit it with a particle which has been accelerated through at least 13.6 volts, that is, by a particle having an energy of at least 13.6 eV.

We can summarize the behavior of the atom conveniently by drawing an *energy-level diagram* such as the one for the hydrogen atom shown in Figure 12.1. This diagram is simply a vertical scale showing the energy of the electron in the atom. When the electron is at rest far from the nucleus, its energy is zero and so it is in the zero energy level (that is, at $n \to \infty$ on the diagram). If it is far away from the nucleus and is moving (for example, a free electron shooting down a TV tube), then it also has kinetic energy, and so its energy is greater than zero. Hence it will be located in the region of the diagram labeled "free electron" where the electron energies are greater than zero. Because this electron is perfectly free, its associated de Broglie wave cannot be made to resonate. For this reason, the region of the diagram above $E = 0$ is a continuum; the electron can have any energy in this range.

However, an electron which is part of a hydrogen atom will almost always be found in the innermost Bohr orbit, the $n = 1$ energy level. As the energy-level diagram in Figure 12.1 shows, the $n = 1$ level is the lowest possible energy level. Since we know that things in nature, when left to themselves, will settle to the position of lowest energy, it is not surprising to find that the hydrogen atom electron falls to its lowest possible energy level. Just as a rock rolls to the bottom of a hill and remains there, the electron falls to its lowest energy position and usually stays there. However, the stone differs in a major respect from the electron. The electron's energy is quantized.

Unlike the stone, which we can give any energy depending on how high we lift it, the electron cannot accept every energy we might wish to give it. Notice that when the electron in the unexcited hydrogen atom is in the $n = 1$ level its energy is -13.6 eV.

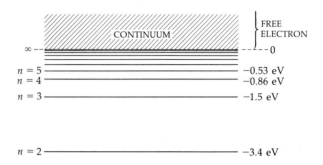

12.1 *The electron in the hydrogen atom can take on only the energies of the levels shown. However, the free electrons can have any and all kinetic energies.*

Since it can fall to no lower level, this is the electron's lowest possible energy. In the next energy level, where $n = 2$, the electron has an energy of -3.4 eV. *The electron cannot have an energy between the values -13.6 and -3.4 eV.* If we try to give it 1.0 eV of energy, thereby raising it from its position in the -13.6-eV level to a position of -12.6 eV on the diagram, we find that the electron is unable to accept this energy. No allowed level exists at -12.6 eV.

The electron has its next highest level at -3.4 eV, an energy that is 10.2 eV (13.6 $-$ 3.4) higher than the lowest level. The electron can never accept less energy than 10.2 eV when it is in the $n = 1$ level. There is no intermediate level to which it can go. As a result, it must rise at least to the $n = 2$ level or not leave the $n = 1$ level at all.

Similarly, the electron cannot possess energies which would place it between any of the other energy levels shown in the diagram. If it is in the $n = 1$ level, it can accept only energy increases of 10.2 eV, 12.1 eV, 12.7 eV, and so on. These are the energies required to lift it to the $n = 2$, $n = 3$, $n = 4$, and higher energy levels. We see, then, that the electron in the hydrogen atom can take certain definite energy increments and no others. This fact greatly influences the atom's behavior as it collides with other atoms, particles, and photons.

From Figure 12.1, notice that 13.6 eV is the energy needed to tear the electron loose from the hydrogen atom and to raise it from the $n = 1$ to the $n \to \infty$ level. This energy, called the *ionization energy* or *ionization potential* of the atom, is the energy required to free an electron from the atom and thereby transform the atom into an ion.

Emission and Absorption of Light

We have seen that a hydrogen atom emits light when its electron falls from a high energy level (n large) to a lower one (n small). A container of hydrogen gas does not ordinarily emit light because the electrons within the atoms have already fallen to the lowest possible energy level, the $n = 1$ level. To cause an atom to emit light, we must knock its electron out of the $n = 1$ level to a higher energy level. Generally, we excite the electron to a higher level by causing some other particle or atom to collide with the hydrogen atom.

Since an energy of at least 10.2 eV is required to excite the atom from the $n = 1$ level to a higher level, collision energies of this order of magnitude are necessary. We know from our study of gases that the energy of a gas molecule is about kT at temperature T. When $T = 300°$K, this energy is about 1/40 eV. We need energies 400 times larger than this to excite a hydrogen atom. Thermal energies are thus not large enough for this purpose except at very high temperatures. Such high temperatures can be attained in the hot sparks that occur in electric arcs.

We can achieve such energies more easily by striking the atom with electrons or ions which have fallen through potential differences of 10 or 20 volts. This is not easy to do in air, because the particles collide after traveling very short distances, and hence do not fall through such large potential differences between collisions. However, if potential differences of several thousand volts (easily obtained from a neon sign transformer) are placed across a tube of hydrogen gas at reduced pressures, the ions and electrons easily acquire 20 eV of energy between collisions. As a result, many of the particles, ions, and electrons in the tube have enough energy to ionize and excite other atoms upon impact. The light emitted as the electrons fall back into the lower energy levels causes the whole tube to glow. In this age of the ever-prevalent neon (and other gas) sign, this phenomenon is familiar to us all.

As we saw in Chapter 10, the spectrum of radiation emitted by hydrogen atoms is a series of spectral lines. The shortest-wavelength series, the Lyman series, is emitted as the electron falls from higher levels to the $n = 1$ level. It is obvious from Figure 12.1 that the energy lost as the electron falls to the $n = 1$ level will be much larger than when it falls to the $n = 2$ or higher levels. This is reflected in the fact that the Lyman series consists of high-energy photons, i.e., short-wavelength radiation. (Recall that the photon energy is $h\nu = hc/\lambda$.) This series lies in the long x-ray and ultraviolet region. The Balmer series, on the other hand, is emitted as the electron falls to the $n = 2$ level. Since the energy differences between the $n = 2$ and higher levels are small in comparison to the $n = 1$ transitions, the Balmer series photons have lower energies (i.e., longer wavelengths).

We also see from Figure 12.1 that as n becomes larger, the energy levels come closer together. The lines in each series bunch together as the series limit is approached. For example, when the electron falls from the $n = 10$ to $n = 2$ level, the emitted photon has about the same energy as when the electron falls from the $n = 1{,}000$ to $n = 2$ level. Even though these two spectral lines will be very close together, $1{,}000 - 10 = 990$ lines fall between them! This is why the series becomes a blur as the series limit is approached. The series limit line is emitted as the electron falls from the $n \to \infty$ to $n = 2$ level in the Balmer series or to the $n = 1$ level in the Lyman series.

The reverse of emission is also possible. If we send a beam of electromagnetic radiation through a tube of hydrogen as in Figure 12.2, the light quanta or photons in the beam will strike the atoms. In most cases the photon will not be able to give its energy to the electron it strikes because it does not have the appropriate energy to throw the electron to one of the higher levels. Almost all the electrons in the atoms will be in the $n = 1$ level since we assume the gas to be initially unexcited. Therefore, the incident photon must have an energy $13.6 - 3.4 = 10.2$ eV to excite the electron up to the $n = 2$

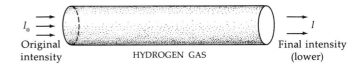

12.2 What wavelengths of electromagnetic radiation will be absorbed by the gas?

level. This energy corresponds to a wavelength of 1216 Å, which is in the ultraviolet. Photons of this wavelength will be *absorbed* strongly.

Similarly, if a photon having the energy $13.6 - 1.5 = 12.1$ eV (which is about 980 Å) strikes one of the atoms, it can excite the electron to the $n = 3$ level. Photons of this wavelength will also be strongly absorbed. (It is convenient to remember that a 1-eV photon has a wavelength of 12,400 Å. Since energy and wavelength are *inversely* proportional, a 2-eV photon corresponds to $12{,}400 \div 2$ Å, while 3 eV corresponds to $12{,}400 \div 3$ Å and so on.) Clearly, photons which have just enough energy to throw the $n = 1$ electron to the higher levels will be strongly absorbed. The absorbed photons have exactly the same energy as those emitted when the electron falls from the higher levels to the $n = 1$ level. These photons are in the Lyman series, and so we see that *hydrogen atoms absorb strongly at the wavelengths of the lines in the Lyman series.* All other wavelengths will pass through the gas with little loss in intensity.

Absorption is a useful way of finding the energy levels in an atom. For example, in the case of hydrogen gas, the electrons are in the $n = 1$ level. We measure the wavelengths of light absorbed by the atoms and find them to be 1216 Å, 980 Å, and so on. These wavelengths correspond to photons with energies 10.2, 12.1 eV, and so on, so we know there are energy levels at 10.2 eV, 12.1 eV, etc., above the $n = 1$ level. Furthermore, the series limit absorption line at 912 Å corresponds to 13.6 eV. Therefore we know that the $n = 1$ level is 13.6 eV below the zero level, the level at which the electron is free from the atom. From these absorption data we can construct directly the energy-level scheme of the hydrogen atom as shown in Figure 12.1.

The Helium Atom and Electron Spin

Let us now extend our discussion to the second atom in the periodic table, helium. Its nucleus has two neutrons and two protons. Since the protons provide a charge $+2e$ to the atom, two electrons must provide a balancing charge of $-2e$. It is reasonable to think that one of these electrons must behave much like the single electron in the hydrogen atom. However, the second electron will repel the first and so we can expect its behavior to change. Before we examine this complication, let us first consider the behavior of the helium *ion*, an atom which has lost one of its electrons. The helium ion differs in two basic respects from the hydrogen atom: the atom is not neutral (this will not concern us now) and the doubly charged helium nucleus attracts the single electron more strongly than did the singly charged hydrogen nucleus. Because of this stronger attractive force, the first Bohr orbit ($n = 1$) has a smaller radius than in the hydrogen atom.

In addition, the electron falls down a larger hill than in hydrogen when it falls from outside the atom to the $n = 1$ level, so higher-energy photons are emitted as the electron falls to the $n = 1$ level. Thus the equivalent of the Lyman series in ionized helium is at shorter wavelengths than it is in hydrogen. We can surmise from this that the equivalent of the Lyman series moves to shorter and shorter wavelengths as we proceed to atoms of higher nuclear charge. This is true, and the series is in the 0.1-Å range for the heavy elements. These photons are in the region of x-ray wavelengths; we see that *x-rays are the radiation emitted when some of the innermost electrons of the heavy elements go to even lower levels.*

In helium under normal conditions, the two electrons in each helium atom will have fallen to the lowest energy level, the $n = 1$ Bohr orbit. (In this case, too much knowledge complicates the situation; you will soon know enough so it will not be at all obvious that both electrons should be in the $n = 1$ level.) We can learn something about the behavior of these two electrons within the atom by examining how the atom behaves in a magnetic field.

You will recall that the electron in the Bohr model for hydrogen goes around the nucleus in a circular path, as shown in Figure 12.3a. In effect, it is a current flowing around a circular loop and so it generates a magnetic field. From Bohr's calculation, we can compute the magnetic field, and its value can be confirmed by experiment. But in addition to this magnetic effect, the electron itself is found to behave like a bar magnet. This magnetic behavior is thought to result from motion of the charge *within* the electron, due to the spinning of the electron about an axis through its center.* For this reason, the electron is said to have a *spin*, and this spin is usually represented by an arrow pointing along the electron magnet from south to north pole. For example, the arrow for the spin of the electron in Figure 12.3b would be pointing up, as shown in part c.

How spin affects the helium atom is shown in Figure 12.3d and e. Notice in (d) that the electrons traverse the Bohr orbit in opposite directions. As a result, the magnetic properties of the atom due to the orbital motion of the two electrons are canceled out; their two current loops give equal but opposite magnetic fields. Moreover, as shown in (e), the spins of the two electrons have opposite orientations. If one electron has its spin upward, the other has its spin downward. If we try to place the second electron in orbit so that its spin will be in the same direction as the first electron, we will find that the electron successfully resists the attempt.

For *all atoms* in nature, when the second electron is placed in the $n = 1$ Bohr orbit, its spin is antiparallel to the spin of the first electron. We shall soon see that this is an extremely important property of electrons.

---

* This picture is useful, but is not accepted literally; we do not know enough to propose an accurate picture.

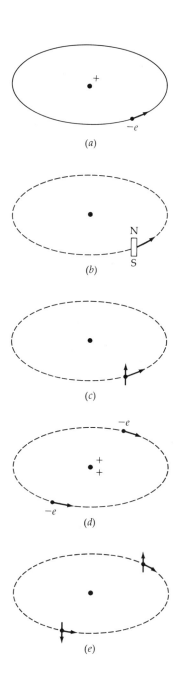

12.3 *The spin magnets are antiparallel for the two electrons. Although the simple Bohr picture leads us to believe the electrons in d and e should collide, they act like waves and no particle-like collision occurs.*

First, however, let us examine the light emitted by helium atoms. The spectrum is more complicated than that of the hydrogen atom, because the repulsion force between the two electrons gives more complex energy levels. In spite of this complication, the series of lines still result from the photons thrown out as the electrons fall from outer to inner orbits. The energy needed to tear an electron loose from helium, the ionization energy, is 24.6 eV, considerably higher than the 13.6 eV for hydrogen. We should expect this, since the nucleus of helium is doubly charged and therefore more work is required to tear an electron from near the nucleus.

Lithium and the Exclusion Principle

The third in the series of elements is lithium. Its nucleus contains three protons and so the neutral atom must have three electrons. What is the fate of the third electron? We expect two of the electrons to be in the first Bohr orbit. Since the nucleus has a charge $+3e$, these electrons should be even more tightly bound in the atom than those in the helium atom. If the third electron is in the $n = 1$ level as well, it too should be tightly bound to the nucleus. Since the ionization energy for helium is 24.6 eV, we would expect the lithium atom to have at least this high an ionization energy if all three electrons are in the innermost orbit. We would also expect them to fall to this orbit, since things in nature tend to fall to the lowest possible energy when left to themselves.

However, when the ionization energy of lithium is measured, it is found to be 5.4 eV. Only 5.4 eV of energy is needed to tear an electron out of the lithium atom instead of the expected 25 eV or more. This observed low ionization energy is more like the energy we would need to tear an electron out of the second Bohr orbit rather than the first. In fact, to pull a second electron loose from the lithium atom requires much more energy. This indicates

ATOMS; THE PERIODIC TABLE 271

AMERICAN INSTITUTE OF PHYSICS

WOLFGANG PAULI (1900–1958)

Pauli, born in Vienna, Austria, was a child prodigy. He wrote a masterful exposition of the theory of relativity while a 19-year-old student at Munich, and his Ph.D. thesis explored the quantum theory. He joined the Zurich group of physicists in the 1920's and spent the rest of his life there except for five wartime years in America. His exclusion principle did much to clarify and unify the many new discoveries being made in the field of quantum mechanics, but it was by no means his only important contribution to physics. Pauli was an indifferent lecturer and a notoriously poor experimenter; however, he was brilliant and witty in discussions and correspondence, and he made many of his most important suggestions in such informal situations. He was awarded the Nobel Prize in 1945.

that there is a first energy level but that only two, not three, of the electrons have settled into it. Some startling new feature of nature prevents the third electron from settling to the lowest energy level. It is like saying that if we tried to drop three balls off a table, only two of them would fall. This is preposterous in the case of the balls, but electrons in atoms appear to behave this way.

Wolfgang Pauli (1900–1958) first pointed out this unexpected feature of electron behavior in 1925; he was investigating the relation of the Bohr atomic theory to the structure of the elements at the time. Although the case of lithium is only a single piece of evidence for what has become known as the *Pauli exclusion principle,* many other situations show its effect clearly. Since the Pauli exclusion principle is closely linked with the wave properties of matter and wave resonance, we shall need a terminology to describe the resonance of waves.

You will recall that a string fastened at both ends can resonate in many ways. For example, it can resonate in one segment with nodes at the ends and an antinode at the center, or it can resonate in two segments, or three, or four, and so on, as shown in Figure 12.4. Each form shown in the figure represents a resonance mode of the string. We characterize each mode by a quantum number; in this case it is denoted by n.

In more complicated situations, such as the resonance of a flat plate (a cymbal for example) or a drumhead or the flat sheet on a tympan, it is not possible to characterize the resonance of the sheet with only one quantum number. For example, consider the rectangular drumhead shown schematically in Figure 12.5. Its simplest mode of motion is shown in (a): the whole drumhead vibrates up and down as one unit. However in (b), (c), and (d) we see increasingly complicated motion. In these cases, two integers (the m and n values given) are needed to describe the resonance patterns. The motion of a drumhead is therefore characterized by two quantum numbers, n and m.

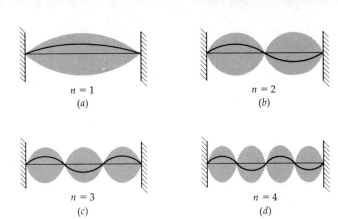

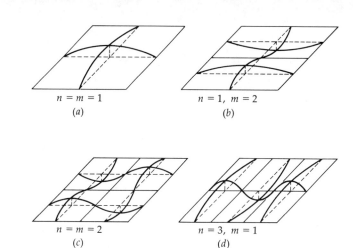

12.4 *A string can resonate in many modes. The modes shown here are characterized by the quantum numbers n = 1, 2, 3, and 4.*

12.5 *Resonance modes for a rectangular drumhead.*

Similarly, if we work out the resonance behavior of an electron near the nucleus of an atom, the solution (obtained from Schrödinger's equation) provides many possible resonance modes, each one characterized by a set of quantum numbers. We call each resonance mode a *state* of the atom. The Bohr energy levels and the states of the atom are not the same; in fact, many *states* have identical energies. The number of states for a given Bohr energy level increases as n, the energy-level quantum number, increases.

For example, the first Bohr energy level, or $n = 1$ orbit (although we now know that we cannot think of them as distinct orbits), has an energy of -13.6 eV. This energy level actually has two states, which we have already discussed. The two states are identical except for their electron spins, which are opposite. We showed this schematically in Figure 12.3e, but of course the electron resonance pattern is much more complex than a simple circular orbit. (It turns out that, for the $n = 1$ case, the circular orbit must be replaced by a fuzzy spherical shell with the same average radius.) Thus there is only one resonance form for $n = 1$, but, since the electron spin can be either up or down, there are two distinct possibilities and therefore two distinct states.

Similarly, the wave solution shows that eight states are associated with the $n = 2$ energy level. There are only four distinct resonance forms; but since the electron spin can be either up or down, the number of distinct states is eight. Another set of states corresponds to the $n = 3$ energy level, and so on for $n = 4$ and higher. Thus the simple, circular orbits of the Bohr model must be replaced by a whole set of orbits for each energy level. Nonetheless, the energy of the electron in all the orbits of a given energy level is the same.

Let us review the concept of atomic states. According to de Broglie's wave concept, the electron can resonate in many patterns about the nucleus. An electron can resonate in each resonance pattern in two ways, with spin up or spin down. Each kind of resonance is called a state. This brings us to Pauli's exclusion principle.

> *Only one electron can exist in each state of an atom or molecule. Two electrons cannot occupy the same state.*

We can now interpret the situation of the lithium atom. The first Bohr orbit contains only two states, each having the same energy, -13.6 eV. Two of the

Group →	I	II	III	IV	V	VI	VII	0
Period 1	HYDROGEN H 13.6							HELIUM He 24.6
2	LITHIUM Li 5.4	BERYLLIUM Be 9.3	BORON B 8.3	CARBON C 11.3	NITROGEN N 14.5	OXYGEN O 13.6	FLUORINE F 17.4	NEON Ne 21.6
3	SODIUM Na 5.1	MAGNESIUM Mg 7.6	ALUMINUM Al 6.0	SILICON Si 8.1	PHOSPHORUS P 11.0	SULFUR S 10.4	CHLORINE Cl 13.0	ARGON Ar 15.8

12.6 *Electronic structure and ionization energies in eV.*

lithium electrons occupy these two states and thus fill the $n = 1$ energy level. According to Pauli, a third electron cannot fall to this level because there are no more empty states to accommodate it. The third electron must therefore occupy a much higher energy level, the $n = 2$ level corresponding to the second Bohr orbit. Notice that this electron is easily pulled loose from the lithium atom; its ionization energy is only 5.4 eV. Lithium atoms frequently lose this electron in chemical reactions.

The Periodic Table of the Elements

As we have seen, the electronic structure of hydrogen, helium, and lithium atoms can be explained by use of the Bohr energy-level scheme, the Pauli exclusion principle, and the number of states available in a given Bohr energy level. Since the $n = 1$ level has only two states, Pauli's principle tells us that it can hold only two electrons. For the lithium atom, which has three electrons, the third electron must go into the $n = 2$ level since the two states in the $n = 1$ level are filled. As a result, the third electron in a lithium atom is relatively easy to remove from the atom.

The next atom in the series of elements is beryllium (Be). Since its nucleus contains four protons, the atom must have four electrons. Two of these electrons fill the $n = 1$ level (or <u>shell</u> as we shall call it). Since the $n = 2$ shell (or level) has eight states, the additional two electrons are in this shell. Figure 12.6 shows the shells of beryllium and the nearby elements. Below each diagram is shown the energy required to tear a single electron loose from the atom. The additional plus charge on the beryllium nucleus apparently makes it more difficult to remove an electron from beryllium than from lithium.

If we go on to boron (B) in Figure 12.6, we see that a third electron has been added to the $n = 2$ shell. Carbon (C) has four electrons in this shell, while nitrogen (N), oxygen (O), fluorine (F), and neon (Ne) have five, six, seven, and eight electrons respectively in the $n = 2$ shell. Finally in neon, all the states in this shell are filled. As a result, the next element, sodium (Na), must add its extra electron to the $n = 3$ level or shell. Notice that this electron, being farther from the nucleus, is relatively easily removed. Thus, the outer electron in sodium should behave much like the outer electron in lithium, since the energy needed to remove each is about the same. In fact, these two atoms do behave similarly in chemical reactions.

Since the third shell has only eight states, it is filled when the element argon (Ar) is reached. The intermediate elements, sodium (Na), magnesium (Mg), aluminum (Al), silicon (Si), phosphorus (P), sulfur (S), chlorine (Cl), and argon (Ar), form a _period_ similar to the elements in the period above them. Just as lithium and sodium are alike in their chemical behavior, so also are other pairs: Be and Mg, B and Al, C and Si, N and P, O and S, F and Cl, and Ne and Ar. For example, both lithium and sodium react explosively when dropped in water; neon and argon are both gases and generally do not react with anything. We see, then, that the vertical columns, or _groups_, of elements contain atoms with similar chemical properties, which is not surprising in view of their similar outer electron-shell structures.

Long before the concept of atoms was accepted, chemists noticed the similarities between these elements. Using these similar chemical properties as a guide, the Russian chemist Dmitri Mendeleev (1834–1907) organized the elements known at the time (1869) into a table called the periodic table or chart. Its more modern form is shown in Table 12.1. The group and period labeling system is the same one that we used in Figure 12.6 for the electronic structure table. Of course Mendeleev was unable to interpret his table in terms of electronic structure. Nevertheless, he found certain gaps in his table and predicted that they represented elements that had not yet been discovered. Using these gaps as guides, chemists of the latter 1800s searched diligently for these missing elements, and soon the list of elements that exist naturally on earth was complete.

EDGAR FAHS SMITH COLLECTION

DMITRI MENDELEEV (1834–1907)

Mendeleev was born in Siberia, the youngest of 14 children. His father, the local school principal, went blind a few years after Mendeleev's birth, whereupon his mother set up a prosperous glass factory to support the family. In 1849, his father died and the factory burned down; his mother took Dmitri to Moscow and eventually managed to obtain his entrance into the university in St. Petersburg. Finishing at the top of his class, he did graduate work in France and Germany before returning to St. Petersburg, where he became a professor of chemistry. In 1892, he became Director of the Bureau of Weights and Measures. His work on the chemistry of atoms was not widely accepted until experiments demonstrated the existence of elements that his periodic chart had predicted. Thereafter, he received widespread acclaim, but the Russian Imperial Academy of Science never acknowledged his accomplishments. Bold and impetuous, he once went up alone in a balloon to photograph an eclipse, though he knew nothing about ballooning. He was an outspoken populist in Tsarist Russia, which may explain his lack of official recognition there.

Table 12.1 Periodic Table of the Elements
  (C^{12} = 12.0000)

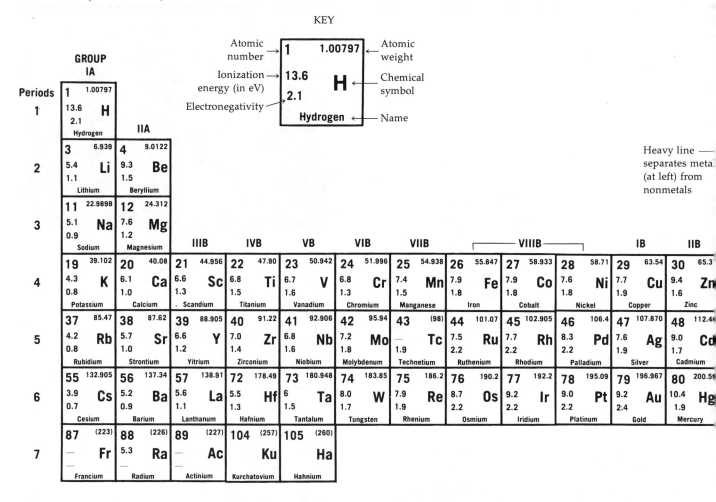

276   CHAPTER 12

IIIA	IVA	VA	VIA	VIIA	VIIIA or 0
					2 4.0026 24.6 He — Helium
10.811 5 B — Boron	6 12.01115 11.3 C 2.5 Carbon	7 14.0067 14.5 N 3.0 Nitrogen	8 15.9994 13.6 O 3.5 Oxygen	9 18.9984 17.4 F 4.0 Fluorine	10 20.183 21.6 Ne — Neon
26.9815 Al 8.1 1.8 Aluminum	14 28.086 8.1 Si 1.8 Silicon	15 30.9738 11.0 P 2.1 Phosphorus	16 32.064 10.4 S 2.5 Sulfur	17 35.453 13.0 Cl 3.0 Chlorine	18 39.948 15.8 Ar — Argon
69.72 Ga 8.1 1.8 Gallium	32 72.59 10 Ge 2.0 Germanium	33 74.922 9.8 As 2.4 Arsenic	34 78.96 9.8 Se 2.4 Selenium	35 79.909 11.8 Br 2.8 Bromine	36 83.80 14.0 Kr — Krypton
114.82 In 7.3 1.8 Indium	50 118.69 7.3 Sn 1.8 Tin	51 121.75 8.6 Sb 1.9 Antimony	52 127.60 9.0 Te 2.1 Tellurium	53 126.904 10.4 I 2.5 Iodine	54 131.30 12.1 Xe — Xenon
204.37 Tl 7.4 1.8 Thallium	82 207.19 7.4 Pb 1.8 Lead	83 208.980 8 Bi 1.9 Bismuth	84 (210) — Po 2.0 Polonium	85 (210) — At 2.2 Astatine	86 (222) 10.7 Rn — Radon

Heavy line separates metals (at left) from nonmetals

| 162.50
Dy
1.2
Dysprosium | 67   164.930
— Ho
1.2
Holmium | 68   167.26
— Er
1.2
Erbium | 69   168.934
— Tm
1.2
Thulium | 70   173.04
6.2 Yb
1.1
Ytterbium | 71   174.97
5.0 Lu
1.2
Lutetium |
| (249)
Cf
Californium | 99   (254)
— Es
Einsteinium | 100   (253)
— Fm
Fermium | 101   (256)
— Md
Mendelevium | 102   (254)
— No
Nobelium | 103   (257)
— Lw
Lawrencium |

Periodic table of the elements. The meanings of the various numbers accompanying each element are given in the key. (You will learn about electronegativities in Chapter 13.) A few features of the table which you should notice are as follows:

The elements in the same <u>group</u> (the vertical columns) have the same number of electrons in their outer shells. Thus the elements in Group IA have one electron in the outer shell. Those in Group VIIIA (or 0) have filled outer shells. The elements of the A groups are called the <u>representative</u> elements. The elements in the center of the periodic table, the B groups, are called the <u>transition</u> elements. Within each period (the horizontal rows) the transition elements are similar to each other.

All the elements in a given period (or row) have the same number of electron shells. For example, all the elements in the fourth period have four electron shells. Potassium has only one electron in its fourth shell, whereas that same shell is completely filled in the krypton atom. As indicated, the two rows at the bottom of the table are portions of the sixth and seventh periods. They are called the <u>rare earths</u>.

The heavy line near the right side of the table separates the metals (at left) from the nonmetals; some of the borderline elements have both metallic and nonmetallic properties. All the elements with atomic number greater than 92 are so radioactive that they no longer exist naturally on earth.

ATOMS; THE PERIODIC TABLE

You will notice that, starting with the fourth period, there are elements in the center of the periodic chart whose group numbers are followed by B. All of these elements are metals and have roughly similar properties. They are atoms for which the shells beyond those shown in Fig. 12.6 are being filled. There are 18 available states in these shells, and so a large number of different atoms are formed before the shell is completed. Adding or subtracting one electron from such a large number in a shell has little influence on the chemical behavior of the atom. Only those atoms which have only one or two electrons in the new shell, or which need none or only one or two electrons to fill the shell, are much changed if one electron is added to or taken from the shell. Therefore, only the atoms in Groups IA, IIA, VIA, VIIA, and O are distinctive in their chemical reactions for the fourth and higher periods. The central section of the periodic table for the fourth and higher periods contains atoms which are roughly similar, although they can be distinguished chemically.

One important exception to this statement is represented by the elements with atomic numbers 26, 27, and 28. These are iron, cobalt, and nickel, the ferromagnetic metals we encountered in Chapter 7. Because of energy considerations involving the fuzzy nature of the electron orbits, the outermost electrons in these atoms do not properly cancel each other's magnetic effects. You will recall that usually the second electron added to an atom will cancel the magnetic effects of the spin and orbital motion of the first electron. This does not happen in these three atoms, and so inordinately large magnetic effects are observed when these atoms are placed together in a solid. In spite of these unusual magnetic properties of iron and cobalt and nickel, the atoms still react chemically in ways similar to their neighbors in the periodic table.

Spectra of Complex Atoms

Most atoms emit light of many different wavelengths. The _emission spectra_ of these atoms have many sharp spectral lines, which is not surprising, since there are many electrons in different energy levels within the atom. These energy levels are not the simple Bohr levels, because electrostatic and magnetic forces between the electrons change the energies of the various electrons. As a result, the spectral lines emitted by atoms of one element differ from those of every other element. The spectrum of iron, for example, has thousands of times more lines than the hydrogen spectrum. Chemists and metallurgists use these distinctive spectral lines to "fingerprint" each element. For example, the light given off by a material vaporized in a hot electric arc is easily analyzed to determine what elements are present in the substance.

In spite of the complexity of most atomic spectra, certain features are easy to understand. For example, the two electrons in the $n = 1$ shell are relatively undisturbed by the electrons in the outer regions of the heavier atoms. Thus the energy of an electron in the $n = 1$ shell is essentially that of an electron in the field of the highly charged nucleus. As we mentioned before, the energy differences between the inner shells of these heavy atoms are much larger than the 10.2-eV difference between the $n = 1$ and $n = 2$ levels in hydrogen. For the atoms near the bottom of the periodic table, this energy difference exceeds 30,000 eV. Because of this large energy difference between levels, the photons emitted as an electron falls from one to the other have wavelengths of tenths of angstroms. These transitions therefore produce x-rays.

At the other extreme is the behavior of the single electron in the outer shells of lithium, sodium, potassium, rubidium, cesium, and francium, the atoms in Group IA. Because the inner electrons act as though they are more or less fused with the nucleus, leaving a central unbalanced charge of simply $+e$, the lone electron acts like the electron of a hydrogen atom. Although this view is oversimplified, the portion of the spectral lines that comes from the lone electron is similar to the corresponding portion of the hydrogen spectrum. The spectral lines due to transitions of the other electrons are much more complicated.

The Laser

We have been discussing the behavior of individual atoms and the light emitted from them. In almost all light sources, the atoms act independently; the emission of a photon by one atom is not coordinated with the emission by other atoms. As a result, the light beam is a complex mixture of electromagnetic waves from the various atoms. Of course, these waves are not all in phase with each other; they sometimes cancel and sometimes add. Clearly, the light beam will be far more intense and coherent if all the atoms can be made to emit their waves in phase. One light source comes close to achieving this; it is called a _laser_.

Many types of lasers are available; they all operate on the principle from which they get their name: "Light Amplification by Stimulated Emission of Radiation." We shall describe a helium-neon gas laser, which gives a low-intensity, continuous light beam. (More intense beams are achieved by newer lasers which give out pulses of energy.) Figure 12.7 shows the basic outline of the helium-neon gas laser and the narrow, pencil-thin beam it gives off.

The heart of the laser is a glass tube containing helium and neon gas at relatively low pressure. At the ends of the tube are flat glass plates accurately parallel to each other. Each plate is coated like a mirror, but one end plate is coated lightly enough so that about 1 percent of the light striking it from inside leaks through the mirror and leaves the tube.

The two gases in a laser are each chosen to perform special functions. In this case, the helium acts as an energy source for the neon. It is excited by means of a very high frequency electrical discharge in the tube. Basically, the helium acts just as mercury vapor does in an ordinary mercury arc lamp. The high voltage causes a gas discharge so that many energetic ions and electrons shoot around inside the tube. The helium gives off the usual helium spectrum as the electrons fall back to the $n = 1$ level. However, the light from this effect is a small fraction of the total light emitted from the tube.

The electrons in the helium atoms are thrown up to all the possible energy levels and states of the atom

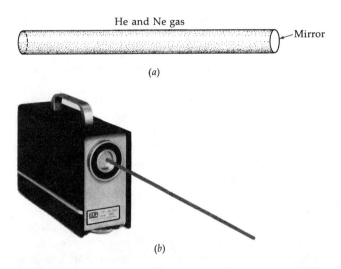

12.7 Although the laser shown in (b) has an output of only 0.0005 watt, its pencil-thin beam shines brightly from the building at the right in (c), more than two miles from the camera. Compare its brightness with that of the high-intensity lamps on the Oakland Bay Bridge. (Courtesy of Electro Optics Associates.)

by collisions inside the tube. One state, which has an energy 20.6 eV higher than the $n = 1$ state, is called a *metastable state*. In such a state, the electron is reluctant to fall to the lower states, and so the atom remains in this state for an abnormally long time. However, the energy-level structure of a neon atom is such that when the neon and helium atoms collide, the excited helium atom gives up its energy to the neon atom. Actually, the excited neon atom has an energy slightly greater than that of the helium atom; this excess energy comes from the kinetic energy it acquires upon collision.

What has happened is this: the helium atoms are excited in the gas discharge; some are in a metastable state and wander around until they collide with a neon atom. During the collision, the neon atom takes the energy from the helium atom and is then in an excited state. This state is rather stable in the neon atom and so the electron in it does not fall to a lower state immediately. In time, the helium atoms excite many neon atoms which then simply wait to radiate their energies. When one eventually falls it falls to a level 1.96 eV lower and emits a pulse of electromagnetic radiation with wavelength 6328 Å, in the red.

On its way through the tube, the pulse passes other excited neon atoms and subjects them to an oscillating electromagnetic field having the same frequency as the light which the excited atom should emit. It is found that the oscillating field effectively causes each excited neon atom to fall to its lower state, so that the photon it sends out is exactly in phase with the radiation which *stimulated* it to emit the photon. Thus the two waves reinforce each other.

As these waves are reflected back and forth between the ends of the tube, they are vastly augmented by identical waves which they stimulate other neon atoms to emit. Consequently, there exists in the tube a strong coherent wave with wavelength 6328 Å. A small portion of this radiation exits through the slightly leaky mirror at one end of the tube, and this is the laser beam. The neon atoms have added their radiation together so that the waves all reinforce each other to produce a strong, coherent beam. Because the beam reflects back and forth many times between the parallel mirrors, the light rays come straight out from the end of the tube. All those which diverge from the axis of the tube are lost out from the sides during the many trips back and forth.

Unlike light from a bulb, however, the energy of a laser does not spread out in space. Instead it flows out like a pencil-thin cylinder of light, and maintains its strength over long distances, as the photograph in Figure 12.7c shows. We can reflect such a beam off the moon and measure it when it returns to the earth.

The laser in Figure 12.7b and c is a very low power source, despite its brightness. However, it is possible to design extremely powerful lasers which send out intense radiation. These devices will vaporize a coin or set wood on fire. The total energy emitted by the laser concentrates on the object its narrow beam strikes; this precise concentration makes the laser extremely useful in industry and medicine. Other uses of the laser, such as communication over large distances, depend on the coherence and single-frequency properties of the laser beam. Since its invention in 1955, the laser has achieved wide use in many different areas.

BIBLIOGRAPHY

1. OLDENBERG, O.: *Introduction to Atomic and Nuclear Physics*, 3d ed., McGraw-Hill, New York, 1961. Chapters 9–11 give more detail concerning the topics treated here.

2. SEMAT, H. and WHITE, H.: *Atomic Age Physics*, Holt, New York, 1959. This textbook was used in the first nationally televised physics course. Chapters 1–10 deal with the atom and its structure.

3   HOLTON, G. and ROLLER, D.: *Foundations of Modern Physical Science*, Addison-Wesley, Reading, Mass., 1958. Chapters 21–24 give detailed historical information about the discoveries that revealed the structure of atoms and molecules.

4   PIERCE, J. R.: *Quantum Electronics* (paperback), Anchor Science, Doubleday, Garden City, N.Y., 1966. An introductory treatment of atoms in solid materials and how they form the basis for transistors, lasers, and other solid-state electronic devices.

5   BOOTH, V: *The Structure of Atoms* (paperback), Macmillan, New York, 1964. A readable account of the structure of atoms and how the original ideas about them were conceived.

6   ASIMOV, I.: *A Short History of Chemistry* (paperback), Anchor Science, Doubleday, Garden City, N.Y., 1966. The title well describes this book written by a famous popularizer of science.

SUMMARY

In the Bohr model of the hydrogen atom, the electron is restricted to definite allowed orbits; these orbits were shown by de Broglie to be the resonances of the electron wave. However, these resonance positions are like the maxima in an interference pattern; the electron is not restricted to them any more than the photons are restricted to the point where maximum light intensity occurs. Despite this deficiency of the Bohr theory, the energy levels given by Bohr are correct. Expressed in electron volts, the energies allowed for the electron in the hydrogen atom are $E_n = 13.6/n^2$ eV, where n is the Bohr orbit number. (We now call it the quantum number describing the energy state of the atom.)

An energy-level diagram is a vertical scale showing the allowed energies of the system. In the case of hydrogen, the zero level corresponds to $n \to \infty$, that is, to the ionized atom. As the electron comes closer to the atom, it loses energy, and so it is below the zero level on the energy-level diagram. Levels exist at the values $-13.6/n^2$, where n is an integer. Positive energy values in this case represent the electron torn free from the atom but possessing kinetic energy. These positive energies are not quantized.

Light is emitted by an atom when it falls from a higher state to a lower state on the energy-level diagram. This coincides with Bohr's postulate about emission of light by the hydrogen atom. Since his model gives the proper energy-level diagram for the atom, it successfully predicts the emission spectra.

The hydrogen atom can absorb only those photons which have the proper energy to excite the atom to an allowed energy level. Since hydrogen atoms are normally in their lowest energy level, they can only absorb the wavelengths which will raise them to the $n = 2$ and higher levels. These are the same wavelengths emitted by the atom as it falls to the $n = 1$ state, namely, the Lyman series of lines.

For atoms with large Z, the large nuclear charge holds the innermost electrons strongly. As a result, the $n = 1$ and $n = 2$ levels in these atoms lie at energies approaching $-100{,}000$ eV. When the inner electron in such an atom falls from one level to another, it emits a very high-energy (short-wavelength) photon in the x-ray region.

Placement of electrons in a multielectron atom is summarized by the Pauli exclusion principle, which states that no two electrons can exist in the same resonance state within an atom. There are two such states for the $n = 1$ energy level, eight states for $n = 2$, and so on. Hence helium has its two electrons in the $n = 1$ level, but lithium, which has three electrons, must have one of them in the $n = 2$ level. Using the Pauli principle, we can build up the shell-type electronic structure of multielectron atoms.

The energy needed to tear the outermost electron loose from an atom is called the ionization energy. Since the helium nucleus is doubly charged, more energy is needed to tear loose the $n = 1$ electrons than for hydrogen. Thus helium has a higher ionization energy than hydrogen. In the case of lithium, the third electron is in the $n = 2$ level, and so its ionization energy is less than that of helium. Similar reasoning explains the ionization energies of the rest of the atoms in the periodic table.

Since the outermost electrons of an atom are responsible for the chemical behavior of the atom, and since these electrons are arranged in a definite energy structure according to the exclusion principle, the atoms show chemical properties which repeat with increasing Z. Atoms such as helium, neon, and argon, which have filled outer shells, all behave similarly chemically. Similar examples of properties which repeat periodically formed the basis for the periodic table, constructed first by Mendeleev. Gaps in the table provided guidelines for the discovery of elements then unknown.

The laser sends out a narrow, pencil-thin beam of light which is highly coherent. By means of stimulated emission, the light waves emitted by the atoms within the laser are all in phase. As a result, the beam is very intense. It is used for such diverse purposes as long-range signaling, eye surgery, and welding metals.

TOPICAL CHECKLIST

1. Why is Bohr's idea of distinct electron orbits unacceptable?
2. What major feature of the hydrogen atom does Bohr's theory correctly predict?
3. Summarize the energy levels of the hydrogen atom by giving the equation for E_n.
4. What is the meaning of the negative sign appended to the electron energy levels in hydrogen?
5. Draw the energy-level diagram for hydrogen and explain its meaning.
6. What is meant by the following terms in respect to the energy-level diagram: ionization energy; continuum; Balmer series; Lyman series; series limit?
7. Normally, hydrogen atoms in nature are found in which energy level?
8. Discuss the effect of the energy-level structure on the energies which a hydrogen electron can accept.
9. Relate the light emission and absorption properties of the hydrogen atom to its energy-level structure. Why are not the Balmer series of wavelengths absorbed?
10. A 1-eV photon corresponds to what wavelength? How can this be used to convert quickly between energy and wavelength?
11. How can absorption data be used to construct the hydrogen atom's energy-level diagram?
12. In helium atoms, the two electrons are normally found in which energy level?
13. What would you predict about the spectrum of ionized helium?
14. What do we mean by the spin of the electron? How are the spins of the two electrons in a helium atom related?
15. Discuss the origin of x-rays.
16. Why do the ionization energies of hydrogen and helium differ? How do they differ?
17. How can we infer from the ionization energy of lithium that a third electron cannot fall to the $n = 1$ energy level?
18. What do we mean by an atomic state? How does a state differ from an energy level?
19. Discuss the meaning and importance of the Pauli exclusion principle.
20. How does the Pauli exclusion principle aid us in interpreting the periodic table?
21. Describe the rationale used by Mendeleev in setting up the periodic table.
22. How do iron, cobalt, and nickel fail to conform to the other members of their group?

23 How do the spectra of complex atoms differ from that for hydrogen? How are they similar?

24 How is it possible to use the spectrum of an unknown substance to identify the atoms in the substance?

25 Describe the atomic basis for the operation of a (helium-neon) laser. How does its emitted light differ from that of a light bulb or a mercury arc lamp? "Laser" is an acronym for what?

26 Laser light is coherent, and it gives a pencil-thin beam. Explain how each feature originates in the laser.

QUESTIONS

1 What determines how many electrons exist outside the nucleus in a normal atom? How do the chemists differentiate between various types of atoms?

2 Helium, neon, and argon are all gases. Given a sealed tube of gas, how could you tell which of these three gases it contained? What if it contained a mixture of them?

3 One sometimes hears a physicist or chemist say "Atoms emit the wavelengths they absorb." Is this statement correct?

4 In the early 1900s, when Rutherford was trying to discover the nature of alpha particles, he sealed some radium in a vacuum tube. After a year or so, he noticed that the tube contained a gas. What was the gas? (*Hint:* Radium decays to lead finally and, in so doing, emits alpha particles.)

5 Why do excited cesium atoms give off short x-rays, while lithium, which is in the same group, does not?

6 The Pauli exclusion principle also applies separately to protons and neutrons. Can you guess from this why the highest-energy neutrons in a high-atomic-weight nucleus usually have more energy than the protons?

7 Helium was discovered on the sun before it was known to exist on the earth. When the light coming from the sun is sent through a spectroscope, the wavelengths corresponding to the Lyman series in the hydrogen spectrum appear as dark lines in an almost continuous band. Explain why these dark lines exist and use the explanation to guess how the existence of helium was learned.

8 X-ray photons in the wavelength range greater than about 0.1 Å, when traversing matter, are stopped chiefly by a process similar to the photoelectric effect. They strike an electron and disappear, giving all their energy to the electron. Why is lead so much more effective in stopping x-rays in this range than is water or carbon?

9 Even though the charge on the potassium nucleus is more than six times larger than the charge on the lithium nucleus, the ionization energies for the two atoms are roughly the same. Why aren't they vastly different?

10 When Marie and Pierre Curie carried out their famous studies on the radioactive mineral pitchblende, they found it contained two different types of radioactive atoms. One behaved somewhat like barium in chemical reactions while the chemical properties of the other resembled those of bismuth. Knowing that no element below bismuth in the periodic table is radioactive, give an educated guess as to what these new radioactive elements were.

11 Discuss the first three periods of the periodic table which would result if the electron spin did not allow a second electron to exist in each state.

PROBLEMS

1. Find the energy and wavelength of a photon emitted as a transition from the $n = 7$ to the $n = 6$ state occurs in hydrogen.

2. The excited cobalt 60 nucleus emits a gamma ray of energy 1.3 MeV which is used for both industrial and medical x-ray purposes. What is the wavelength of the gamma ray? Compare the origin of a gamma ray to that of an x-ray.

3. Suppose electrons that have fallen through 100 volts strike the atoms in a gas. What is the shortest possible wavelength of radiation the gas will give off under these conditions?

4. A tube filled with neon gas has two electrodes at its two ends, and a potential difference of 100 volts is placed across them. When red light shines upon the tube, no current flows between the electrodes. However, as the wavelength of light is slowly decreased, current begins to flow at $\lambda = 570$ Å and continues for all shorter wavelengths. Explain this behavior and point out what can be calculated from the data.

5. Sodium atoms emit a brilliant yellow light (5798 Å) which gives sodium arc lamps their characteristic yellow color. In addition, sodium vapor strongly absorbs light of this wavelength. From these data, what can you deduce about the energy-level diagram for sodium atoms?

Chapter 13

THE CHEMISTRY OF INORGANIC SUBSTANCES

In nature, atoms rarely exist alone, but are bound together in combinations called molecules. The area of science that investigates the joining together of atoms to form molecules is called chemistry. In this chapter we will begin our study of chemistry. We will learn that four basic bonding forces bind atoms into molecules and cause these molecules to aggregate into liquids and solids. For the present, we will be concerned with inorganic chemistry, the study of molecules that are not usually produced by living organisms. The chemistry of organic molecules, the molecules that compose living things and their decay products, will be the subject for the following chapter.

The Hydrogen Molecule

When two or more atoms are joined together strongly, the combination is called a _molecule_. Sometimes the atoms of a particular molecule are all of the same element. An example of this is the nitrogen molecules of the air. Each of these molecules is composed of two nitrogen atoms bound together by a strong bond, a chemical bond. Often, however, a molecule is a combination of two or more unlike atoms. A typical example is the water molecule; it is composed of one oxygen atom joined to two hydrogen atoms. Another example is table salt, whose chemical name, sodium chloride, is descriptive of its chemical composition, a sodium atom joined with a chlorine atom. These latter molecules, composed of more than one type of atom, are called _compounds_. We often use this designation interchangeably for a single molecule and for a large number of the same molecules taken together. Let us now begin our study of chemistry by considering one of the simplest of all molecules, the hydrogen molecule. The vast majority of atoms in a tube of hydrogen gas are bound together in pairs, forming hydrogen molecules. Since this form of hydrogen is more prevalent than the individual atoms, we might expect that the two atoms in a molecule possess less energy than when they are separated. That is, the atoms would tend to form molecules if they could lose energy by combining into pairs. We can test this hypothesis by measuring the potential energy of the two hydrogen atoms as they are brought close together. Figure 13.1 shows such measurements in a graph which relates the potential energy of the atoms to the interatomic separation distance r.

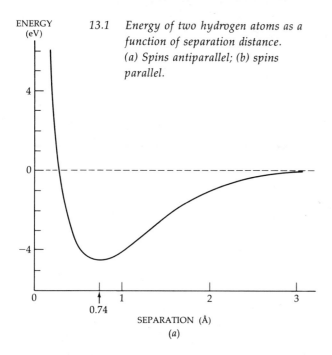
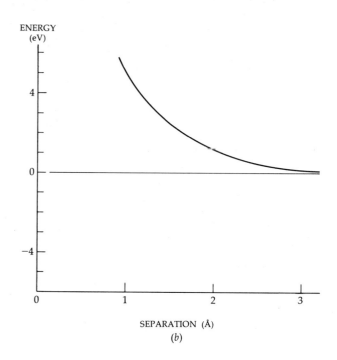

13.1 Energy of two hydrogen atoms as a function of separation distance. (a) Spins antiparallel; (b) spins parallel.

From the bead-on-a-wire analogy, it is clear that a bead placed on the curve of Figure 13.1a will slide down the curve to its lowest point. This tells us that the two hydrogen atoms attract each other at all separations greater than 0.74 Å. However, when very close together, for r less than 0.74 Å, they repel. Thus, if left to themselves, they will come to equilibrium at a separation of 0.74 Å. This distance between atom centers in the hydrogen molecule is shown schematically in Figure 13.2.

Part of the attractive force between the hydrogen atoms is electrical in origin. However, most of the binding force can only be explained by the Pauli exclusion principle and calculations from Schrödinger's equation. We can see the exclusion principle's tremendous influence by measuring the energy required to push two hydrogen atoms toward each other when the spins of their electrons are parallel. This is shown in Figure 13.1b. Notice that the atoms do not attract at all; they strongly resist being pushed together.

If Bohr's model of the hydrogen atom were correct, we could draw a simple picture showing the two electrons of the hydrogen molecule circulating about the two nuclei in definite orbits. But we know such a picture is untenable. The uncertainty principle tells us that we cannot locate the positions of the electrons in the molecule exactly. The best we can do is to state that an electron is more likely to be found in certain regions than in others. Bohr's concept of orbits is of little aid when we try to locate the electrons in a molecule. To describe locations of the electrons, one must ask what the wave theory (and Schrödinger's equation) tells us about them.

The wave theory result is shown in Figure 13.3; the likelihood of finding an electron in a given position is proportional to the darkness of the fuzzy cloud. The Bohr orbits are lost completely as a result of the uncertainty introduced by the wave nature of the electrons. Notice that the electrons are most often found in the region between the two nuclei, and so the two atoms are apparently sharing their two electrons. If the two electrons had parallel spins, the electrons would avoid the region between the nuclei. Electrons that are being forced into the same state will repel each other, just as Pauli's exclusion principle predicts.

We can picture what is happening in the hydrogen atom with a diagram showing Bohr orbits, even though we know that these orbits are not correct

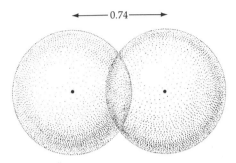

13.2 Although the individual hydrogen atoms have radii of about 0.5 Å, their distance between centers in the hydrogen molecule is 0.74 Å.

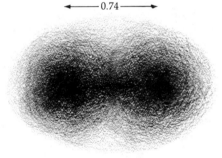

13.3 The two electrons in the hydrogen molecule are most likely to be found where the cloud is darkest. Notice how the most likely electron locations are concentrated in the region between the nuclei so that the two atoms apparently share the two electrons.

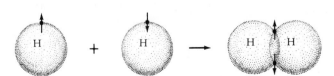

13.4 In combining to form the H_2 molecule, the two atoms have in effect filled their $n = 1$ shells by sharing their two electrons. (Why must the spins be antiparallel?)

in detail. Two hydrogen atoms automatically form a hydrogen molecule whenever two low-energy hydrogen atoms come close enough so that their attraction forces overpower their thermal motion:

$$H + H \rightarrow H_2$$

where H_2 is the symbol for two hydrogen atoms bound together. We call this process of combination a *chemical reaction*. It is illustrated schematically in Figure 13.4.

We know that the $n = 1$ shell of an atom is filled when it has two electrons with antiparallel spins. In Figure 13.4, the two hydrogen atoms have joined together in such a way that they share their two electrons. In effect, each atom has a filled $n = 1$ shell. If the electrons did not have antiparallel spins, they would not be able to enter the $n = 1$ shell together because of the exclusion principle. We conclude from this that two hydrogen atoms which can complete their electronic shell by sharing their outer electrons will bind together to do so. Two such hydrogen atoms are bound together strongly enough that an energy of about 4.5 eV (see Figure 13.1a) is required to tear them apart. We call such a bond, in which electrons are shared by two atoms, a *covalent bond*.

The Covalent Bond

Two atoms which share one or more electrons equally form a covalent bond; the hydrogen molecule is an example. If you know that nitrogen, oxygen, and chlorine molecules are represented by the chemical formulas N_2, O_2, and Cl_2, you may guess that these *diatomic* molecules (i.e., two-atom molecules) are also held together by covalent bonds. This is in fact true.

For example, the chlorine atom has two electrons in the $n = 1$ shell, eight in the $n = 2$ shell, and seven in the $n = 3$ shell. Since the inner shells do not influence chemical behavior greatly (because they are filled and also because they are shielded by the atom's exposed outer shell), we will consider only the outer shell, the $n = 3$ shell in chlorine. The $n = 3$ shell has seven electrons and needs one more to be filled. The atom obtains this electron by sharing one of its electrons with another chlorine atom, as shown in Figure 13.5. The reaction is

$$Cl + Cl \rightarrow Cl_2$$

Again, a covalent bond binds the two atoms together. The reaction in the upper part of Figure 13.5 is shown more simply in the lower part of the figure, where the electrons are represented by dots. A third way to picture the reaction is

$$Cl + Cl \rightarrow Cl\text{—}Cl$$

where the bar between the two chlorine atoms represents two shared electrons.

The case of oxygen molecules is more complicated. Oxygen atoms have only six electrons in their outer shells and require two more to fill them. Consequently, when two oxygen atoms bind together they share *two pairs* of electrons in order to complete their shells.

The reaction is represented in the following ways:

$$O + O \rightarrow O_2$$

or

$$\ddot{\text{:}O\text{:}} + \ddot{\text{:}O\text{:}} \rightarrow \ddot{\text{:}O\text{::}O\text{:}}$$

or

$$O + O \rightarrow O\text{=}O$$

Of course all these forms are purely schematic. In the last two, we show the *two pairs* of shared electrons by two pairs of dots or two bars. A covalent bond which involves two pairs of electrons is called a <u>double bond</u>. Chemists generally use the bar notation to represent covalent bonds. One bar represents one pair of shared electrons, a single covalent bond. Two bars and three bars represent two and three mutually shared electron pairs and are double and triple covalent bonds.

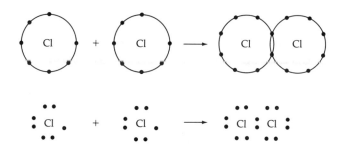

13.5 It is sometimes convenient to show the electron pair bond, the covalent bond, by the symbolism used in the lower part of the figure.

Perhaps the most common covalent bond is that between carbon and hydrogen atoms. Since carbon has an atomic number $Z = 6$, it has two electrons in the $n = 1$ shell and four in the $n = 2$ shell. Hence it requires four electrons more to fill its outer shell. Four hydrogen atoms can provide these electrons. The resulting compound, a gas called *methane*, can be given by three alternative formulas:

$$CH_4 \quad \text{or} \quad H\text{:}\overset{\cdot\cdot}{\underset{\cdot\cdot}{C}}\text{:}H \quad \text{or} \quad H\text{—}\underset{|}{\overset{|}{C}}\text{—}H$$

(Methane gas is formed in nature by the decay of vegetable matter and makes up a large fraction of the gases in natural gas. It is also called marsh gas.) Notice that both the hydrogen and carbon shells have been filled by forming covalent bonds. The notation CH_4 for methane tells us that four hydrogen atoms are combined with one carbon atom in the same molecule. This is an example of a more general representation. As an example of it, $X_n Y_m Z_p$ represents a molecule composed of n X-type atoms, m Y-type, and p Z-type. We shall use this notation frequently.

The Ionic Bond

Several times during the previous chapters we pointed out that some atoms hold their outer electrons more strongly than others. For example, we saw in Figure 12.6 that the helium ionization energy (the energy needed to tear loose an electron from the atom) is 24.6 eV while it is only 5.4 eV for lithium. Clearly the atoms in the periodic table differ widely in the strength with which they retain their outermost electron. As you might expect, this ionization energy is an important factor in determining how one atom interacts with another.

Another factor that is just as important as ionization energy in directing atomic interactions is called the *electron affinity* of the atom. The electron affinity is the energy with which a neutral atom attracts and binds an extra electron to itself if such an electron is available. For example, we already know that a chlorine atom needs only one more electron to fill its outer shell, so it has a large electron affinity. All the atoms of Group VII in the periodic table (fluorine, chlorine, bromine, iodine) need one electron to fill their outer shell and they readily acquire any available electron to do so. In the process, the atom itself becomes charged; it becomes a *negative ion*. Typical electron-affinity energies for the Group VII atoms are near 3.7 eV.

In contrast, the Group O atoms (helium, neon, etc.), which have filled shells, have nearly zero electron affinities. They have no need for an extra electron. With the exception of hydrogen, the Group I atoms (lithium, sodium, potassium), which have one lone electron in their outer shells, also have little need for another electron. Their electron affinities are of the order of 0.5 eV, much smaller than those of the Group VII elements. This trend persists through the various groups of the periodic table.

We are now ready to discuss the behavior of two atoms when they are brought close together. In the case of two hydrogen atoms, the atoms share their electrons equally. This is reasonable, since each has the same ionization energy and the same electron affinity. They are equally reluctant to lose an electron and both need one more electron. As a result, they form a covalent bond and share their outer (and only) electrons equally.

The situation is quite different, however, when a sodium atom is brought close to a chlorine atom. Since the chlorine atom has a high affinity for an extra electron and the sodium atom binds its outer electron only loosely, the chlorine atom can steal the sodium atom's electron. In the process, the sodium atom becomes a *positive ion* while the chlorine atom becomes a *negative* ion. Other atoms react similarly, and in each case the reaction permits the formation of *ionic bonds*.

Ionic bonds are characteristic of molecules that are held together by strong electrostatic forces. These forces are found between oppositely charged ions. Let us look more closely at our example, the reaction of a sodium atom (Na) with a chlorine atom (Cl) to form the molecule NaCl, sodium chloride (familiar to us as table salt). This molecule consists of a positive sodium ion and a negative chloride ion.

Since the sodium ion has lost the single electron in its $n = 3$ shell, its charge is $+e$. The chloride ion is a chlorine atom which captured the electron to fill its outer shell; its charge is $-e$. The resulting oppositely charged ions are held together by the electrostatic attraction of the positive sodium ion for the negative chloride ion. This reaction may be written

$$Na + Cl \rightarrow NaCl$$

or

$$Na + Cl \rightarrow Na^+ + Cl^-$$

The ionic nature of the reaction is pictured in the shell diagrams of Figure 13.6.

Another ionic reaction, common but more complicated, occurs when a rubber band is placed in the same drawer as the household silverware. The

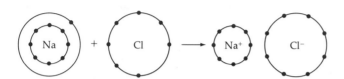

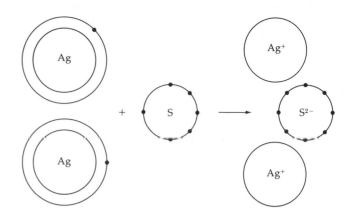

13.6 The chlorine atom steals the outer electron from the sodium atom. The resultant ions are then held together by the attractive force between the ions. This is an ionic bond.

13.7 The two silver atoms lose their outer electrons to the sulfur atom, thereby forming two silver ions and a doubly charged sulfide ion. These charges then hold the ions together in a molecule by means of ionic bonds.

rubber band often contains a small amount of a sulfur compound, which is used to tie the rubber molecules together in a process called vulcanization. Some of the excess sulfur slowly vaporizes (probably not as pure sulfur) and comes into contact with the silver. Silver atoms (Ag) have one loosely held electron in an outer shell, while sulfur (S) needs two electrons to fill its outer shell. As a result, they react as shown in Figure 13.7:

$$2Ag + S \rightarrow Ag_2S$$

Notice that it takes two silver atoms to furnish the electrons needed to complete the shell of each sulfur atom. The product is a brownish-black compound, Ag_2S, silver sulfide, which you see as tarnish on the silverware.

We should note that Ag_2S is not a purely ionic molecule. In NaCl, the sodium atom and its outer electron are almost completely separate. However, the situation is different for the two silver atoms in Ag_2S. While their outer electrons are pulled into the vicinity of the sulfur atom, the attractive force is not strong enough to pull them loose completely from the silver atoms. Because this situation is frequent, chemists often apportion degrees of ionic and covalent character to various chemical bonds. It will be sufficient for us to understand what is meant by these designations.

We might note in passing that sulfur compounds account for a considerable portion of the air pollution in industrialized areas. The sulfur originates in the combustion of low-grade coal and natural gas. Both of these fuels contain sulfur compounds; how much depends upon what area of the earth the fuel comes from. When the fuel is burned, the sulfur is converted to volatile gaseous compounds which enter the atmosphere (chiefly sulfur dioxide, SO_2, and hydrogen sulfide, H_2S). Since they are corrosive, these pollutants are harmful to people who breathe them. They also react destructively with housepaint, metals, and some fabrics. At present, considerable research is being done to find economical means for keeping these noxious gases out of the atmosphere.

As we have seen, both the ionization energy and the electron-affinity energy determine how two atoms will share their electrons. A given atom will be most likely to take an electron from another atom rather than lose one if the given atom's ionization energy and its electron affinity are larger than that of the other atom. Hence a reasonable measure of an atom's tendency to take on an additional electron is the sum of these two energies. With these figures, we could set up an *electronegativity scale** for the atoms of the periodic table; a high electronegativity would indicate those atoms that have large values for the sum of their ionization and electron-affinity energies. Unfortunately, electron-affinity energies are very difficult to measure and often are not available. Therefore, we cannot practically evaluate this sum in many cases. Instead, we set up an arbitrary electronegativity scale as follows. We assign to fluorine, the most electronegative atom, a value of 4.0 on our scale. At the other extreme, the unreactive atom helium is assigned the value zero. All the other atoms are included within this range, and their electronegativities are given in Figure 12.7.

To see the use of the electronegativity scale, consider what it tells us about the bond in the potassium bromide (KBr) molecule. Since potassium has an electronegativity of 0.8 and the value for bromine is 2.8, it is clear that potassium will lose its outer electron to the highly electronegative bromine. Hence an ionic bond must be present in this molecule. An entirely different situation occurs in the BrCl molecule. Since these atoms have nearly identical electronegativities (2.8 and 3.0), they share two electrons nearly equally and are bound by a covalent bond. In general, atoms with widely different electronegativities will form ionic bonds, and when the difference is small, covalent bonds are formed.

Bonding in Solids and Liquids

The forces that bind atoms into molecules are strong enough so that the bonds between the atoms are not broken by ordinary thermal energy. Like the atoms they are composed of, molecules have thermal energy, which makes them vibrate much like balls connected by springs. Moreover, if they are part of a gaseous substance, they shoot from place to place and often collide with each other. Nonetheless, the bonds we have been discussing are stable and will not break easily at normal temperatures.

This is not true, however, for a system as simple as two helium atoms. Both of these atoms have filled outer shells. They have no reason to form ions or to share electrons with other atoms. As a result, only extremely unusual conditions will force helium atoms to react chemically and form molecules. This is true of all the atoms with filled outer shells that make up Group O of the periodic table. Helium, neon, argon, krypton, xenon, and radon (called the *noble*, or *rare*, *gases*) rarely form molecules with each other or with other atoms. However, all these gases will condense to a liquid if they are cooled low enough: 4.2°K in the case of helium. Clearly, even in this extreme case, a small attractive force must draw the atoms together. This force is called the *van der Waals force* after the man who first investigated it extensively.

The van der Waals force exists between all atoms and molecules. It is generally a very weak force and can cause molecules to condense to liquids and solids only at low temperatures. The force arises out of the way the electrons in the two atoms adjust themselves to maximize attraction and minimize repulsion between the charges within the atoms.

In the case of two adjacent helium atoms, the electron-affinity effect is essentially zero. Since the outer shells of the helium atoms are full, the two atoms have no tendency to share their electrons. However, if we use Bohr's crude picture of electrons circling the nucleus, as shown in Figure 13.8, then we will see it is possible for the electrons to mesh in an especially advantageous way. Notice that the two sets of revolving electrons have meshed so that no two electrons are ever very close together. More importantly, one or more electrons is always in the region close to the midpoint be-

---

* Linus Pauling (1901–   ) devised the scale we are using. Now teaching at Stanford University, he won the Nobel Prize for Chemistry in 1954 and the Nobel Peace Prize in 1962.

KAMERLINGH ONNES LABORATORY

JOHANNES D. VAN DER WAALS (1837–1923)

Van der Waals, shown here on the right with H. Kamerlingh Onnes in the latter's laboratory, was born in Leiden in the Netherlands. His theory about the forces between molecules was only one of his many contributions to our understanding of gases, liquids, and solids. Though he was 35 years old before he published his doctoral dissertation, the time was well spent. When it appeared, his dissertation was widely acclaimed and it laid the groundwork for most of his later work. He was appointed to a professorship at the University of Amsterdam in 1877 and later taught physics at Deventer and The Hague while continuing his researches. He won the Nobel Prize in 1910 for his theory which related the pressure and volume of gases and liquids to their temperature.

13.8 *Notice that the electron between the two nuclei causes the two nuclei to be pulled toward the electron. This tends to hold the nuclei together. An instant later, an electron from the other atom will perform the same function. The resulting weak force is the van der Waals force.*

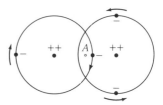

tween the two nuclei, point A in the figure. Since the positive nuclei are both attracted by this centrally located negative charge, the nuclei of the two atoms experience a force which tends to hold the atoms together.

In short, the atoms will arrange themselves in the configuration of lowest energy, since physical systems left to themselves will fall downhill. Although the electrons are circling in their orbits, if they circle as shown, the repulsion between the electrons is minimal. Moreover, the attraction between the nucleus of one atom and the electron of the other will be larger on the average than the repulsion forces. As a result, the net effect is to hold the two atoms together.

The van der Waals force exists between all atoms and molecules. In holding atoms together to form molecules, it is usually almost negligible. However, it is a major force in holding together many types of liquids. In addition to liquid helium, hydrogen, and nitrogen, the molecules of most organic liquids (benzene, gasoline, etc.) are held together by this type of force. As we would expect, these liquids have very low boiling points, a reflection of the fact that the forces holding the molecules together are not large.

In addition to the van der Waals force, ionic bonds, and covalent bonds, a type of bond which we have not yet discussed—the metallic bond—holds materials together. We shall now give examples of each of these basic bonding forces.

Types of Crystalline Solids

You will recall from earlier chapters that crystals consist of ordered groups of atoms or molecules which repeat throughout a relatively large region. These atoms or molecules are held in their ordered array by the four basic bonds. Each type of bond gives rise to a crystal with certain characteristic features, as discussed below.

1 *Ionic crystals.* One of the best-known ionic crystals is ordinary table salt, NaCl, sodium chloride. As we have seen, the chlorine atom takes the lone electron from the sodium atom's outer shell, leaving a negative chlorine ion and a positive sodium ion. (We say that the sodium atom has a <u>valence</u> of +1 since it loses one electron to become a Na$^+$ ion. Similarly, the chlorine atom has a valence of −1 since it gains an electron to become a Cl$^-$ ion.) When a group of NaCl molecules are joined to form a solid, they arrange themselves as shown in Figure 13.9. Because this structure consists of many cubes, it is called a *simple cubic lattice*. Notice that each positive ion is surrounded by six negative ions, while each negative ion is surrounded by six positive ions. The attraction of these opposite charges holds the crystal together. As we can infer from the fact that salt is hard and brittle and not easily melted, the ionic bond is relatively strong.

We can generalize from these considerations in the following way. The atoms of Group I, known as the <u>alkali metals</u>, include lithium (Li), sodium (Na), potassium (K), rubidium (Rb), and cesium (Cs). The alkali metals combine with the atoms of Group VII, the <u>halogens</u>—fluorine (F), chlorine (Cl), bro-

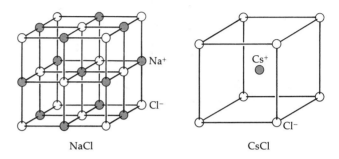

13.9 *The sodium chloride crystal consists of 10^{18} times as many atoms as shown, packed in this cubic lattice structure. Actually the Na$^+$ ions are considerably smaller than the Cl$^-$ ions. Another alkali halide crystal, cesium chloride (CsCl) is called a body-centered cubic structure because the cesium atom is in the center of the cube.*

mine (Br), and iodine (I)—to form ionic molecules and ionic crystals of the NaCl type. Typical examples are sodium iodide (NaI), lithium bromide (LiBr), potassium fluoride (KF), and cesium chloride (CsCl). The alkali metal atoms in Group I are all univalent (that is, they have a valence of +1), while the atoms of Group VII all have a valance of −1, because the Group I elements have a single electron in their outer shells while Group VII elements have a single vacancy.

The alkali metals of Group I also react with elements in Group VI (which lack two electrons in their outer shell) to form ionic compounds such as Na$_2$S and their crystals. However, reactions of the Group IB elements with Group VI elements are more common. For example, to form Ag$_2$S, the two silver atoms (each with a valence of +1) contribute the two electrons needed to fill the outer shell of the sulfur atom, which has a valence of −2. The reaction is

$$2Ag + S \rightarrow 2Ag^+ + S^{2-}$$

where the $^+$ and $^{2-}$ indicate the valence of the ions. The crystal structures of compounds between Group I and Group VI elements are not usually cubic but are of more complex form.

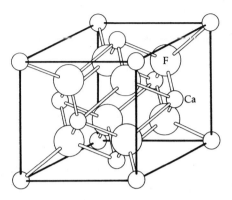

13.10 *The fluorite lattice structure of calcium fluoride, CaF_2, is complicated by the fact that the ions are not equal in number.*

The atoms in Group II are usually divalent (have valences of +2) and react with Group VII elements in a reaction such as

$$Ca + 2Cl \rightarrow Ca^{2+} + 2Cl^-$$

in which calcium chloride, $CaCl_2$, is formed. The calcium atom has two electrons in its outer shell and it gives one to each of two chlorine atoms (which require one electron to fill their shells) to form calcium chloride. This ionic compound is a salt which is widely used to melt ice on roads in winter and to draw water vapor from the air to form a liquid which settles dust on dirt roads in the summer. The crystal structure of a similar ionic compound, calcium fluoride, CaF_2, is shown in Figure 13.10. Because its ions form autonomous groups, CaF_2 does not exist as a molecule, but rather as separate ions in an ionic crystal.

Since the ions in an ionic crystal hold tightly to their electrons, no electrons are free to wander through the crystal. As a result, ionic crystals are nonconductors (or insulators). Even in a strong electric field, the electrons cannot free themselves to carry current. When the crystals are melted by heating them to high temperatures, the ions still maintain their identity and hold the electrons. In this case, however, the ions themselves, free from the lattice, can move as mobile charges in an electric field. As a result, *molten* ionic substances do have movable charges which can carry current, though they are much poorer conductors than metals are.

2 *Covalent crystals.* Perhaps the two best known covalent crystals are diamond and graphite. Both have identical composition: they are pure carbon. The difference between these two crystalline forms is the result of the lattice in which the carbon atoms are held. As you will recall, carbon has four electrons and four vacancies in its outer shell. It therefore tends to form covalent bonds. In graphite (the material in a pencil), the carbon atoms arrange themselves in sheets such as the one shown (in part) in Figure 13.11. If you count the bonds running from each carbon atom you will see that three covalent bonds have been drawn for each. However, carbon needs four covalent bonds to fill its outer shell, so one of the bonds shown for each carbon atom should be a double bond.

We have not drawn it because the double bonds in the graphite layer structure shift about. They are often called *delocalized* double bonds. Another example of this is benzene, which we will study in Chapter 14. For now we will note that the electrons in the layer can move around as the double bonds shift from place to place. As a result, graphite can conduct a current along the layer planes, although poorly.

Van der Waals forces, shown by the dashed vertical lines in Figure 13.11, hold the graphite sheets together. As might be expected, these layers or sheets slide over each other relatively easily. Because of this property, graphite is a slippery, powdery substance that can be used as a lubricant.

13.11 The graphite crystal consists of parallel planes. Van der Waals forces hold the planes together, but only weakly.

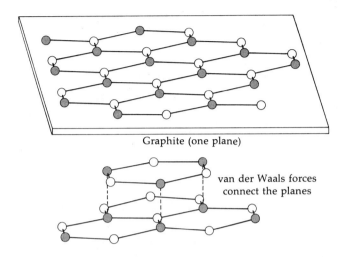

13.12 The diamond structure has a repeating set of tetrahedrons with carbon atoms at the tips of the tetrahedrons. Each heavy bar represents a covalent bond.

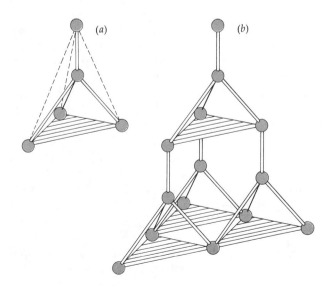

Diamond, of course, has completely different physical properties from graphite even though both are pure carbon. This difference is due to its different lattice structure, which is shown in Figure 13.12. In part (a), we see how one carbon atom bonds to other atoms with single covalent bonds. To obtain the four electrons it needs in its outer shell, the carbon atom shares four pairs of electrons with four other carbon atoms. The dashed lines show that the four outer carbon atoms form a tetrahedron, a geometrical figure with four triangular faces. Tetrahedral bonds are characteristic of carbon compounds.

The diamond lattice is built up from tetrahedrons like the one in Figure 13.12a; a small portion of the diamond lattice is shown in (b). Unlike the graphite lattice, the carbon atoms in diamond are held firmly in a three-dimensional lattice by the strong covalent bonds. The electrons, too, are all held tightly in place. As a result, diamond does not conduct electricity.

To form this diamond lattice, molten carbon is subjected to extremely high pressures and then cooled. Natural diamonds apparently were formed under the extreme conditions present when the earth first solidified. Recently it has been possible to duplicate these conditions in the laboratory, and small synthetic diamonds are now available. They are used mainly for industrial cutting and grinding (diamond is one of the hardest substances known). Large, perfect diamonds have also been synthesized, but they are not yet produced commercially.

In addition to carbon, two other elements in Group IV, silicon and germanium, also form covalent crystals. Although the atoms of pure silicon and germanium bond covalently like diamond, they do not hold their electrons nearly as tightly as diamond. As a result, both silicon and germanium conduct electricity slightly at high temperatures, germanium more effectively than silicon. When the thermal energy of the germanium atoms is high, some of the electrons break loose and move through the crystal. Although germanium has a low conductivity, tin, the next element in Group IV, has a comparatively high conductivity even though it also crystallizes in the diamond lattice form. We notice that as the atoms become larger, their hold on the electrons in a crystal is less pronounced.

When a tiny amount of an element from Group III or Group V is added to germanium or silicon, the resulting substances show a large increase in electrical conductivity. They are called <u>semiconductors</u>. Since the Group V atoms have five electrons in their outer shell, they have one more electron than is necessary to form the covalent diamondlike structure. Hence an atom of this group will contribute an extra electron if it is placed as an impurity in germanium. The extra electron breaks loose easily and then moves freely through the lattice. As a result, the slightly impure crystal becomes a semiconductor. A similar situation occurs with a Group III impurity, but now there is one electron too few. Although the process is not as easy to describe, this too leads to charge motion through the lattice, and the material is a semiconductor. Both kinds of semiconductor are widely used in transistors and other solid-state electronic devices.

3 *Metal crystals.* Nearly all the familiar metals exist in crystalline form. Mercury, a liquid at room temperature, is the only common exception; even it crystallizes at $-39°C$. Metal atoms have one, two, or three electrons in their outer shell. When the atoms are packed in a solid, they do not have enough outer electrons to fill their outer shells by electron sharing. Hence they cannot undergo covalent bonding. Nor can they form ionic bonds, because no atom is present to accept their extra electrons. Consequently, no particularly strong forces act on the outer electrons other than the attraction of their own nuclei and inner electron shells. Let us see what happens to these outermost electrons.

When a group of sodium atoms is brought together as shown in Figure 13.13, the atoms will approach

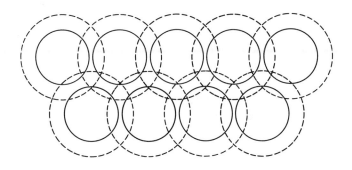

13.13   *In metals, the outer electrons overlap other atoms enough so that they are essentially free from the atom.*

until the filled $n = 2$ shells of the atoms begin to repel each other. The lone $n = 3$ electron does not prevent the atoms from coming even closer than the third Bohr orbit. The valence electron (i.e., the outer electron) in sodium is essentially lost among the neighbor atoms. It wanders about freely and takes up the lowest-energy positions in the electric field of the atoms. It loses energy in the process, and this lost energy constitutes the basic binding energy of the metallic crystal. For the most part, this binding energy is small, and so metallic crystals are less stable than ionic or covalent crystals. For example, sodium metal melts at $97°C$ while the ionic crystal NaCl melts at $801°C$.

Since the outer electrons of the metal atoms are essentially free, they can migrate easily through the crystalline metal. As a result, they move under the action of any imposed electric field and generate a current. It is for this reason that metals are good conductors of electricity.

4 *Molecular crystals.* Many substances are crystalline aggregates of molecules. In these so-called

molecular crystals, the molecules maintain their identity even in the crystal. The forces holding the molecules together are van der Waals forces or some similar nonionic electrical force. Typical molecular crystals are ice, solid carbon dioxide (dry ice), many solid waxes, and iodine crystals. Many common gases also fall into this category. For example, liquid nitrogen gas, N_2, crystallizes at $-210\,°C$, while liquid oxygen crystallizes at $-183\,°C$. As their low melting points indicate, these crystals are held together by very weak van der Waals forces.

The water molecule is of special interest, since ice is one of the most strongly bound molecular crystals. Even in liquid water the molecules are partly bound. The water molecule is typical of those molecules that have a hydrogen and oxygen atom joined together by a covalent bond. (Among this group are the alcohols, which will be discussed in the next chapter.)

The interaction of two water molecules is complex, since the electron positions become distorted as the molecules approach. However, we can see the essential feature of the interaction in Figure 13.14. Notice that, since each hydrogen atom shares its solitary electron with the oxygen atom in order to fill the outer oxygen shell, the electron is confined chiefly to the region between the hydrogen and oxygen atom. The hydrogen nucleus is thus exposed, since it stands essentially alone on the outside of the H_2O molecule. (Hydrogen is the only atom whose nucleus can be denuded of all its negative charge in this way, since it is the only atom with a single electron.)

In effect, the charge of the water molecule is separated into two positive charges and a double negative charge, as Figure 13.14b shows. Molecules such as this are called *dipolar molecules* or simply *polar molecules*. Many molecules have slightly separated charges (which we call *dipoles*), but the oxygen-hydrogen combination is unique in the amount of charge separation and the exposure of the proton, the hydrogen nucleus.

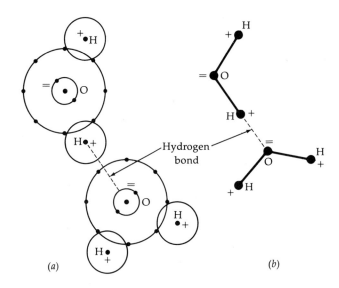

13.14 *The hydrogen bond between two water molecules is represented in two schematic ways.*

Clearly, adjacent water molecules will attract each other; the proton attracts the negative, oxygen portion of a neighboring molecule. This attraction between the molecules is called a *hydrogen bond* and is shown by the dashed lines in Figure 13.14. Because the exposed proton can come close to the oxygen atom, the bond is fairly strong. As a result, hydrogen bonds hold water molecules together in an ice crystal so that the crystal is more stable than most van der Waals crystals, but not as stable as a covalent or ionic crystal. Even after the ice melts, hydrogen bonds hold small groups of the water molecules together in the liquid.

Because of hydrogen-bonding, the molecules in an ice crystal are not packed to the best advantage, so the ice crystal has a larger volume than it could have. As the ice is melted and the water is heated from 0 to 4°C, the volume contracts, because these highly ordered but loosely packed molecular aggregates break up. Water is one of the few substances that contract when heated, and this happens only between 0 and 4°C, as the thermal energy given to the molecules breaks the hydrogen bonds apart.

THE CHEMISTRY OF INORGANIC SUBSTANCES

Ionic Solutions in Water

Because of its highly polar nature, water has the ability to dissolve ionic compounds. For example, we taste salt only because it dissolves in water. In a solution of salt water, the water is the _solvent_, the dissolving agent, and the salt is the _solute_, the substance dissolved. The charged portions of the water molecule counterbalance the attractive forces of the sodium ions and chloride ions for each other, as shown in Figure 13.15. (For simplicity we show the water molecules as ellipsoids with opposite charges on the two ends.)

As we see from Figure 13.15, the ions Na⁺ and Cl⁻ are freed from the crystal by opposite portions of the water molecule. Once in solution, the electric fields due to the $+e$ charge on the Na⁺ ion and the $-e$ charge on the Cl⁻ ion cause more water molecules to position themselves around the ions. The water molecules tend to cancel the charge on the ions by grouping around them, effectively reducing the electric field around the ion by a factor of 80. We call this factor by which the solvent reduces the ion's electric field the _dielectric constant_ of the solvent. Because of the high dielectric constant of water, the Na⁺ and the Cl⁻ ions in solution do not attract each other sufficiently to rejoin.

For all ionic substances that dissolve in water, the water molecules orient themselves to minimize the attraction between ions. As a result, the ions escape from each other and the substance dissolves. Other highly polar substances are also capable of dissolving ionic material, but to a much lesser extent than water. Water, because of its strong hydrogen-bonding capabilities, has the highest dielectric constant of any common liquid. As we shall see in the next chapter, alcohols have fairly high dielectric constants (ranging up to 20 and 30, as compared with water at 80) and act somewhat like water.

When ions are present in a solution, the solution will conduct current. This is reasonable, since the

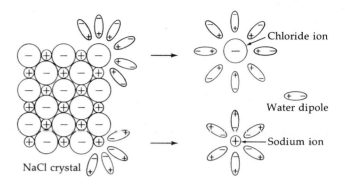

13.15 *The polar water molecules counterbalance the attraction between the Na⁺ and Cl⁻ ions and allow them to pull apart. Then, by effectively neutralizing the electric fields around the Na⁺ and Cl⁻ ions, the water molecules keep the ions from rejoining.*

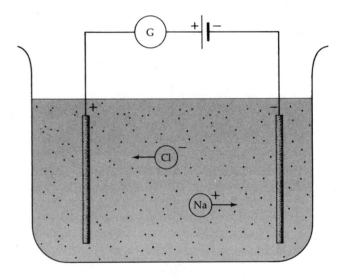

13.16 *The ions dissolved in the water will cause the solution to be a conductor.*

ions act much like charged balls in the liquid. Under the action of a potential difference, they will move through the solution. A typical situation is illustrated in Figure 13.16 where the ions involved are Na^+ and Cl^-. Although the ions will tend to pull several water molecules along with them due to the effect shown in Figure 13.15, the water molecules only slow the progress of the ions toward the electrodes. You may be surprised to learn that *pure* water is a very poor conductor of electricity. Only when ionic substances are dissolved in it does it become a conductor. The fact that pure water is not a conductor is further evidence that the water molecule is stable and does not break into ions. Clearly, electrical conductivity of solutions and liquids is a simple way to test for the presence of ions.

13.17 *Typical acids. Formic acid is the first in a series of several organic acids that includes acetic and citric acid.*

Acids

Acids play an important role in our daily lives as well as in science and industry. Most people know that the citric acid in lemons accounts for their sour taste. In fact, this sour taste is characteristic of all acids. You may also know that the very weak acid, oxalic acid, is still strong enough to remove various stains. And mothers frequently (and accurately) warn their children that the carbonic acid in carbonated beverages may be bad for their teeth.

These acids we have mentioned are what we term <u>weak acids</u>, as opposed to the corrosive <u>strong acids</u>, such as nitric acid, hydrochloric acid, and sulfuric acid. None of the acids we have mentioned is a good conductor of electricity when perfectly pure. We therefore conclude that their molecules are stable if water is not mixed with them. However, when water is added to them, the weak acids ionize (dissolve into ions) slightly, while the strong acids become almost completely ionized.

The composition of hydrochloric acid is shown in Figure 13.17. Notice that the bond within this molecule is partly covalent. When acids like hydrochloric are added to water, they ionize in the following way:

$HCl \rightarrow H^+ + Cl^-$ Hydrochloric
$HNO_3 \rightarrow H^+ + NO_3^-$ Nitric
$H_2SO_4 \rightarrow H^+ + HSO_4^- \rightarrow 2H^+ + SO_4^{2-}$ Sulfuric

These three acids are strong acids and so ionize almost completely. However, these reactions show only a portion of the story. Actually, the water molecules are clustered about the ions. The hydrogen ion, H^+, is bound tightly to a water molecule and should more properly be written as H_3O^+, which is called the hydronium ion. On the other hand, the sulfate ion, SO_4^{2-}, and the nitrate ion, NO_3^-, are stable against dissolution by water.

Formic acid, the second acid in Figure 13.17, is an organic acid. These acids contain only carbon, hydrogen, and oxygen, and are products of living organisms in nature. (They can also be made synthetically, of course.) Formic acid is found in nature in some plants and in ants; when placed in water it ionizes according to the reaction

$$HCOOH \rightleftharpoons H^+ + HCOO^- \quad \text{Formic}$$

which can be written more descriptively as

$$H-\overset{\displaystyle O}{\underset{\displaystyle O-H}{C}} \rightleftharpoons H^+ + H-\overset{\displaystyle O}{\underset{\displaystyle O^-}{C}}$$

Acetic acid, found in vinegar, reacts as

$$CH_3COOH \rightleftharpoons H^+ + CH_3COO^- \quad \text{Acetic}$$

Citric and other weak acids undergo this same partial ionization.

It is customary in the case of weak acids to show arrows pointing in both directions, as we have done above, to emphasize the fact that only a portion of the acid molecules ionize. The ions frequently recombine to form the acid even after ionization. Eventually, an equilibrium is set up, with equal numbers of molecules dissociating into ions and ions combining into molecules. Since far fewer ions are available, the weak acids are not as reactive as the highly ionized acids; this is the origin of their designation as "weak."

Bases and Hydroxides

When any of the alkali metals (lithium, sodium, potassium, rubidium, and cesium) is added to water, a violent reaction takes place. With sodium, for example, we have

$$2Na + 2H_2O \rightarrow 2NaOH + H_2 \uparrow$$

where $H_2 \uparrow$ means that the product H_2 is a gas and escapes. Instantaneously, the water ionizes the sodium hydroxide, NaOH, and so the solution contains Na^+ and OH^- ions.

This is just one way of preparing hydroxides, also called *bases*. The OH^- group has a valence of -1, and so other typical hydroxides are

KOH	Potassium hydroxide
Ca(OH)$_2$	Calcium hydroxide
NH$_4$OH	Ammonium hydroxide

Hydroxides of the alkali metals and ammonium hydroxide are caustic reagents which can cause damage and burns in their concentrated solutions. Ordinary cleaning ammonia is a dilute solution of ammonium hydroxide in water.

Water solutions of the hydroxides react readily with solutions of acids. For example, if equal numbers of NaOH (sodium hydroxide) and HCl (hydrochloric acid) molecules are added to water, both types of molecules ionize:

$$Na^+ + OH^- + H^+ + Cl^- \rightarrow H_2O + Na^+ + Cl^-$$

In other words, the H^+ and OH^- ions combine to form water molecules. The only ions left in the water are Na^+ and Cl^-, and so the product is table salt dissolved in water. If the water is evaporated away, white powdery table salt, NaCl, will remain. We say that the HCl has been *neutralized* by the NaOH.

There are many acid-base reactions in which an acid neutralized by a base gives a salt. Several typical ones are

$$KOH + HNO_3 \rightarrow H_2O + KNO_3$$
$$\text{Potassium nitrate}$$

$$Ca(OH)_2 + H_2SO_4 \rightarrow 2H_2O + CaSO_4$$
$$\text{Calcium sulfate, or gypsum}$$

$$NaOH + H_2SO_4 \rightarrow H_2O + NaHSO_4$$
$$\text{Sodium bisulfate}$$

$$2NaOH + H_2CO_3 \rightarrow 2H_2O + Na_2CO_3$$
Sodium carbonate, or washing soda

You can think of similar reactions. For example, NaOH can be used to neutralize any acid.

To test for acidity you can moisten a piece of specially treated paper, *litmus paper*, with the solution. If the solution is acidic, the paper will turn red. If the solution has a base dissolved in it, the paper will turn blue.

Minerals in Water

It is very difficult to obtain perfectly pure water from natural sources. Rainwater, when carefully collected as it falls, is usually quite pure. The water which forms the raindrops was purified when it evaporated into the air, and most of the dissolved salts, acids, and bases were left behind. However, even rainwater is slightly acidic because some of the carbon dioxide gas (CO_2) always present in the air dissolves in the moisture droplets and forms carbonic acid, a weak acid, in the following reaction:

$$H_2O + CO_2 \rightarrow H_2CO_3 \quad \text{Carbonic acid}$$

Lake water and other water in sources exposed to air usually contain even more carbonic acid. Although the waters of the earth are only slightly acid, over long periods of time they can dissolve almost any mineral. For example, carbonic acid reacts with calcium carbonate (limestone) in the following way:

$$H_2CO_3 + CaCO_3 \rightarrow Ca(HCO_3)_2 \rightarrow Ca^{2+} + 2HCO_3^-$$
Calcium bicarbonate

Since the calcium bicarbonate is very soluble in water, it simply washes away. Many huge underground caves were carved out as the limestone dissolved in this way.

Many similar reactions take place as water flows over and through the earth. As a result, most natural water is a complex solution of many ionic salts. The residues left in teakettles and pots by supposedly clean tap water are salts left behind as the water evaporates. For the most part, the dissolving of the earth by water in streams and lakes and oceans is a very slow process and has been going on for hundreds of millions of years. It has dissolved out many of the deep caverns in the earth and helped level mountains and hills as erosion has taken place (although the major work was done by the friction of wind and water).

There are many more examples of the reactions between metals, nonmetals, acids, and bases. Most of the reactions we have discussed are in the field of *inorganic chemistry*. Inorganic chemistry is the chemistry of all compounds except those of carbon. Carbon compounds are so numerous that they are studied separately in *organic chemistry*, which is the subject of our next chapter.

A Summary of Chemical Names and Symbols

The chemical symbols for elements consist of one or two letters; the first is always capitalized. Most symbols are derived from the English name for the element: calcium, Ca; strontium, Sr; barium, Ba; oxygen, O; sulfur, S. Some symbols come from their Latin names: sodium, *natrium*, Na; potassium, *kalium*, K; mercury, *hydragarum*, Hg; silver, *argentum*, Ag. Tungsten receives its symbol, W, from its German name, wolfram.

Let us consider compounds having just two elements (called binary compounds). To write the formula and the name of a binary compound, we need to know the symbol for each element (from the periodic table, Figure 12.7). We must also know how many electrons each atom can give up or take; in other words, the valence of each atom. Elements with positive valences always combine with elements of negative valence so that the total positive and negative valences are balanced out. For example, sodium chloride is NaCl because one Na (+1) matches one Cl (−1). Calcium chloride is $CaCl_2$ because one Ca (+2) matches two Cl's. Ferric oxide is Fe_2O_3 because two ferric irons (2 × 3 = +6) match three oxides (3 × −2 = −6).

Some elements can give up different numbers of electrons depending on the reaction to which they are subjected. Thus they can have different valences. For example, copper can have valences of +1 and +2. The name for the element in combination is given by suffixes: *-ic* for the higher valence, *-ous* for the lower valence. Copper reacts with chlorine in different ways to form

Cuprous chloride   CuCl

and

Cupric chloride $CuCl_2$

A more modern way of showing these valences is by Roman numeral; for example:

Copper(I) chloride   CuCl
Copper(II) chloride $CuCl_2$

Some more examples are:

KBr	Potassium bromide	K_2O	Potassium oxide
Fe_2S_3	Iron(III) sulfide or ferric sulfide	FeS	Iron(II) sulfide or ferrous sulfide
$CoCl_3$	Cobalt(III) chloride or cobaltic chloride	$CoCl_2$	Cobalt(II) chloride or cobaltous chloride
$SnBr_4$	Tin(IV) bromide or stannic bromide	$SnBr_2$	Tin(II) bromide or stannous bromide

The table on the following page shows the symbol, valence, and name in combination for some common elements. If you want to write chemical formulas, you should memorize the valences.

Some ions consist of two or more elements which act as a unit in reactions. For example, we have encountered the sulfate ion, SO_4^{2-}, and the carbonate ion, CO_3^{2-} in this chapter. Most of the ions shown in the following table have negative valences and contain oxygen. The one exception is the ammonium ion, NH_4^+, which has a positive valence.

Ion	Symbol	Valence
Hydroxide	OH^-	−1
Nitrate	NO_3^-	−1
Nitrite	NO_2^-	−1
Acetate	CH_3COO^-	−1
Sulfate	SO_4^{2-}	−2
Sulfite	SO_3^{2-}	−2
Carbonate	CO_3^{2-}	−2
Phosphate	PO_4^{3-}	−3
Ammonium	NH_4^+	+1

Some compounds with these ions are:

$CaCO_3$	Calcium carbonate
$Ca(OH)_2$	Calcium hydroxide
NH_4Cl	Ammonium chloride
$FeSO_4$	Iron(II) sulfate or ferrous sulfate
$Fe_2(SO_4)_3$	Iron(III) sulfate or ferric sulfate
KNO_3	Potassium nitrate
NH_4OH	Ammonium hydroxide
H_2SO_4	Hydrogen sulfate or sulfuric acid

Hydrogen can combine with one other element or negative ions like the ones above to form an acid.

Element	Symbol	Valence	Name in combination
Hydrogen	H	+1	Hydrogen
Lithium	Li	+1	Lithium
Sodium	Na	+1	Sodium
Potassium	K	+1	Potassium
Silver	Ag	+1	Silver
Copper	Cu	+1	Copper(I) or cuprous
		+2	Copper(II) or cupric
Mercury	Hg	+1	Mercury(I) or mercurous
		+2	Mercury(II) or mercuric
Magnesium	Mg	+2	Magnesium
Calcium	Ca	+2	Calcium
Strontium	Sr	+2	Strontium
Barium	Ba	+2	Barium
Zinc	Zn	+2	Zinc
Cadmium	Cd	+2	Cadmium
Chromium	Cr	+2	Chromium(II) or chromous
		+3	Chromium(III) or chromic
Manganese	Mn	+2	Manganese(II) or manganous
		+3	Manganese(III) or manganic
		+4	Manganese(IV) or manganese*
Iron	Fe	+2	Iron(II) or ferrous
		+3	Iron(III) or ferric
Cobalt	Co	+2	Cobalt(II) or cobaltous
		+3	Cobalt(III) or cobaltic
Nickel	Ni	+2	Nickel(II) or nickelous
		+3	Nickel(III) or nickelic
Tin	Sn	+2	Tin(II) or stannous
		+4	Tin(IV) or stannic
Lead	Pb	+2	Lead(II) or plumbous
		+4	Lead(IV) or plumbic*
Aluminum	Al	+3	Aluminum
Carbon	C	†	Carbon*
Silicon	Si	†	Silicon*
Nitrogen	N	†	Nitrogen*
Phosphorus	P	†	Phosphorus*
Fluorine	F	−1	Fluoride
Chlorine	Cl	−1	Chloride
Bromine	Br	−1	Bromide
Iodine	I	−1	Iodide
Oxygen	O	−2	Oxide
Sulfur	S	−2	Sulfide

* Compounds of elements with high valences are sometimes named by adding a prefix to the second word instead of the *-ic* or *-ous* suffix to the first. For example:

$MnCl_2$ Mangan*ous* chloride
$MnCl_3$ Mangan*ic* chloride
$MnCl_4$ Manganese *tetra*chloride

CO_2 Carbon *di*oxide
CCl_4 Carbon *tetra*chloride
PbO_2 Plumb*ic* oxide or lead *di*oxide

† Many different valences, all named differently.

When it combines with one element, the acid is named with the prefix *hydro-* and the suffix *-ic*:

    HCl   *Hydro*chloric acid
    HF   *Hydro*fluoric acid

In the latter case, the name of the acid depends on the ion with which the hydrogen combines:

NO_3^-	Nitrate	HNO_3	Nitric
NO_2^-	Nitrite	HNO_2	Nitrous
CH_3COO^-	Acetate	CH_3COOH*	Acetic
SO_4^{2-}	Sulfate	H_2SO_4	Sulfuric
SO_3^{2-}	Sulfite	H_2SO_3	Sulfurous
CO_3^{2-}	Carbonate	H_2CO_3	Carbonic
PO_4^{3-}	Phosphate	H_3PO_4	Phosphoric

Bases are combinations of the OH^- ion with an element having positive valence; they are called hydroxides. For example:

    NaOH   Sodium hydroxide
    $Ca(OH)_2$ Calcium hydroxide

When elements and compounds combine in chemical reactions, the starting materials are called *reactants*, the final materials *products*. We can summarize a chemical reaction by an equation. For example,

$$6HCl + 2Al \rightarrow 3H_2 + 2AlCl_3$$

The number in front of each compound or element is the number of molecules of that substance. Notice that the total number of atoms of each element on the left must equal the number of atoms of that element on the right. In this equation, six hydrogen atoms, six chlorine atoms and two aluminum atoms appear on each side.

---
* Organic acids (discussed in Chapter 14) are written with the "acid hydrogen" last, while inorganic acids are written with the "acid hydrogen" first.

BIBLIOGRAPHY

1. ASIMOV, I.: *A Short History of Chemistry* (paperback), Anchor Science, Doubleday, Garden City, N.Y., 1965. An interesting account written by a famous popularizer of science.

2. MCANALLY, J.: *Chemistry* (paperback), Merrill, Columbus, Ohio, 1966. This text goes into more detail than we have been able to, but is still brief and uncomplicated enough to be useful.

3. WEEKS, M. E.: *Discovery of the Elements*, Journal of Chemical Education, 1956. An authoritative yet readable history of the early scientists and how they discovered the elements in the periodic table.

4. LEICESTER, H., and H. KLICKSTEIN: *Source Book in Chemistry*, McGraw-Hill, New York, 1952. Contains the original writings of many famous early chemists. They are selected for the general audience, not just for chemists.

5. BAXTER, J., and L. STEINER: *Modern Chemistry*, (paperback), vols. 1, 2, Prentice-Hall, Englewood Cliffs, N.J., 1960. A good place to look if you want to explore the field of chemistry in more detail (written for the Continental Classroom TV series).

6. HOLDEN, A., and P. SINGER: *Crystals and Crystal Growing* (paperback), Anchor Science, Doubleday, Garden City, N.Y., 1960. This is a "how to" book as well as a delightful discussion of crystals in general. It even gives instructions for growing crystals at home.

SUMMARY

Several types of forces or bonds cause atoms to bind together. The covalent bond is a sharing of electrons by two or more atoms in such a way that both atoms effectively have filled outer shells; the H_2 molecule is an example. Both the diamond and graphite crystal forms of carbon are also covalently bonded. We represent a sharing of two electrons in a covalent bond by the symbolism H—H, where the bar represents the two shared electrons. An oxygen molecule, O_2, has four shared electrons and we represent them as O=O. This is called a double covalent bond.

Both the ionization energy of an atom (the energy needed to tear an electron loose) and its electron-affinity energy (the energy lost when an electron is captured) are important in determining its chemical behavior. Those atoms that lack only one electron to have a filled outer shell (F, Cl, Br, I) have large electron affinities and high ionization energies. At the other side of the periodic table, those atoms (Li, Na, K) that have a single outer-shell electron have small electron affinities and small ionization energies. Consequently, sodium easily loses an electron to a chlorine atom. The result is two charged atoms called ions; they are represented as Na^+ and Cl^-. In ionic crystals such as table salt, NaCl, the atoms are held together by the attraction between the oppositely charged ions. Such a bond is called an ionic bond, and is usually a very strong bond. There are intermediate cases in which the bond combines ionic and covalent characteristics.

The van der Waals force is rather weak but is still strong enough to bind many molecules together into liquids. This force is important when the electrons in adjacent atoms or molecules arrange or distort their normal motion patterns slightly to minimize coulomb repulsion forces and maximize attraction forces. Because of van der Waals forces, even the unreactive (or noble) gas element atoms (He, Ne, etc.) can be obtained as liquids at low temperatures. The molecules in liquids such as gasoline, oil, and many plastics are held together by this force.

Metals are solids or liquids in which atoms with loosely held outer electrons are packed together. The outer electrons become detached from the parent atom and are then free to carry charge through the solid or liquid. Moreover, the loose electrons have lower energies than when they were a part of the original atom. This energy, lost when the atoms were brought together to form the liquid or solid, contributes much of the binding energy of metallic substances.

Some molecules, such as water, have an exposed positively charged proton because of the way the hydrogen atom is bound into the molecule. Moreover, the same molecule can have a negatively charged atom in it—the oxygen atom in the case of water. As a result, a hydrogen bond forms between the exposed proton and the charged oxygen atom. This bond is intermediate in strength between the van der Waals bond and the true ionic and covalent bonds. It is responsible for the distinctive properties of water—for example, the way water contracts between 0 and 4°C.

Water molecules congregate around ions in such a way that the charge on the ion is partly compensated. As a result, ions in molecules normally held together by ionic forces are able to escape from each other. This allows the substance to dissolve, giving rise to free ions in the water solution. These ions can then react with any other ions present in the water solution.

Acids are substances which, when dissolved in water, give rise to hydrogen ions. The compensating negative ion depends on the acid involved. In the case of HCl, it is Cl^-, while for HNO_3 it is NO_3^-. Those acids that ionize nearly completely in water are called strong acids, while their opposites are termed weak. Bases, on the other hand, release OH^- ions when dissolved in water. A typical example is sodium hydroxide, NaOH. When a solution of a base is neutralized by adding acid to it, the H^+ and OH^- ions join to form water while the remnant ions join to form compounds known as salts.

TOPICAL CHECKLIST

1   Draw the potential energy diagram for two hydrogen atoms as a function of their separation distance r. What is the significance of the minimum in it?

2. How does electron spin influence whether or not two hydrogen atoms bond together? Describe the electron behavior in the hydrogen molecule.

3. What is a covalent bond? Give examples of it.

4. What does the formula P_2S_5 tell us about the number of phosphorous and sulfur atoms in the molecule?

5. Explain the type of bond in a molecule such as Cl_2 or Br_2.

6. Give two alternative ways of representing symbolically a single covalent bond; a double covalent bond. What is the physical situation corresponding to each type of bond?

7. What is an ion? How does an ionic bond differ from a covalent bond? Give examples of ionic molecules.

8. What are the origins of sulfur-containing air pollutants?

9. What is a van der Waals bond? Give examples of it.

10. List the four types of bonds between molecules in a series running from strongest to weakest.

11. Describe a common ionic crystal such as NaCl, pointing out how the crystal is held together. What is meant by electron affinity?

12. Explain the meaning of valence. What are the normal valences of the following atoms: Li, Cl, K, S, Ca, F, Na, He, H, Br?

13. Are ionic crystalline materials conductors or insulators? Explain.

14. Discuss covalent crystals in terms of the structure of diamond.

15. Graphite is both a covalent and a van der Waals crystal; explain.

16. Although graphite and diamond are both pure carbon, one is a conductor while the other is not; explain.

17. When a Group V impurity is added to a germanium crystal, the crystal shows a much higher electrical conductivity. Explain why.

18. Discuss the origin of the binding force within metals. Why do Group I atoms form electrically conducting crystals?

19. What is a molecular crystal and why is it given this name? What forces hold the crystal together?

20. What is the hydrogen bond? Why is it unique to hydrogen and not found in lithium and sodium?

21. Discuss how water molecules achieve the solution of ionic crystals. Why are liquids of lower dielectric constant unable to dissolve ionic substances?

22. What is an acid? Distinguish between weak and strong acids. What is the hydronium ion?

23. What is meant by the terms hydroxide and base? How do they react with acids? What is a salt?

24. Discuss the composition of water found in nature. How does it react with limestone?

QUESTIONS

1. Draw a curve similar to Figure 13.1 showing the energy of a helium atom as it is brought close to another helium atom. In drawing this, remember that liquid helium does exist.

2. When sodium is added to water, a gas is given off in a violent reaction. This gas can be ignited with a match so that it combines with the oxygen in the air, that is, it burns. What is the reaction by which it burns?

3. Why do many soft drinks "fizz"? Carbonated soft drinks lose some of their acidity as they sit exposed in a glass. Why?

4. When a dilute water solution of silver nitrate, $AgNO_3$, is added to dilute hydrocloric acid,

HCl, a solid precipitate settles out of the solution. What is the reaction and what is the precipitate?

5 One test for the purity of water is to measure the resistance of the water between two electrodes immersed in the water. Explain the principle behind this test.

6 Sulfur dioxide is a colorless, suffocating gas found in smog and as a residue of many industrial processes. Show its bond structure.

7 Copper and iron are among the atoms in the center of the periodic table which can give up different numbers of electrons depending on the reaction to which they are subjected. As a result, they have two possible valences: +1 and +2 for copper, and +2 and +3 for iron. The +1 and +2 compounds of copper are called cuprous and cupric while the +2 and +3 of iron are called ferrous and ferric. Write the formulas for the following: (a) ferrous chloride; (b) cupric chloride; (c) cupric nitrate; (d) ferrous sulphate; (e) ferric nitrate. Another terminology frequently used is iron (II) chloride and iron (III) chloride for ferrous and ferric respectively.

8 In studying the moon and the planets, scientists look for evidence of erosion. If a structure similar to the Grand Canyon could be found on the surface of Mars, what would this imply? What sort of processes on the moon can obliterate a footprint?

9 The substance Carborundum is used as an industrial cutting material since it has many of the same properties as diamond. Its formula is silicon carbide, SiC, and it crystallizes in the diamond lattice with each atom of one type surrounded by four atoms of the other type. Draw a small portion of its lattice and explain why the Si atom will form a lattice like this with carbon.

10 Much of the earth's crust is composed of silicon compounds. What chemical properties would you predict for this element?

11 A common ingredient of antiacids is milk of magnesium, $Mg(OH)_2$. Why is this effective in reducing stomach acidity? Why is it suicidal to use NaOH instead? What about baking soda, $NaHCO_3$? (Note: the structure of $Mg(OH)_2$ is the following: H—O—Mg—O—H.)

12 Before a person has an x-ray taken of his digestive tract, he is required to drink a milky suspension of a barium compound such as $BaSO_4$ in water. What is the purpose of this? Why won't $Mg(OH)_2$ work just as well?

13 Most medicine cabinets have first-aid instructions. Invariably they prescribe the drinking of baking soda (sodium bicarbonate, $NaHCO_3$) dissolved in water for one who has swallowed acid *or* base (alkali). How can the same chemical counteract these two opposite types of chemical poisoning?

PROBLEMS

1 The ionization energy of sodium is 5.1 eV while the electron affinity of a chlorine atom is 3.6 eV. An energy of 6.0 eV is required to tear a sodium chloride molecule apart. Using these figures, explain why 4.5 eV of energy is given off when a Na atom combines with a Cl atom to form NaCl.

2 If the atomic weight (or molecular weight, the atomic weights of every atom in the molecule added together) of a substance is M, then in M grams of the substance there will be 6.0×10^{23} atoms (or molecules). Find the mass of a hydrogen atom. The atomic weight of hydrogen is 1.00 atomic mass unit. (The number 6.0×10^{23} is called Avogadro's number after the man who proposed that such a constant should exist. In his time, however, its value was unknown.)

3   Find the mass of one molecule of nitrogen in nitrogen gas (see Problem 2).

4   Suppose a carbon atom is approximated by a cube 2 Å on an edge. How many carbon atoms would exist in a cubic centimeter of a diamond? (One cubic centimeter is a cube 1 cm on an edge.)

5   If the answer to Question 4 is taken to be 1×10^{23}, what would be the mass in grams of the 1-cm cube composed of carbon, assuming the carbon is C^{12}? Compare this with the densities (mass per unit volume) of diamond and graphite which are 2.25 and 2.0 grams per cubic centimeter respectively.

6   In carrying out the following reaction

$$H + Cl \rightarrow HCl$$

how many grams of chlorine will be required to combine with 2 g of hydrogen? The atomic masses of hydrogen and chlorine are 1.0 and 35.5, respectively. How many grams of HCl are formed?

7   When hydrogen gas is burned in oxygen to form water, the reaction is

$$2H_2 + O_2 \rightarrow 2H_2O$$

How many grams of water will be formed when 3 g of hydrogen is burned? (Atomic weights: hydrogen = 1.0; oxygen = 16.0.)

8   How many grams of sodium hydroxide are necessary to neutralize 1 g of HCl?

9   The energy needed to break an H_2 molecule apart, the "bond strength," is 4.7 eV. What wavelengths of light shining into a tube of hydrogen gas will be capable of breaking the molecule apart?

Chapter 14

THE MOLECULES OF LIVING THINGS: ORGANIC CHEMISTRY

Everything on earth which is or once was part of a living organism is basically composed of hydrogen, carbon, and oxygen. These atoms combine to form organic molecules in living things and in their decay products. Organic molecules are molecules that contain carbon. They range from the simple molecules found in alcohol and natural gas to the complex molecules in proteins and vitamins, and the intricate polymer molecules found in muscle and flesh. We will start our study with the simplest organic molecules, the aliphatic hydrocarbons.

Hydrocarbon-Chain Molecules

Compounds composed only of hydrogen and carbon are called _hydrocarbons_. Let us first consider the wide variety of compounds that can be made simply by connecting carbon atoms together in uncomplicated chains. You will recall that carbon forms covalent bonds to obtain the necessary four extra electrons to fill its outer shell. The four covalent bonds always arrange themselves about the carbon atom to form a tetrahedron; the methane molecule shown in Figure 14.1a is an example.

Notice that the carbon atom is at the center of the tetrahedron and the four hydrogen atoms are at the corners.

Methane, CH_4, is the first of a series of molecules called _paraffin_ molecules. In Figure 14.1b, the covalent bonds of methane are indicated by bars. Other members of this series are similarly diagrammed in Figure 14.2. Of course these are only schematic diagrams; the atoms do not actually lie in a plane or along a straight line. The molecular models in Figure 14.3 give a more realistic picture.

Table 14.1 lists the boiling points and melting points of the paraffin molecules. In the order listed in the table, these molecules are found in nature in the following forms:

1 _Natural gas_ consists mostly of methane, with some ethane and propane. Both propane and butane are used as bottled gas for cooking and cigarette lighters.

2 _Gasoline_ consists mostly of the liquids pentane, hexane, heptane, octane, and nonane. Since pentane has a boiling point only 20°F above room temperature, it makes gasoline extremely volatile (easily vaporized). The exact composition of gasoline

14.1 *The carbon atom is at the center of a tetrahedron formed by the hydrogen atoms, to which it is joined by four covalent bonds.*

14.2 *Several members of the straight-chain or normal paraffin family.*

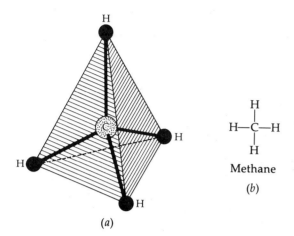

is adjusted to match the season and the region in which it is sold, to provide higher volatility and quicker starts.

3 *Paraffin wax*, a white crystal, is a mixture of the longer-chain hydrocarbons, with the carbon-atom chains straightened out and lying side by side in the crystal. Notice that after hexadecane, the paraffins are solid at room temperature (about 20°C).

We know that the lower, more volatile members of this series burn readily. When a paraffin (or any hydrocarbon) is burned in oxygen, carbon dioxide and water are formed; pentane is a typical case:

$$C_5H_{12} + 8O_2 \rightarrow 5CO_2 + 6H_2O$$
Pentane   Carbon dioxide

If automobiles burned their gasoline perfectly and if gasoline were pure hydrocarbon, the end products would be water and a relatively harmless gas. However, automobiles do not burn gasoline completely; instead, the following reaction occurs:

$$2C_5H_{12} + 11O_2 \rightarrow 10CO + 12H_2O$$

14.3 *Molecular models showing some of the various configurations an n-butane molecule can take on. The dark interior atoms are the carbons; the light, outer atoms are the hydrogens.*

Table 14.1 Some Normal Paraffins

Formula	Name	Boiling Point, °C	Melting Point, °C
CH_4	Methane	−161	−184
C_2H_6	Ethane	− 88	−172
C_3H_8	Propane	− 43	−188
C_4H_{10}	Butane	+ 0.6	−135
C_5H_{12}	Pentane	36	−148
C_6H_{14}	Hexane	69	− 94
C_7H_{16}	Heptane	98	− 90
C_8H_{18}	Octane	126	− 98
C_9H_{20}	Nonane	150	− 51
$C_{10}H_{22}$	Decane	174	− 32
$C_{13}H_{28}$	Tridecane	234	− 6
$C_{16}H_{34}$	Hexadecane	288	18
$C_{19}H_{40}$	Nondecane	330	32
C_NH_{2N+2}	Polymethylene*	—	~130

* N is an integer of the order of a thousand or more. This is also called polyethylene.

ORGANIC CHEMISTRY

One product, CO, is poisonous carbon monoxide; so-called side reactions also occur. In addition, gasoline contains certain additives (tetraethyl lead, for example) which make some of its combustion products corrosive and otherwise dangerous to health. These combustion products and carbon monoxide are major sources of harmful air pollution in urban areas.

Hydrocarbons also react with elements other than oxygen. For example, the paraffins react with chlorine, as these reactions of methane show:

$CH_4 + 3Cl_2 \rightarrow 3HCl + HCCl_3$ Chloroform
$CH_4 + 4Cl_2 \rightarrow 4HCl + CCl_4$ Carbon tetrachloride

Since the Cl atom has a high electron affinity, it steals the H atom from the carbon.

Nomenclature

There are literally hundreds of thousands of carbon compounds with different structures or arrangements of the atoms. For this reason, a system of naming these compounds is necessary. Here we will mention a few rules of this nomenclature, or system of names.

The number of carbon atoms in a group or molecule is commonly given by a prefix. For example from Table 14.1, it is clear that the following prefixes are associated with the indicated numbers:

Prefix	Number of carbon atoms
meth-	1
eth-	2
prop-	3
but-	4
pent-	5
hex-	6
hept-	7
oct-	8
non-	9
dec-	10

When joined with the suffix *-ane*, these prefixes designate a paraffin hydrocarbon. Hence ethane is C_2H_6.

Some molecules, such as butane, can exist in two different molecular forms. Therefore, another designating prefix is necessary. All of the molecules shown in Figure 14.2 are in straight chains. They are called <u>normal paraffins</u>, because in chemical nomenclature, the word "normal" always means a straight chain. However, consider the two following forms of butane below, both of which have the same formula, C_4H_{10}:

         n-Butane   Isobutane

Where *n-* stands for normal, and the prefix *iso-* means that the chain is not a straight chain. Alternate structures such as these are called <u>isomers</u> (from Greek *isos*, equal, and *meros*, part). Both forms have the same composition; they differ in chemical properties because the arrangement of their atoms is different.

Another important rule of nomenclature is best illustrated by an example. When propane reacts with chlorine to obtain monochloropropane, the prefix <u>mono-</u> tells us that only one chlorine has been added. However, the resultant compound may be either of the following:

All other positions for the chlorine are equivalent to one or the other of these. To distinguish these two compounds, the one on the left is called 1-chloropropane and the one on the right is 2-chloropropane.

The position of an atom is given by counting from one end of the molecule or from some unique point on the molecule. Another example would be

$$\begin{array}{c} \text{Cl} \quad \text{H} \quad \text{H} \quad \text{Cl} \quad \text{H} \\ | \quad | \quad | \quad | \quad | \\ \text{H}-\text{C}-\text{C}-\text{C}-\text{C}-\text{C}-\text{H} \\ | \quad | \quad | \quad | \quad | \\ \text{H} \quad \text{H} \quad \text{H} \quad \text{Cl} \quad \text{H} \end{array}$$

which can be called 1,4,4-trichloro-n-pentane, or alternatively 2,2,5-trichloro-n-pentane.

We will point out other features of nomenclature as we continue. The rules above are valuable in naming complex compounds. Unfortunately, most of the simple compounds were named before a universal nomenclature was widely accepted. As a result, many organic molecules have two names, the traditional name and the one corresponding to the universal nomenclature system.

Alcohols

The most important alcohols differ from the paraffin hydrocarbons only in that one H atom is replaced by an OH group. The OH group is typical of all alcohols. For example, we have

$$\begin{array}{cc} \begin{array}{c} \text{H} \\ | \\ \text{H}-\text{C}-\text{OH} \\ | \\ \text{H} \end{array} & \begin{array}{c} \text{H} \quad \text{H} \\ | \quad | \\ \text{H}-\text{C}-\text{C}-\text{OH} \\ | \quad | \\ \text{H} \quad \text{H} \end{array} \\ \text{Methanol} & \text{Ethanol} \\ \text{(methyl alcohol)} & \text{(ethyl alcohol)} \end{array}$$

Methyl alcohol is commonly called wood alcohol and is used as rubbing alcohol; it is very poisonous. Ethanol, also called grain alcohol, is the alcohol in alcoholic beverages.

The traditional and scientific names for alcohols are fairly transparent. For example propyl alcohol can be of two types, 1-propanol or 2-propanol. Similarly, normal hexanol can also be designated as 1-hexaonal. The alcohols are characterized by the -ol endings.

Alcohols such as methanol and ethanol are perfectly miscible with water; that is, they form solutions with water in all proportions. This is not surprising, since water itself has the typical —OH alcohol group. The alcohols form hydrogen bonds with themselves and with water. For example, two methanol molecules may be hydrogen-bonded:

$$\text{H}_3\text{C} \underset{\text{H}}{\overset{\text{O}}{\diagdown}} \text{H} \cdots \text{O}-\text{CH}_3$$

Like water, they have high dielectric constants. You will recall that the dielectric constant of water is 80; for methanol it is 30 and for ethanol, 24. In spite of these rather high dielectric constants, alcohols cannot dissolve ionic substances such as sodium chloride. They are not polar enough to free the ions.

In the long-chain alcohols, the hydrocarbon portion of the molecule causes the substances to be less like water and more like an oil or wax. For example, 1-decanol (for which you should be able to write the structure) is an oily liquid that is not at all soluble in water. It does, however, mix in all proportions with methanol.

Of course, a single molecule can have more than one OH group. These molecules are also alcohols, but the common ones have different names. For example, the compound

$$\begin{array}{c} \text{H} \quad \text{H} \quad \text{H} \\ | \quad | \quad | \\ \text{H}-\text{C}-\text{C}-\text{C}-\text{H} \\ | \quad | \quad | \\ \text{OH} \quad \text{OH} \quad \text{OH} \end{array}$$

is called glycerol. Solutions of this liquid in water are commonly called glycerin. Pure glycerol is an extremely viscous liquid, because its molecules are held together tightly by hydrogen bonds. It is widely used in toothpaste, lotions, and creams as the moisturizing agent, since it evaporates very slowly.

Organic Acids and Esters

An interesting transformation in organic chemistry is the reaction of a foul-smelling organic acid with an alcohol to form a sweet-smelling *ester*. This reaction is the organic counterpart of the reaction of an acid with a base to form a salt. We mentioned the organic acids in the last chapter; we list a few of them again in Table 14.2. Notice that they are characterized by the

$-C\begin{smallmatrix}O\\ \\OH\end{smallmatrix}$ group, the *carboxyl* group.

A typical reaction between an acid and an alcohol to form an ester is the following:

$$HO-\overset{O}{\underset{O-H}{C}} + [H-O-CH_3] \rightarrow HC\overset{O}{\underset{O-CH_3}{}} + H_2O$$

Formic acid   Methanol   Methyl formate

As indicated, the OH⁻ and H⁺ groups split away from the two molecules to form water.* In a similar way, butyric acid (which causes the rancid odor of bad butter) plus ethanol gives sweet-smelling ethyl butyrate. You are familiar with some esters of acetic acid, since ethyl acetate gives apples their

---

* You might think that the esterification would proceed by taking only the hydrogen from the acid:

$$HO-\overset{O}{\underset{OH}{C}} + [H-O]-CH_3$$

The end result is the same in either case, but a radioactive tracer will easily show which process is correct. If a radioactive isotope of oxygen O^{18} is used in the alcohol, then the water molecule formed in the reaction will be radioactive if this alternative is correct. Since this does not occur, the oxygen in the water molecule must come from the acid. Thus the reaction given in the text is the one that actually occurs.

Table 14.2   *Organic Acids*

Name	Structure	Origin in nature
Formic	$HC\begin{smallmatrix}O\\ \\OH\end{smallmatrix}$	Ants and nettles
Acetic	$CH_3C\begin{smallmatrix}O\\ \\OH\end{smallmatrix}$	Vinegar
Butyric	$C_3H_7C\begin{smallmatrix}O\\ \\OH\end{smallmatrix}$	Spoiled butter
Stearic	$C_{17}H_{35}C\begin{smallmatrix}O\\ \\OH\end{smallmatrix}$	Tallow or suet

fragrance, and amyl acetate (more properly, 1-pentanol acetate) is "banana oil," and gives the characteristic banana flavor and odor. Although these esters are edible in the small quantities found in fruit, others, such as the odiferous material in model airplane cement, are quite harmful even if only inhaled.

Benzene and Related Compounds

So far we have been dealing with linear carbon molecules, which are generally called *aliphatic* hydrocarbons. There are other molecules which exist as rings. Benzene is the best known of these ring molecules, but before we look at its structure, we will discuss a somewhat simpler ring compound called cyclohexane.

As its name implies, cyclohexane is a hexane molecule with its ends tied together. To form the tie point, each end of the hexane molecule is missing one hydrogen atom, and so the formula of cyclohexane is C_6H_{12}. Its structure is shown in Figure 14.4. Notice that the molecule is nonplanar and is somewhat flexible; it can be snapped from form (*a*) to form (b) and vice versa. Cyclohexane is a liquid having properties intermediate between hexane and benzene.

The benzene molecule has six fewer hydrogen atoms than cyclohexane. To fulfill the requirements of their outer shells, the carbon atoms are joined by alternate single and double covalent bonds as shown in Figure 14.5. Since the bond angles for single and double bonds are different, the benzene molecule does not have the same configuration as cyclohexane; it is in fact a planar molecule. Benzene is a volatile liquid, which boils at 80°C. It is used widely as a solvent for organic materials.

Because of the shared nature of the electrons in ring compounds such as benzene, the double-bond and single-bond electrons often cannot decide which bond they belong to. As a result, the molecule shown in Figure 14.5 actually takes on two configurations which alternate rapidly. The second configuration is the same molecule with single and double bonds reversed. This alternation is called _molecular resonance_.

The molecular resonance of unsaturated ring compounds (an _unsaturated_ compound is one which has some double or triple bonds) causes electrons to move freely around the ring. This does not lead to conductivity, because the electron is still not free to escape from the molecule. However, we can find the energy levels of the "free" electron by considering the resonance of the electron's de Broglie wave going around the molecule, much as we did for the Bohr orbits of hydrogen. The energy differences between the resonance levels cause benzene to absorb light in the ultraviolet. In more complex ring molecules consisting of several benzene rings, such as the one shown in Figure 14.6, these energy levels are close enough together so that light absorption occurs in the visible spectrum. These compounds often have brilliant colors and are used as dyes.

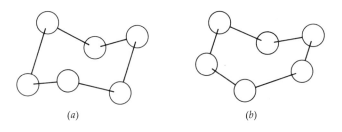

14.4 *Cyclohexane. Each carbon atom also has two hydrogens attached, but these are not shown. The molecules are somewhat flexible, and thermal motion can snap form (a) into form (b) and vice versa.*

14.5 *In the benzene molecule, the double bonds move around the molecule, so the bonds may reverse at any moment, as the diagrams show.*

ORGANIC CHEMISTRY 315

14.6 The dye algol red. The hydrogen and carbon atoms of the benzene ring are not labelled.

14.7 Several common ring compounds. The atoms of the ring are not shown.

Many chemicals based upon benzene are widely known and used. A few of these are shown in Figure 14.7. While important in themselves, these compounds are also used as starting points in the synthesis of drugs and other chemicals. Most unsaturated ring compounds come directly or indirectly from tar, petroleum, and coal. However, organic chemistry has progressed to the stage where these compounds, originally found in nature, can often be synthesized more economically in the laboratory.

Amines and Amides

Although the amide group of molecules is not a pure hydrocarbon group, it plays an important role in polymers (such as nylon) and proteins. As its name suggests, it is related to the ammonia molecule, NH_3, as shown.

If one of the hydrogens is replaced by a methyl group, CH_3, we have the simplest *amine*, methylamine. The amines are characterized by the

$$-C-N\begin{matrix}H\\H\end{matrix}$$

combination.

Amines are foul-smelling compounds, responsible for the odor of decaying fish. They are extremely important, however, since they react with organic acids to form chemicals that are vital to fiber-making and many other industrial processes. A typical reaction is that between methylamine and acetic acid:

$$CH_3-\overset{O}{\underset{OH}{C}} + H-\underset{H}{\overset{CH_3}{N}} \rightarrow CH_3-\overset{O}{\underset{\underset{H}{N}}{C}}CH_3 + H_2O$$

Acetic acid   Methyl amine   Methyl acetamide

We call the salt of an organic acid and an amine an *amide*. It is characterized by the group (the amide group)

$$-\overset{O}{\underset{}{C}}-\overset{H}{\underset{}{N}}-$$

The exposed proton in the amide group can form strong hydrogen bonds.

Polymers

A *polymer* is a giant molecule, formed from a small molecule called a *monomer* (Greek *monos*, single, and *meros*, share or part) that repeats itself thousands of times. Rubber, proteins, and other polymers are found in nature. Recently, scientists have learned how to synthesize polymers in the laboratory and thus create new materials such as polyethylene and nylon out of raw materials like natural gas, coal, and petroleum.

Polyethylene is one compound of the class of polymers called the *vinyls*. The vinyls are all based on the following molecule:

$$\overset{H}{\underset{H}{C}}=\overset{R}{\underset{H}{C}}$$

which repeats itself over and over. R can symbolize any atom or molecule. In the case of polyethylene, it represents the hydrogen atom. If R is a benzene ring, this monomer is styrene and its polymer is polystyrene. If R is a chlorine atom, then the monomers will form polyvinyl chloride (one of the most common of the vinyls). If R is an acetate group, the polymer is polyvinyl acetate.

For a vinyl monomer to form a polymer, its double bond must be broken by heat, by collision with high-energy photons or electrons, or by use of special initiating compounds. When the double bond is broken, the molecule becomes a *radical*: both its carbon atoms need another electron. The radical reacts with molecules of the same type in a chain reaction until the molecule has become thousands of carbon atoms long:

$$-\overset{H}{\underset{H}{C}}-\overset{R}{\underset{H}{C}}- + \overset{H}{\underset{H}{C}}=\overset{R}{\underset{H}{C}} \rightarrow -\overset{H}{\underset{H}{C}}-\overset{R}{\underset{H}{C}}-\overset{H}{\underset{H}{C}}-\overset{R}{\underset{H}{C}}-$$

Radical + Monomer → Polymer radical

The chain terminates only when an impurity is encountered. We generally ignore the impurity when writing the structure of the final polymer molecule:

$$H-\overset{H}{\underset{H}{C}}-\overset{R}{\underset{H}{C}}{\left(\overset{H}{\underset{H}{C}}-\overset{R}{\underset{H}{C}}\right)}_n \overset{H}{\underset{H}{C}}-\overset{R}{\underset{H}{C}}-H$$

The central section repeats itself n times, where n is usually a thousand or more.

When R is a hydrogen atom, the molecule is just a very long paraffin chain, polyethylene. For example, if n is 2, the molecule is octane; if n is 3, it is decane. In commercial polyethylene, n is not the same for all molecules, since the length of the molecule is determined by when an impurity terminates the molecule's growth.

Like the vinyls, many other molecules polymerize in long chains. In synthetic rubber, the forces between adjacent molecules are so small that the molecular chains constantly change shapes and orientations as they undergo thermal vibrations and rotations. If the molecules were not tied together in a process called vulcanization, the rubber molecules would flow as a viscous liquid.

ORGANIC CHEMISTRY

In contrast, Plexiglas (or Lucite) and polystyrene are hard and glasslike. Because these molecules are more bulky and bound together by stronger forces than the rubbers, they have less thermal motion. Even though they are technically very viscous liquids, they look like solids.

The polymers we have discussed so far are <u>addition</u> polymers: they are made by addition of a monomer to a radical. Most of them are unsuitable for making fabrics and other fibrous materials. The van der Waals forces between their chains are too small to make a material which does not melt at high temperatures and in which the chains can be stretched out parallel to each other to form strong, thin filaments. To make such materials we need chains that will hydrogen-bond to each other. Polyvinyl alcohol, whose R group is OH, was tried as a fiber, but it has a fatal drawback—it dissolves in water.

The first satisfactory fiber-forming synthetic polymers were rayon and nylon (actually, rayon is wood pulp, a natural polymer, which has been modified chemically). Nylon was first synthesized in a laboratory in 1929, but years of refinement were required to produce a nylon fiber that would make an acceptable material for clothing.

Nylon is produced in a <u>condensation reaction</u>. One of the most common kinds of nylon, nylon 66, is made in the following reaction:

$$H_2N-(CH_2)_6-NH_2 + HOOC-(CH_2)_4-COOH \rightarrow$$

Hexamethylene diamine + Adipic acid

$$H_2N-(CH_2)_6-NH-CO-(CH_2)_4-COOH + H_2O$$

Notice that water is separated out so the two molecules can combine to double their size. Since the new ends have an acid and an amine group, this new molecule can react with an adipic acid molecule and a diamine molecule over and over until all the reactants are gone. The final polymer, nylon 66, is

$$H{-}\!\left[\!{-}N(H){-}(CH_2)_6{-}N(H){-}C(=\!O){-}(CH_2)_4{-}C(=\!O)\!{-}\right]_n\!\!{-}OH$$

where n in a typical condensation polymer may be a few hundred.

In the nylon polymer, many hydrogen atoms share their electron with nitrogen atoms, leaving a proton hanging on the chain. As in water, the proton forms strong hydrogen bonds with nitrogen and oxygen, so the nylon chains make crystals with high melting points, and thus good fiber-forming materials.

Nylon 66 is one of many condensation polymers. Others are encountered daily as synthetic fibers, films, and molded plastic objects. Polymers are a subject of intense industrial development, and we will see even greater use of these materials in the future. You may wish to read more about them in reference 6.

Proteins

There are two giant molecules that occur in all living systems, and only in living systems. They are <u>proteins</u> and <u>nucleic acids</u>. These two polymers are so characteristic of living things that their detection on Mars or some distant planet would virtually prove the existence of life there. Proteins are the major constituent of enzymes, muscle, tendons, hemoglobin, and many other portions of living things. They carry out the routine functions of various parts of a living body. The nucleic acids tell cells how to form, and carry the information needed to duplicate living organisms. An understanding of both molecules is essential to an understanding of biology.

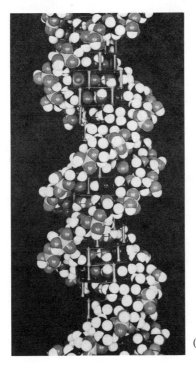

14.8 *The drawing of a protein alpha helix (a) has been simplified by including only atoms in the carbon-nitrogen backbone of the helix. The photograph (b) shows a double helix, two alpha-helixes joined together (see Figure 14.11). Only very small portions of the chains are shown. [Photo (b) Courtesy of the Klinger Scientific Apparatus Corp. and E. Leybold's Nachfolger.]*

Protein molecules are incredibly complex biological polymers composed of amino acid molecules. <u>Amino acids</u> are compounds having the following general structure:

$$\begin{array}{c} H \quad R \quad O \\ | \quad | \quad \| \\ N-C-C \\ | \quad | \quad \backslash \\ H \quad H \quad O-H \end{array}$$

The R groups vary greatly; many have rings and oxygen and nitrogen atoms. Hence each amino acid can be quite a complicated molecule in itself. About twenty different amino acids are important in proteins.

A protein is a chain of amino acids attached end-to-end by the following condensation reaction:

$$\begin{array}{c} H \quad R_1 \quad O \\ | \quad | \quad \| \\ N-C-C \\ | \quad | \quad \backslash \\ H \quad H \quad O-H \end{array} + \begin{array}{c} H \quad R_2 \quad O \\ | \quad | \quad \| \\ N-C-C \\ | \quad | \quad \backslash \\ H \quad H \quad O-H \end{array} \rightarrow$$

$$\begin{array}{c} H \quad R_1 \quad O \quad H \quad R_2 \quad O \\ | \quad | \quad \| \quad | \quad | \quad \| \\ N-C-C-N-C-C \\ | \quad | \quad \quad \quad | \quad \backslash \\ H \quad H \quad \quad H \quad H \quad O-H \end{array} + H_2O$$

By successive additions, chains thousands of units long are built up to form the protein molecule, which is also described as a <u>polypeptide</u> chain. The amazing thing about these protein chains is that *each has a very precise sequence of amino acids*. Since the 20 major amino acids are usually abbreviated by their first three letters, a typical protein chain is written as

Val-His-Leu-Thr-Pro-Glu-Glu-Lys- · · ·

Every protein that has the same function in a living body or organism has exactly the same sequence. A protein having another function will have a different, but equally precise, sequence. These discoveries about the sequential structure of proteins, and the synthesis of some proteins in the laboratory, are among the great triumphs of biochemistry.

Once we find the chemical structure of the molecule, we seek to know how these molecules orient themselves in their biological environment. This difficult task has been achieved by x-ray crystal structure analysis, which reveals that the chain most often exists as a helix coil, the so-called alpha helix structure. In Figure 14.8 we see two representations of this helix. Part (*a*) is only schematic; a more realistic representation is shown in (*b*).

The chain is stabilized in the helical form by hydrogen bonds. These hydrogen bonds can be broken by heating or by the addition of ionic solutions in water. When the hydrogen bonds are broken, the helix is destroyed and the chain contracts. It is not hard to visualize from this how a protein chain in a muscle or tendon can lead the structure of which it is a part to move. We do not yet know the exact chemical reactions by which the protein carries out its assigned functions. However, such problems involving protein behavior are the subject of much research effort, because protein behavior, along with the action of nucleic acids, is fundamental to the understanding of all biological behavior.

Nucleic Acids

Proteins, the workhorses of the living organism, have structures tailored to match their particular function. However, they cannot reproduce themselves. This is the function of the nucleic acids DNA (deoxyribonucleic acid) and RNA (ribonucleic acid). In addition to producing proteins (or more correctly, directing the course of their production), the nucleic acids must also reproduce themselves in order to provide continuity in living things. Like the proteins, the nucleic acids are long-chain molecules composed of an ordered sequence of units. These sequences carry all the genetic information necessary for the reproduction of every feature of a living body. If their sequences become faulty, mutations and birth defects occur.

Unlike the proteins, the DNA nucleic acid is composed of only four basic units. A typical unit is shown in Figure 14.9. RNA is identical to DNA except that —OH groups appear on both of the internal carbon atoms in the sugar section of the molecule. The types of units in DNA differ only in

14.9 *One of the four units from which DNA is composed.*

14.10 *A small section of a nucleic acid chain. The S stands for a sugar while B stands for one of four base groups.*

the structure of their base group, shown on the left in Figure 14.9. In the sugar portion of the molecule the four carbons and the oxygen actually form a ring. The long oxygen bonds do not literally exist that way in the molecule, but it is customary to show the ring nature in this way in order to draw the structure more easily.

To form the DNA molecule from units such as that shown in Figure 14.9, a reaction takes place between the phosphate group on one unit and the OH of the sugar on another unit. The resulting structure is shown schematically in Figure 14.10. Notice that the base units hang off the main chain. These base groups form hydrogen bonds with base groups of another chain. In 1953, James Watson and Francis Crick found that these chains exist in double helixes, and thus proposed a mechanism for the duplication of DNA. (For a very human and informative account of this discovery read the small book *The Double Helix*, reference 1.) The arrangement of two DNA molecules in the double helix is shown in Figure 14.11.

Each base unit in one chain is paired in a hydrogen

14.11 *The double helix. Hanging on the inside portions of the two helical DNA molecules are the base groups. These pair together in a definite pattern and hydrogen-bond the two chains together, as shown by the horizontal dashed lines.*

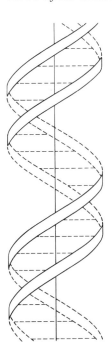

14.12 *A schematic diagram showing that in order for two DNA chains to form a twin-chain helix, their functional groups must exist in precise, complimentary sequence so that they can pair with each other.*

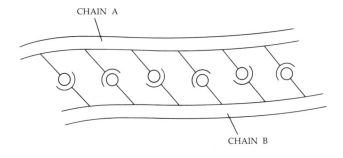

bond to a different but complementary base unit in the other chain. Only if the base groups are ordered in perfectly complementary sequences will two chains pair together. This is illustrated schematically in Figure 14.12. This pairing provides the key for distinguishing one DNA sequence structure from another. However, this matching process is not enough; the DNA molecules must also be able to form new DNA molecules in their exact image so that cells can reproduce and grow.

We presently believe that new DNA molecules are created as the double helix unwinds. Starting from one end, one of the chains has one or two of its end hydrogen bonds broken, presumably by thermal motion, as shown in Figure 14.13a. If a similar unit happens to be floating freely in the surrounding liquid, it is likely to form a hydrogen bond with the detached chain before the original chain can re-form the bond. If the first two chains unwind a bit more, the next portion of each chain will also be replaced by a free unit from the surrounding solution. These newly attached, adjacent groups can now react together to begin a new chain. After several such steps, the situation will be as shown in Figure 14.13b. As the original chains continue to unwind, previously unpolymerized units attach themselves to each one, and these new pairs react with each other to form an exact replica of the original chain. In this way the double helix unwinds and simultaneously duplicates the two DNA molecules that composed it. At the end, there are two double helixes, each identical to the original.

By this duplication of the units on the DNA chain, the information that controls the life processes is passed on. The DNA molecules direct the proper synthesis of other molecules in the living organism. Which molecule it produces depends on the sequence structure within the DNA molecule. The RNA molecules apparently act as messengers in carrying out the construction details, but we are not yet sure how the reproduction process is carried out. You may be interested in pursuing this fascinating subject further in references 2 and 3.

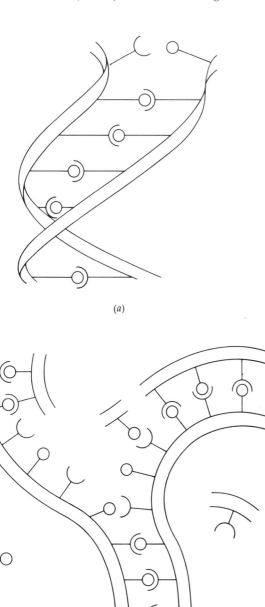

14.13 *As the DNA pair which forms the original double helix splits apart, two identical chains are formed by addition of units from the surrounding solutions.*

Mutations from Cell Damage

The reproduction of all living cells is determined by the information stored in the precise sequences along the DNA chains. Any disruption of the sequence structure may lead to improperly assembled cells within the living organism. Fortunately, a single damaged DNA chain is not sufficient to disrupt the growth process since to duplicate, the chains must cooperate in pairs. In spite of this, genetic accidents do happen.

Any outside influence that causes several DNA molecules to rupture can lead to a faulty sequence structure. The broken chain may re-form with a chain fragment that is different from its original partner. This new DNA sequence structure may not function normally; it may be capable of directing cell growth, but the cells will be abnormal. An organism with faulty DNA molecules and their resultant cells may show visible abnormalities. Many birth defects probably have this origin.

A DNA molecule can be damaged in several ways. If a photon in the x-ray region strikes a chain, it has enough energy to break the chain's bonds. For this reason, exposure to x-rays should be kept as low as possible for those of child-bearing age. However, an x-ray examination of an arm is far less dangerous than an x-ray beam directed near the body's reproductive system.

Although we can control many of the sources of harmful radiation, this danger cannot be completely removed. Particles (chiefly beta- and x-rays) are constantly coursing through our bodies. These come from slightly radioactive materials in our buildings and in the ground beneath our feet, as well as from outer space in the form of cosmic rays. Thus genetic damage can never be avoided completely. Fortunately, these so-called "background sources" of radiation are not intense enough to damage living things seriously.

At present, a great amount of discussion and controversy is centered on the similar dangers associated with nuclear power plants. Although the radiation from such a plant will always be far less than background, any radiation at all may be harmful. We have two alternatives to nuclear power, we can either restrict our consumption of power; or we can build coal and gas power plants which will contribute further to air pollution and the depletion of natural resources. Clearly, there is no perfect solution to the problem.

Photosynthesis and the Energy of Life

All living organisms consume energy in order to sustain life as well as to do useful work. Let us consider the source of our energy and its mode of reaching us. As we shall see, the sun provides us with the energy upon which we exist.

All forms of energy on earth, except nuclear energy, can be traced back to the sun. The sun's heat evaporates water from the earth; this water returns to the earth as rain and as it flows to the sea, furnishes us with energy to operate our huge hydroelectric generating stations. Most of our energy, however, comes from the foods we eat and the coal, oil, and gas we burn. Since coal, oil, and gas are products of plant decay, and since even the animal flesh we eat originates in the plants the animals consume, our major energy supply comes from the plants of the earth. From where does this energy come?

Although we do not know all the details of how a plant grows, we have a general idea of the basic reaction leading to plant growth. Plants take carbon dioxide (CO_2) from the air and use the energy of sunlight to combine it with water. This process forms the sugar, starch, and cellulose (called carbohydrates) that constitute the plant; it also generates oxygen. Written as an equation, this process, known as *photosynthesis*, is

Sunlight + carbon dioxide + water →
$$\text{carbohydrates} + \text{oxygen}.$$

Photosynthesis begins when carbon dioxide from the air diffuses into the leaves of the plant. The water for the reaction is usually carried up from the ground through the roots and stem to the leaves. Sunlight by itself cannot cause these two ingredients to react. However, in the presence of a substance called chlorophyll (which gives the leaves their green color), the leaves absorb sunlight which then supplies the energy for the reaction of the CO_2 and water to form carbohydrates.

Carbohydrates are compounds of carbon, oxygen, and hydrogen. They derive their name from their formula $C_n(H_2O)_x$ where n and x are integers. In spite of this, their structure is not a simple combination of water and carbon atoms.

When we (and other animals) consume plants, our bodies effectively burn the plants by combining their substance with the oxygen we breathe. Thus, our bodies use the energy given to the plants by the sun, and generate CO_2 and water, which are expelled mainly from our lungs. Thus our bodies reverse the reaction of photosynthesis. The carbon dioxide and water are then reused to grow new plants in the energy-giving light of the sun.

Biochemistry is as exciting today as the study of atoms and molecules was in the time of de Broglie's discovery. Rapid advances in this very complex field may shortly enable us to alter the biological building-blocks of life itself. Faced with that possibility, we will soon have to make difficult and serious decisions about the morality and legality of influencing the development of life.

BIBLIOGRAPHY

1. WATSON, J.: *The Double Helix,* Signet Books, New York, 1969. The very human story of the discovery of the double helix, written by the codiscoverer. Watson won the Nobel prize for his work on this subject.

2. STEINER, R. and EDELHOCH, H.: *Molecules and Life,* Van Nostrand, Princeton, N.J., 1965. This small book is written for the intelligent layman.

3. COHEN, D.: *Biological Role of the Nucleic Acids,* American Elsevier, New York, 1965. A concise, lucid exposition of the nucleic acids and their functions in living things.

4. MELVILLE, SIR H.: *Big Molecules,* Macmillan, New York, 1958. Written by a famous British chemist for the nonscientist who is scientifically inclined.

5. ASIMOV, I.: *The World of Carbon,* Abelard-Schuman, New York, 1958. A wide-ranging look at the reactions and uses of this very important element. Written for the nonscientist by a famous popularizer of science.

6. TRELOAR, L.: *Introduction to Polymer Science* (paperback), Springer Verlag, New York, 1970. One of a series of texts written to acquaint British high school students with various areas of science.

SUMMARY

The paraffin series of hydrocarbon molecules are linear chains of carbon atoms with hydrogens attached. Their general formula is C_nH_{2n+2}. They are characterized by the suffix -ane, as in methane, ethane, and propane. These molecules occur in natural gas, gasoline, oils, and waxes; the heavier molecules are found in the more solid materials.

In naming these compounds, straight chains are given the prefix *n-*, for normal. Branched chains can be designated by the prefix iso-. Molecules having the same composition but different structures are called isomers. The position of a substituted atom on a molecule is designated by numbering positions along the molecule starting at its end or some other distinctive point.

Alcohols are characterized by the —OH group and the suffix -ol. Typical examples are methanol and ethanol. Because of their exposed proton, alcohols form hydrogen bonds. The lower alcohols have high dielectric constants. However, the very long-chain alcohols behave more like their analogous paraffin molecules.

Organic acids are characterized by the —COOH group, the carboxyl group. They react with alcohols to separate a water molecule out. The resultant compound is called an ester. For example, when butanol is reacted with acetic acid, a sweet-smelling ester called butyl acetate is formed. Most esters have pleasant odors.

Benzene is typical of a whole series of unsaturated ring molecules containing delocalized double covalent bonds. Because of the rapid shifting of these bonds, some of the electrons of the molecule move more or less freely around each ring.

Amine molecules are characterized by the group —C—NH$_2$. They react with organic acids to form amides. Amides are characterized by the amide group, —CO—NH— . This group, like the —OH group, forms hydrogen bonds since it too has an exposed proton. The hydrogen bond holds protein molecules in distinctive configurations.

When a monomer is polymerized, a long-chain molecule is formed. Polymers range from oils to rubbers to glassy plastics and to synthetic fibers. Protein molecules are also polymers.

Protein and nucleic acids are found in living things. They have a precise, complicated sequence structure along the chain. The function of the molecule within the living organism is determined by its chain sequence structure and its three-dimensional shape. Protein molecules are the workhorses of the organism. The nucleic acids direct the synthesis of proteins and duplicate themselves precisely as pairs of chains (double helixes) unwind.

When a nucleic acid chain is damaged by radiation or any other means, it can lead to the production of defective molecules within the organism. These faulty molecules can result in birth defects and cancer.

Life on earth receives virtually all of its energy from the sun. This energy reaches us as the energy of the photons in the sun's radiation. Plants absorb this energy in the chlorophyll of their leaves. With the sun's energy, plants combine water and CO_2 from the air to form carbohydrates in the process of photosynthesis. When the plants (or the oil, gas, and coal formed from them) are consumed or burned, the products are CO_2 and water. Thus the sun furnishes the energy needed to generate the life cycle.

TOPICAL CHECKLIST

1. What is a paraffin molecule? Give a few examples.
2. Describe the bonding characteristics of the carbon atom in organic molecules.
3. Write the chemical reaction that describes what happens when methane undergoes complete combustion in oxygen.
4. How are the normal paraffins named? Give the structural formula for octane.
5. What is the difference between butane and isobutane? Write the structural formula for 1,3-dichloro normal pentane.
6. What is an alcohol? Distinguish between the properties of methyl and ethyl alcohol.
7. In what respect do alcohols behave like water?
8. What is a typical structure for an organic acid? Discuss the reaction of an organic acid with an alcohol.
9. How can an ester be formed from an alcohol?
10. Give the structural formula for benzene. How does the word "resonance" apply to such a molecule?
11. Give the formula for ammonia. What is an amine? An amide?
12. Discuss the hydrogen-bonding capabilities of the amide groups.
13. What is the structure of a vinyl monomer? How does it react to form a polymer?
14. Discuss the nature of the molecule in a polymer such as polyethylene. What determines whether a polymer is a liquid, rubber, or glass?
15. How are condensation polymers made? What types of reactants are used?
16. What is a protein molecule? Discuss the arrangement of the amino acids within it.
17. Discuss the function of nucleic acids in determining the structure and growth of protein molecules.
18. What is the importance of the double helix?
19. Distinguish between DNA and RNA molecules. What appears to be the function of each?
20. What is the major energy source for life on earth? What is photosynthesis?

QUESTIONS

1. When natural gas is burned in the burner of a gas stove or in a bunsen burner, the flame is cone-shaped with the tip of the cone farthest from the metal burner. The flame is yellow-red around the edges and at the tip of the flame, but blue at its center. If you were to guess from these data alone, which part of the flame would you think was hottest? Why?

2. Since plants take carbon dioxide (CO_2) from the air and give off oxygen in photosynthesis, why isn't the CO_2 eventually depleted from the atmosphere and replaced by oxygen?

3. Describe the changes brought about since 1900 by our present-day ability to synthesize organic compounds.

4. Sodium hydroxide is a white solid. When it is dissolved in water, the water becomes quite hot (we say the reaction is *exothermic*). How can one interpret this heating effect in terms of the strengths of the various bonds involved?

5. When methyl alcohol is added to an equal volume of water, heat is given off. What can you conclude from this?

6. The boiling point of ethane is $-88°C$ and that for methanol is $65°C$. Try to justify this difference.

7. Benzene has a boiling point of $80°C$, and the boiling points of toluene and xylene are $111°C$ and $140°C$ respectively. Why is this variation reasonable?

8. A few polymer chains have special groups along them which dissolve in water to form ions. Some of these same chains can be dissolved in a solvent that does not allow these groups to ionize. In one solvent, the chains stretch out greatly, while in the other they do not. Explain.

9. Discuss the problem of urban air pollution, paying particular attention to the sources of harmful pollutants and feasible ways of minimizing pollution.

10. Some antiperspirants warn on their labels that garments contacting the chemical should be dry-cleaned or washed before ironing. If one ignores this warning, the garment develops a fishy smell. What would you think is the cause of this odor?

11. Explain why exposure to high-energy radiation can cause deformities in a person's offspring.

12. Many cells in the body are being replaced continuously. Sometimes these cells start to grow abnormally and, if their rate of replication is large, a cancer develops. Discuss the possible origins of cancer.

13. The human body is somewhat like a fire in that we "burn up energy." Discuss the nature of the "fire" in terms of the fuel used, its original energy source, and the products of combustion.

14. Many of the dyes formerly used in cloth faded when exposed to strong sunlight for long periods. What causes this? How would you expect the effect to vary with average smog concentration and height above sea level?

Chapter 15

THE EARTH'S ATMOSPHERE AND WEATHER

Early History of the Earth

We believe that the planet earth coalesced from the aggregation of galactic dust, as outlined in Chapter 1. Particles in a wispy cloud broke loose from the streamers that were forming the sun and, drawn by gravitational forces, formed the earth. As the particles fell into the embryonic aggregate, they acquired large kinetic energies. This resulted in high thermal energies for the atoms as they coalesced, but we are uncertain about the exact temperature of the young earth. It is possible that it was a molten sphere at this formative stage of its existence, although many geologists believe that the earth was a solid even then.

Although the galactic dust that formed the earth consisted mainly of hydrogen and helium atoms, these elements constitute only a small fraction of the present-day earth. Since they are the least massive of all atoms, the gravitational forces pulling them to the earth were too small to capture them in the earth's gravitational field. Nearly all the hydrogen found on earth is combined with other atoms in molecules. These more massive molecules are strongly attracted by the earth's gravitational force, and so they and their constituent hydrogen atoms became bound to our planet. However, helium atoms, being nonreactive, were not locked into large molecules and were, for the most part, lost from the earth. Even so, we find helium on earth because helium atoms are being formed continuously during the process of radioactive decay of the very heavy elements. You will recall that an alpha particle is a helium nucleus; hence, each time an alpha particle is emitted by a radioactive atom, a helium atom is formed. Although this process is extremely slow, it has created sizable amounts of helium in the 4.5 billion years of the earth's existence.

Since the atmosphere of the earth is composed primarily of nitrogen (78 percent) and oxygen (21 percent) in diatomic molecular form, it is apparent that molecules of molecular weight 28 and 32 cannot escape from the earth's gravitational field. Therefore, we assume that stable atoms with atomic weight larger than 28 exist on earth in approximately the same concentrations as in the cloud from which the earth was formed. In fact, substantial concentrations of even the lighter atoms are found on earth, but this is partly because of their ability to react with other atoms to form massive molecules. The main constituents of the earth are silicon, iron, oxygen, magnesium, and aluminum, but they are not uniformly distributed throughout the earth.

Most of the atoms that constituted the earth four billion years ago are still here today. However, there were also some radioactive atoms that are no longer found here. All those with short half-lives have long since decayed and disappeared. Today, only those radioactive elements created in the decay series of the very long half-life elements uranium, thorium, and actinium are found in measurable quantities on earth. Even though they are present only in small amounts, they contribute a substantial portion of the heat energy of the entire planet. Perhaps as much as 80 percent of the heat energy that flows into space from the earth's hot core is replenished by the energy produced in radioactive decay processes within the earth.

Indeed, those who support the theory of an originally solid earth attribute its eventual high interior temperatures to the heat generated by the many radioactive species present at earliest times. But whether the source of the energy was gravitational or nuclear, a large fraction of the earth's interior was molten about four billion years ago, and still is today. If we proceed from the picture of a molten earth, we can explain its present stratification in the following way.

Because of heat radiation, the molten earth began to cool; the outer, exposed layer of the earth's sphere cooled most rapidly. Since the material of the earth could not conduct heat from the core to the surface as fast as the surface was losing heat, a temperature difference was set up. The surface became considerably cooler than the interior, and crystals began to form from the molten mass. Among the first were crystalline forms of silicon-oxygen compounds called silicates. These crystals are lighter than molten silicon and iron, and so they rose to the surface of the earth and eventually formed a solid layer upon it. This layer, ranging from 3 to 20 miles thick, is called the _crust_ of the earth. It contains the rocks and other materials we usually think of as constituting the earth. However, this thin crust covers a portion of the earth quite different from the one we know. Indeed, a major portion of the earth below the crust is still molten.

Heat does not readily flow through the earth's crust, and so it acts like an insulating blanket over the molten core. This is why the heat from the sun has little influence on the temperature in deep caves and mines. Temperatures in such cavities are nearly constant all year round; even at these relatively short distances below the surface, summer and winter have little effect.

Because of the insulation of the crust, the surface of the earth and the atmosphere above it cooled to temperatures below the boiling point of water. Then the steam in the atmosphere condensed, and torrential rains fell on the earth as much of the water now in the oceans changed from gas to liquid. These huge water-vapor clouds and torrential rains almost certainly produced violent storms on the earth. The attendant lightning may well have been a primary factor in originating life-sustaining cells at that time.

At present, we have no generally accepted theory for the origin of the simplest forms of living things on earth. However, we are certain that life as we know it could not begin until an atmosphere developed in which elements could form compounds with oxygen rather than hydrogen. Once that atmosphere was created, living organisms somehow appeared on earth and began forming the many features of the earth's surface that we associate with life and its processes of decay.

The Atmospheric Layers

The atmosphere surrounding the earth is mostly composed of nitrogen and oxygen, though traces of other molecules are also present. The average composition of the air is shown in Table 15.1. The

Table 15.1 The Earth's Atmosphere

Gas	Symbol	Concentration (percent)
Nitrogen	N_2	78
Oxygen	O_2	21
Argon	A	0.9
Carbon dioxide	CO_2	0.03
Neon	Ne	0.002
Helium	He	0.0005
Methane	CH_4	0.0002

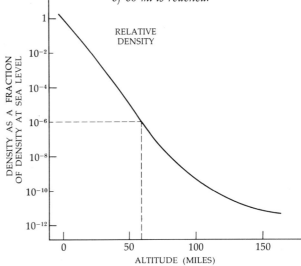

15.1 *Variation of air density with altitude. Notice the density has dropped to one millionth of its value when an altitude of 60 mi is reached.*

water vapor content is not shown, since this varies widely with climatic conditions. You know that the oxygen in the air is necessary to life on earth. The importance of the nitrogen was dramatically illustrated by a tragic accident early in our space program. At that time, pure oxygen was used as the atmosphere in the space capsules. During a preflight test, a spark ignited the oxygen and killed the astronauts in the capsule. Without the nitrogen in the air to dilute the oxygen, life as we know it could not exist.

The percentages in Table 15.1 do not apply to all sections of the atmosphere. Near large cities, the air is laden with contaminants. Moreover, as we go higher above the earth, the relative proportions of the various molecules change. As we showed in Example 5.4, the average nitrogen molecule at the surface of the earth has enough kinetic energy to ascend to a height of about 9 miles. Lighter molecules such as helium and methane can rise higher. Heavier molecules such as carbon dioxide are less able to rise to the top of the atmosphere. Hence, at high altitudes, the atmosphere becomes relatively rich in light molecules.

Since there is always a wide distribution of energies among the individual molecules of a gas (many have energies less than the average while many have energies far larger), the atmosphere has no defined top. It thins out gradually. For example, in Laramie, Wyoming, which is about 1.3 mi above sea level, the air is already thinned to about 0.8 of its sea-level value. This thinning, or decrease in *density* of the air with altitude, is shown in Figure 15.1. Notice that the density scale is in powers of 10, so the variation is much larger than you see at first glance. Thus the air density drops to one millionth of its sea-level density at an altitude of 60 mi. At 100 mi above the earth's surface, the air is as thin as the vacuum in a good vacuum tube. This is why satellites circling the earth at tremendous speeds are not slowed appreciably by air friction if their orbit is at least this high.

The pull of gravity has another important effect on the molecules of the atmosphere. You will recall that the absolute temperature of a gas is a measure of the average kinetic energy of one of its molecules. But we know that molecules at high altitudes have less kinetic energy than those at sea level because some of it changes to potential energy as they ascend. Hence we predict that the temperature of the air will decrease with height above the earth. This prediction will be confirmed by anyone who has

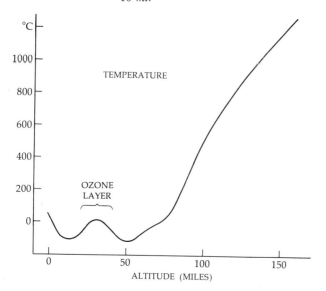

15.2 Variation of air temperature above the earth. Notice the temperature falls rapidly as we ascend through the first 10 mi.

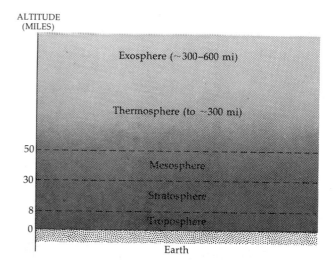

15.3 Layers of the atmosphere. The ionosphere extends from the upper regions of the mesosphere almost to the top of the thermosphere.

flown in unheated aircraft. The measured variation of air temperature with altitude is shown in Figure 15.2. Notice that the temperature drops to about −40°C in the first 10 mi.

However, the temperature increases dramatically with altitude above about 50 mi. At 150 mi above the earth, the temperature of the extremely thin atmosphere is over 1000°C. This high temperature results from the bombardment of the upper atmosphere by the direct, unfiltered rays of the sun. We know that a large quantity of high-energy light radiation reaches us from the sun. This radiation is in the form of ultraviolet and x-rays, since the photon energy increases with decreasing wavelength. These photon energies are large enough to tear apart the molecules which absorb them. When a photon is absorbed, it imparts its energy to the absorbing molecule. Thus, the molecules at the upper edge of the atmosphere have exceedingly high kinetic energies; this means, from our definition of temperature, that the upper atmosphere is very hot.

The high temperature we assign to the upper edge of the atmosphere is somewhat deceptive. Since the density of the air is so low at these altitudes, the thermal properties of the air are those of a vacuum, and ordinary methods for detecting air temperature do not work. For example, an ordinary thermometer requires hours to reach the temperature of its surroundings under conditions such as these. Space travelers are thus more concerned with the radiation striking them from the sun than they are with the temperature of the few molecules existing at these high altitudes.

Experiments have shown that we can divide the atmosphere into five layers; these are shown in Figure 15.3. Let us examine each of these layers in detail.

THE EXOSPHERE

As you might expect from the prefix "exo-", the exosphere is the outermost region in the atmosphere, starting about 300 mi above the earth. At higher altitudes it gradually blends with interplanetary space. About 150 mi high in the exosphere (or about 450 mi above the earth), there are still 10^6 molecules in each cubic centimeter. But this is 10^{-13} times fewer than the number in a like volume at sea level. In fact, most man-made vacuums have thousands of times more molecules per cubic centimeter than this region at the edge of space.

Closer to the earth, the density of the atmosphere gradually becomes greater. The lower layers of the exosphere are not distinct either; the exosphere blends into the thermosphere.

THE THERMOSPHERE

This layer, extending from about 50 to 300 mi above the earth, is extremely hot (hence the prefix "thermo-," from the Greek *thermos,* hot). As we saw in Figure 15.2, the sun's rays heat this region of the atmosphere strongly. In addition, the high-energy photons from the sun tear many of the molecules of the thermosphere into ions. This is extremely important for radio-wave transmission, as we shall see.

THE MESOSPHERE AND STRATOSPHERE

These adjacent layers cover the region from about 8 to 50 mi in altitude. As we see from Figures 15.1 and 15.2, the air here is very thin and temperatures are below freezing for the most part. The air is ionized near the upper boundary, but only slightly ionized at the bottom. At these altitudes there is still no appreciable amount of water vapor and therefore no clouds, winds, or storms. Occasionally a high-flying aircraft leaves a white trail of water and ice droplets (called a *contrail,* from *condensation trail*) in the lowest parts of the stratosphere and below. These form from the fuel combustion products of its engines. You can estimate the stillness of the air in this region from the way in which the trails persist.

In the region 20 to 40 mi above the earth, ozone, O_3, is found. This molecule forms when an isolated oxygen atom combines with an oxygen molecule. The free oxygen is formed when a high-energy photon collides with an oxygen molecule. Hence, ozone exists in appreciable quantities only at altitudes where the sun's rays are still strong. Moreover, formation of the ozone molecule is a complicated reaction that also involves a third molecule, an oxide of nitrogen. Thus ozone production depends on a very delicate balance between the molecules in this region.

The ozone layer is vital to life on earth, because it effectively absorbs ultraviolet light from the sun and vastly decreases the high-energy radiation striking us at the earth's surface. Since even the small quantities of ultraviolet light that do reach us can burn or tan our skins, we can appreciate how the absence of ozone would affect our lives. Ozone's ability to absorb ultraviolet radiation is indicated by the sudden rise in temperature that is characteristic of the mesosphere (see Figure 15.2).

Some scientists believe that life as we know it will be seriously endangered if aircraft fly unregulated in the stratosphere. They fear that the combustion products left in the contrails of these supersonic jets could upset the delicate balance of the ozone layer, and thus weaken or even destroy it. At present, we do not have enough solid scientific evidence to prove or disprove this theory.

THE TROPOSPHERE

This is the region of the atmosphere with which we are most familiar, since it is the one in which we live. Its thickness varies from about 5 mi at the poles to 10 mi at the equator. Of all the atmosphere's layers, the troposphere is most easily distinguished, because its upper boundary is marked by the maximum height of the cloud cover of the earth. More than half the mass of the earth's atmosphere is found in the bottom $3\frac{1}{2}$ mi of the troposphere. It contains large quantities of water vapor, and so it is here that we find the clouds, together with the storms involving them.

The Ionosphere

As we have seen, radiation from the sun tears apart molecules and produces many ions in the thermosphere. This layer of ions, called the *ionosphere*,

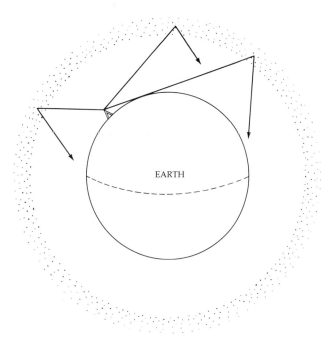

15.4 *The ionosphere layer confines long-wavelength electromagnetic waves to the earth. This makes it possible to detect radio waves from distant stations.*

starts 40 to 50 mi above the earth and ranges almost to the top of the thermosphere. Its structure and height vary according to the time of day or night; at night, ions in the lower, more dense, portion of the layer recombine with the free electrons and so the bottom of the layer disappears. Recombination occurs less frequently in the much less dense upper regions, and so the upper portion of the ionosphere remains nearly unchanged during the course of a single night. In effect, the bottom of the ionized layer is raised as the night progresses.

The ionized layer affects radio transmission as shown in Figure 15.4. Most of the electromagnetic waves sent out by a radio station are reflected from the ionosphere back to earth, so that the waves circle the earth instead of escaping into space. Consequently, we can detect the waves from radio stations half way around the world. However, the ionosphere's reflecting ability depends on the wavelength of the electromagnetic waves; the layer acts as a mirror only if the ions are much less than a wavelength apart. The ions are close enough together to reflect efficiently any radio waves with λ greater than about 100 m (meters). However, TV and FM frequency waves, which have λ shorter than this, are reflected very poorly by the ionized layer. As a result, TV and FM reception is limited by the curvature of the earth to a range of about 50 mi.

Until recently, worldwide radio reception and intercontinental communications depended on the ionosphere. In the past, radio communications have been disrupted, often at critical times, by huge outbursts on the sun. These events, which are easily visible on earth, are called solar flares; they send out large quantities of charged particles which reach the earth about a day later. When they strike the earth's atmosphere, they disrupt the mirror-like properties of the ionosphere and therefore, the transmission of radio waves. With the advent of synchronous satellites (satellites that rotate at the same rate as the earth so their position above the earth remains fixed), we no longer rely on the ionosphere for intercontinental beaming of radio waves. Instead, transmissions are reflected from these communications satellites hovering above the earth. These satellites also reflect TV waves, and so we now have direct TV contact between continents.

Large-Scale Air Motion

Air motion on the earth results largely from the familiar fact that "warm air rises." Since a gas expands when heated, hot air is less dense than the cold air from which it came. As a result, it is lifted upward by the denser surrounding cold air as a piece of wood is pushed upward by the denser water in which it is submerged. Thus, the hot air at the equator rises while the cold air in the arctic regions descends. From this we would expect the air above the earth to flow in the pattern shown in Figure 15.5a. However, other factors, too complicated for us to pursue here, cause the circulation to be more like that shown in Figure 15.5b.

In addition to the circulation of warm air, air movement depends on the rotation of the earth, an effect called the <u>Coriolis effect</u>. To see how it comes

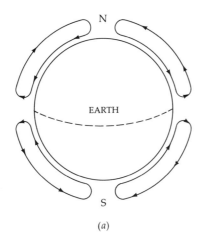
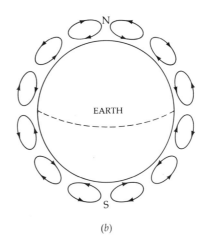

15.5 *Polar cooling and equatorial heating alone would cause the air circulation pattern shown in (a). In practice, (b) is found to be more realistic.*

about, suppose a photographic satellite is fired parallel to the earth's surface from a point high above the North Pole, as shown in Figure 15.6. During its flight, the earth rotates beneath it. Thus, if the satellite is originally aimed at London, by the time it travels the distance from the Pole to London, the rotation of the earth has moved London to the east relative to the satellite's path. Instead of photographing London, the satellite will photograph the middle of the Atlantic Ocean. To anyone watching the process from the earth, the satellite will appear to move westward.

Like the satellite, air masses moving south from the North Pole appear to move westward because of the earth's rotation beneath them. Similar reasoning will show you that air masses moving northward from the South Pole also will seem to drift westward. However, an air mass (or a satellite) starting north from the equator will share the earth's eastward velocity at that point. As the air moves northward, the velocity of the earth rotating under it decreases and finally reaches zero at the pole. But the air mass retains its initial eastward velocity, and therefore it drifts to the east as it moves northward.

15.6 *The Coriolis effect illustrated.*

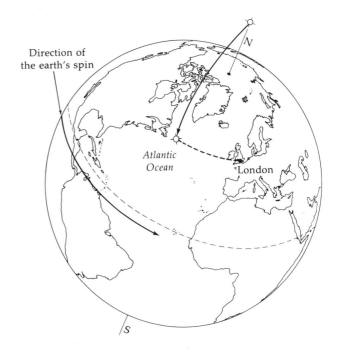

THE ATMOSPHERE AND WEATHER 333

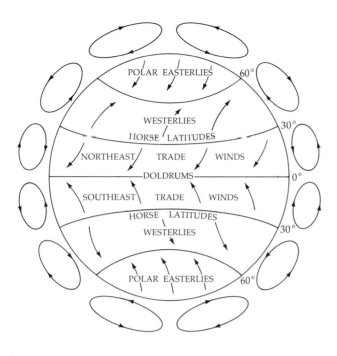

15.7 *Prevailing wind directions on the earth.*

We can summarize the Coriolis effect in the following way. If an air mass is moving either away from or toward the equator, and if we look in the direction the air mass is moving, then the Coriolis effect will deflect the air mass to our right if we are in the Northern Hemisphere; the deflection will be to our left if we are in the Southern Hemisphere.

As a result of the earth's rotation, the prevailing wind patterns on the earth are those shown in Figure 15.7. Most of the United States is in the region of winds coming from the west, or westerlies. Northern Florida lies at 30° latitude while southern Alaska lies at 60° latitude. Thus, the prevailing winds in the United States are from west to east; storms and other weather phenomena usually cross the continent in that direction. For your interest, the common names of other prevailing winds are also shown in Figure 15.7.

Of course, the winds do not always flow in the directions shown in Figure 15.7. Other factors cause localized wind patterns to override the prevailing winds. For example, frontal movements and their associated high- and low-pressure regions seriously influence air movement. The topography of the earth also has an effect, since the earth's surface heats unevenly in the sun's rays. For example, water reflects sunlight differently from tree-covered land, so the two have different temperatures even when the incident sun energy is the same for each. Variations in altitude, such as mountain ranges, also affect the flow of air masses along the earth. In spite of these complications, however, the *overall* wind patterns are those shown in Figure 15.7.

The Earth-Air Boundary

Only a thin layer on the surface of the earth is occupied and extensively utilized by mankind. Our day-to-day existence takes place within a narrow region extending only a fraction of a mile above and below the earth's surface. The atmosphere above this thin layer keeps most solar radiation from penetrating to the region in which we live. High-energy ultraviolet radiation is largely absorbed in the mesosphere, while clouds and dust greatly diminish the heat radiation from the sun. Consequently, even on a clear day, most of the radiation energy reaching the earth's surface from outer space is in the wavelength range near the visible, a range which the atmosphere does not strongly absorb.

The radiation that reaches the earth is largely absorbed by the surface. Much of it becomes heat energy which is then reradiated at longer wavelengths. But these longer wavelengths are in the infrared and are strongly absorbed by the water vapor and carbon dioxide in the atmosphere. Hence, much of this energy remains close to the earth and keeps it warm. In a sense, the atmosphere acts like a blanket around the earth retaining its warmth. This is called the *greenhouse effect* because a similar phenomenon occurs in a greenhouse. After the sun's visible rays come through the glass, they are absorbed and partly reradiated as heat waves. However, glass does not transmit infrared radiation, so the radiant energy is trapped in the greenhouse. In the case of the earth, the CO_2 and H_2O vapor blanket serves the same function as the glass. The greenhouse effect causes cloudless nights to be colder than when the sky is heavily overcast. Can you explain why?

As the earth's surface and the lower atmosphere are warmed by the greenhouse effect, some of the heat is radiated up to the cooler portion of the atmosphere and eventually returns to space. On a yearly average, as much heat is returned to space as has come from the sun; without this balance, the surface of the earth would heat up continuously. During recent years we have begun to realize that the greenhouse effect and other similar effects might bring disaster to the world. Because we burn such huge amounts of fuels containing carbon, we are producing nearly as much CO_2 each year as there is altogether in the atmosphere. The possibility exists that the thickening CO_2 blanket could become too effective, destroying the heat balance and making the surface of the earth noticeably warmer.

Depending on the extent of the change, the effect on man and other forms of life could be serious or even fatal. Unfortunately, it is difficult to test this frightening hypothesis, since the behavior of CO_2—which is consumed by vegetation as well as produced in a multitude of ways—is too complex to analyze easily. We do not yet know how serious the danger is, but it is not as great as the most pessimistic estimates.

Dust, smoke, and other particles can cause similar disturbances in the atmosphere. These materials are, of course, undesirable from a health standpoint. In addition, they hold global significance. For example, volcanic dust thrown into the atmosphere by the volcano Krakatoa, which erupted west of Java in August, 1883, absorbed so much sunlight that corn growth was stunted in the United States in 1884. Equally potent on a regional scale is the smoke released by the belching chimneys of any large industrial city. In cities without stringent smoke controls, the local climate changes measurably.

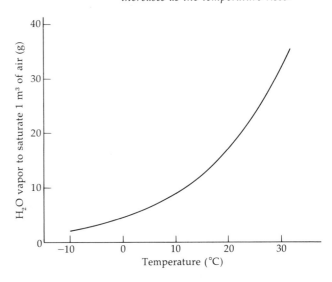

15.8 *The ability of air to hold water vapor increases as the temperature rises.*

Weather and the Troposphere

Weather phenomena are almost always accompanied by characteristic states of the water vapor in the air. These states are linked with the fact that air can hold only a certain amount of water. This limiting concentration of water vapor depends greatly on temperature; hot air can hold much more water per unit volume than cold air. We show this variation in Figure 15.8, where the maximum number of grams of water that one cubic meter (1 m³) of air will hold is plotted against the temperature of the air. For example, at 20°C (64°F) 1 m³ of air can hold almost 18 g of water vapor. When air contains its maximum possible amount of water vapor, we say the air is <u>saturated</u>.

Saturated air is unable to retain its water if it is cooled. For example, in Figure 15.8 we see that saturated air at 10°C can hold only about half as much water as air at 20°C. Therefore, if saturated 20°C air is cooled to 10°C, about half the water vapor in it must drop out by forming dew, fog, rain droplets, or snow. From this, you should be able to explain why fog forms at night after a warm, extremely humid day. We call the temperature at which the moisture begins to fall from the air the <u>dew point</u> of the air.

15.9 · *Notice the difference in the ways the cold and warm fronts progress. Can you explain this difference? Why does rain fall along the fronts?*

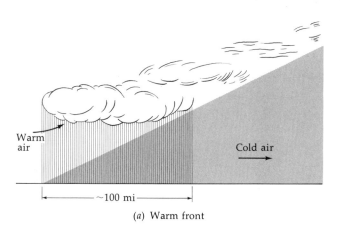
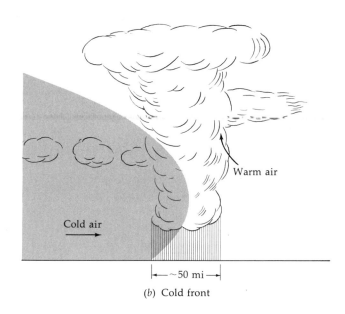

(a) Warm front

(b) Cold front

Relative humidity is a quantitative way of expressing how humid the air is (or how much water it holds). The relative humidity is the ratio of the amount of water in the air to the maximum amount of water that the air can hold at that particular temperature. For example, at 20°C the air can hold about 18 grams of water per cubic meter of air. If the air on a particular day has a temperature of 20°C and has only 6 g water per cubic meter, then it contains only 6/18 or 1/3 as much water vapor as it can hold. We therefore say that the relative humidity is 0.33 or 33 percent. Thus, if the relative humidity is 70 percent, then the air is holding 70 percent as much water as it is capable of holding at that temperature.

Our bodily comfort is quite dependent upon humidity. Dry air (low-humidity air) can take up considerable quantities of water, so evaporation takes place readily when the humidity is low. Since evaporation leads to cooling (as we saw in Chapter 5), our bodies are cooled by evaporation when the humidity is low. When the humidity is high, however, evaporation of more water into the air is not possible; our bodies cannot be cooled by evaporation of sweat, and so we "feel" the heat much more.

The weather is largely the result of the movement of air masses across the surface of the earth. These long-range motions often cause collisions between huge air masses having different temperatures. Two typical situations are shown in Figure 15.9. In (a), a warm air mass is encroaching upon a region of cold air. Since the warm air is expanded and therefore less dense than the cold, it tends to rise above the cold air mass as it displaces the cold air. At the surface at which the two masses touch, the more heavily water-laden warm air is often cooled below its dew point. Thus, the *warm front* (the region of contact of the two masses) is frequently accompanied by heavy clouds and rain.

The opposite situation, the encroachment of a cold air mass upon a mass of warm air, is called a *cold front* and is shown in Figure 15.9b. In this case, the cold air tends to raise the less dense warm air as it pushes the warm air mass away. The rising warm air is cooled as it flows to higher altitudes. This, along with the presence of the cold air mass, cools the air to its dew point, and so precipitation and clouds often accompany a cold front as well.

A moving front frequently brings a cloud line and storms. Texans are accustomed to seeing a "blue Norther" approach from the north; it is a distinct

15.10 In (a), a squall line is seen from a satellite looking down upon it. The appearance of the same line as seen from the earth's surface is shown in (b). (Courtesy of NOAA.)

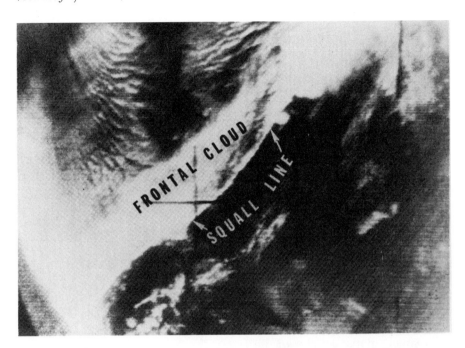

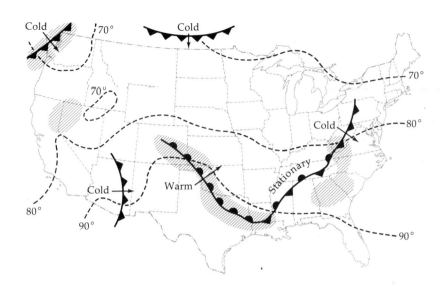

15.11 *A simplified weather map. Note how the cold and warm fronts and their direction of motion are symbolized. Regions of precipitation are shown by shading. Also shown is a stationary front. Since its motion is zero, only the positions of the cold and warm masses can be shown. The dashed lines connect points of equal temperature.*

wall of clouds which moves slowly across the prairie and carries rain and cold weather behind it. A less dramatic example is the <u>squall line</u> in Figure 15.10*a* and *b*.

The frontal patterns and other weather features are represented by weather maps as shown in Figure 15.11. A warm front is represented by the symbol ⌒⌒⌒⌒, where the semicircles point in the direction of motion. A cold front is symbolized by ▲▲▲▲; the triangles point in the direction of its motion. The shading on the map shows areas of precipitation. The dashed lines indicate points of equal temperature and are called <u>isotherms</u>. For example, the isotherm running from Delaware to southern California indicates that the region along it has a temperature of 80°F. Similar lines (not shown) are used to show points of equal barometric pressure and the positions of highs and lows, which we will examine in the next section. A more complete weather map is shown in Figure 15.12.

Localized Air Motions

The motion of fronts generally follows the circulation paths pictured earlier in Figure 15.7. Like most weather phenomena, fronts usually cross the continental United States from west to east. However, this is not correct in detail because of various localized weather effects which disrupt this general pattern.

CYCLONE CIRCULATION

The meteorologist uses the word <u>cyclone</u> in a different way than the layman. To the specialist, a cyclone is any spiraling air current, whether or not it is violent. As we shall see, most cyclone motions are large circulation effects extending over hundreds of miles; rarely are they accompanied by the tornadoes that laymen refer to as cyclones. Let us now see how a large-scale spiral circulation effect comes about.

A cyclone-type motion often originates along the boundary between a cold and a warm air mass by the mechanism shown in Figure 15.13. In (*a*), we see a cold air mass moving to the left adjacent to a warm mass moving toward the right. Friction between the two masses causes the boundary to distort as in (*b*) and (*c*). We say that a wave has developed at the boundary. Eventually, the distortion causes the boundary to break, and then we find the beginnings of a localized circulation as in (*c*).

15.12   *A typical Weather Service map.*

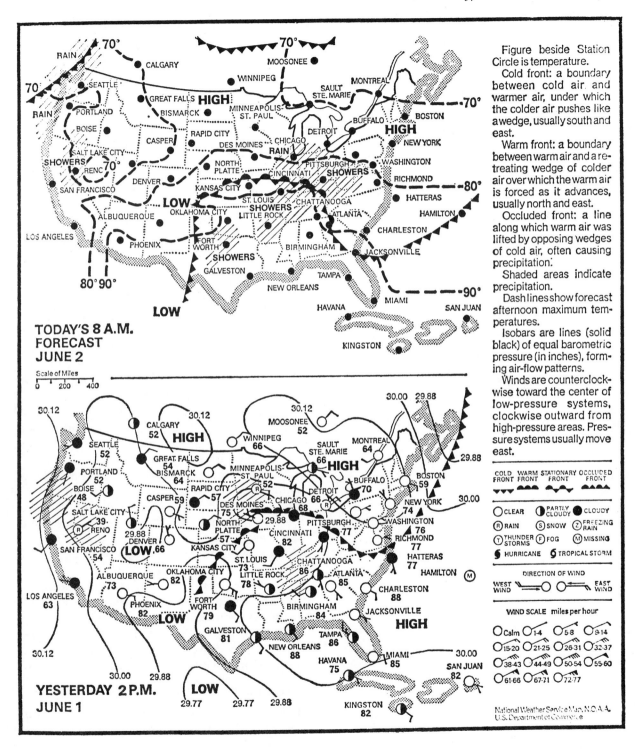

THE ATMOSPHERE AND WEATHER

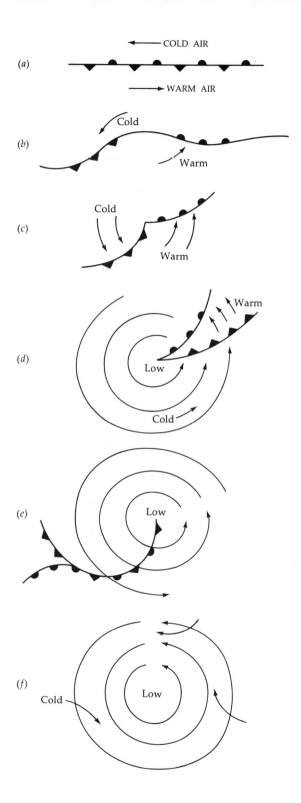

15.13 The process of development for a cyclone motion in the Northern Hemisphere as seen by a viewer looking down upon the earth. In (f), cold air moving into the low is deflected to the right by the Coriolis effect.

In the particular situation shown, the cold front moves more swiftly than the warm front. Eventually it overtakes the warm front, as shown in (d), (e), and (f). During this process, the encroaching cold air pushes the warm air aloft. This provides more room for the cold air and so more air rushes into the spiral from the outer regions. The combined effects give rise to the flow pattern shown in (f). As we saw in our discussion of air movement on the rotating earth, air moving from one region to another in a rotating system is subject to the Coriolis effect. Moreover, any rotating mass has a tendency to resist being drawn into a circular path.

In the situation shown, the various forces involved will create a counterclockwise motion in spirals above the northern hemisphere. Another result of all of these forces is to cause the center of the spiral to be a low-pressure area, or <u>low</u>. In a low, we find warm air rising upward and cold air rushing in to replace it. Both the rising warm air and inrushing cold air behind it lead to stormy conditions at the center. For this reason, stormy weather is usually associated with low-pressure areas.

A similar, but clockwise, rotation (called an <u>anticyclone</u>) can also develop. At its center is a high-pressure area called a <u>high</u>. Highs are characterized by good weather, since the roles of warm and cool air are reversed. Cool dry air settles to the earth at the center of a high and spreads out across the earth from the central region. Since it warms during the process, there is no tendency for clouds and rain to form. Cyclones and anticyclones also occur in the southern hemisphere, of course, but there, the directions of rotation for the highs and lows are reversed, because the Coriolis effect deflects the moving air masses in the opposite direction.

THE HURRICANE (OR TROPICAL CYCLONE)

Every summer, hurricanes move north toward the continental United States from the tropics. They are

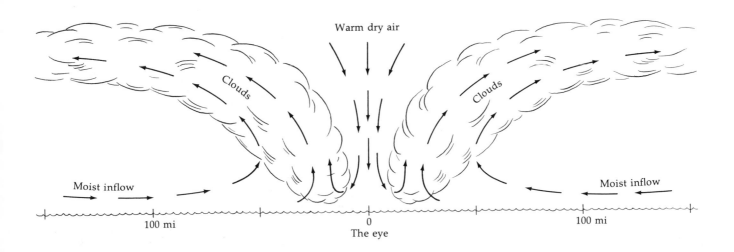

15.14 *The eye of the hurricane is a region of clear, relatively still air.*

typical of the violent storms that move both north and south out of the equatorial zone throughout the world. In the Pacific, they are usually called *typhoons*, whereas in the Indian Ocean they are called cyclones. They are cyclonic motions of high winds, often extending over hundreds of miles. Near the center of a hurricane, wind speeds of close to 200 mi/hr have been recorded.

Invariably, the hurricane originates over the equatorial oceans as the sun heats the moisture-laden air. The air rises as it warms, and the surrounding air begins a cyclonic motion as it rushes in from the sides to fill the void. However, as the rate of rotation of the air mass increases, the lack of sufficient centripetal force to hold the air in this tight circular motion causes a partial vacuum at the center of the rotation. Hence the center, or *eye*, of the hurricane is a region of unusually low pressure—perhaps 10 percent lower than normal. This low pressure causes warm air to flow from the upper atmosphere down into the eye of the hurricane, as shown in Figure 15.14. Notice that for about 10 miles around from the center of the eye, the sky is clear and the winds are low. Just outside the eye, however, the wind speeds may exceed 100 mi/hr. A satellite view of a hurricane is shown in Figure 15.15.

15.15 *The characteristic spiral cloud formation of a hurricane as seen from a manned spaceship. (Courtesy of NOAA.)*

THE ATMOSPHERE AND WEATHER

15.16 *A typical tornado funnel. (Courtesy of NOAA.)*

The hurricane center travels slowly. It is carried along with the prevailing long-range air currents, moving with a speed typical of a front or other weather feature. This is fortunate, for it allows those in its path several days to prepare for its arrival. However, a hurricane's motion is as unpredictable as other weather phenomena, and unexpected changes in path are not uncommon.

THE TORNADO

Like a hurricane, a tornado consists of a high-velocity cyclonic motion about a central eye. However, it is much smaller in size; as you can see from the one shown in Figure 15.16, a tornado's central funnel is only a fraction of a mile in diameter. Exactly how a tornado forms is still a matter of conjecture. The presence of intersecting cold and warm air masses and a swiftly flowing stream of air at high altitudes (a _jetstream_) appear to be necessary. Even though we cannot explain its exact mechanism, we do know the conditions under which a tornado is likely to form. The weather bureau posts tornado warnings when these conditions exist.

Near the funnel of the tornado, wind velocities often exceed 200 mi/hr. In addition, the area in the funnel has a pressure about 10 percent lower than normal. Although this does not seem very large at first glance, it means that the unbalanced force from inside on a window of a tightly closed house will be of the order of a ton. As a result, during the few seconds the funnel takes to pass over a house, the excess pressure inside the house often blows out all the windows and usually the roof and walls as well. Fortunately, the tornado's path is narrow and it usually touches the ground only momentarily, but the destruction within that path is usually total.

A tornado passing over a body of water raises a tower of water within the funnel, and is called a _waterspout_. A common sight in many areas is a _dust devil_, a harmless miniature tornado, which is a swirling pillar of dust that whisks erratically across the terrain.

Seasonal Changes in Weather

Some weather changes result from the yearly motion of the earth around the sun and the orientation of the earth's rotation axis. To refresh your memory, Figure 2.6 is reproduced here as Figure 15.17.

As we see in Figure 15.17, the North Pole of the earth is in continual shadow when the earth reaches the winter solstice (December 21) in its orbit around the sun. At that time of the year, the Northern Hemisphere receives only a small share of the sun's rays, and so winter prevails there while the summer sun warms the Southern Hemisphere. Six months later, the situation is reversed.

A seasonal weather change that does not occur in the continental United States is the _monsoon_, an air-flow pattern most pronounced in Asia. During the summer, the air above the oceans in southeastern Asia is cooler than the air above the land. As a result, the less dense warm air over the land is displaced upward by the cool air moving in from the ocean. The moist sea air is also lifted as it flows inland over the mountainous terrain. But the rising air grows cooler as its kinetic energy is changed to potential energy. As a result, its temperature is lowered

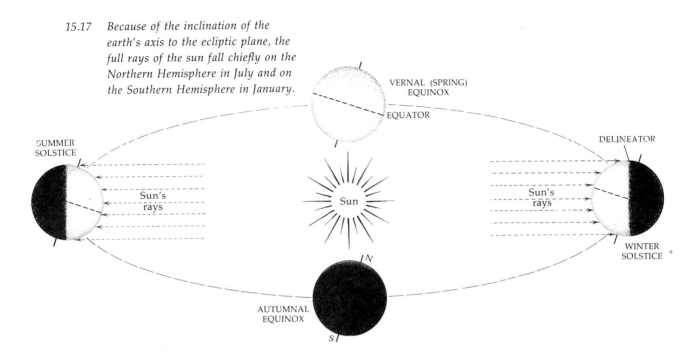

15.17 Because of the inclination of the earth's axis to the ecliptic plane, the full rays of the sun fall chiefly on the Northern Hemisphere in July and on the Southern Hemisphere in January.

below the dew point and precipitation results. During the summer, when this flow pattern occurs, the region is subjected to the rainy season called the <u>summer monsoon</u>. Less frequently mentioned is the <u>winter monsoon</u>, a dry season characterized by a flow of cold, dry air from the mountains to the sea.

The Effect of the Oceans on Climate

You probably know that a large body of water moderates the temperatures of the land around it. In climates where temperatures drop below freezing in winter, the oceans and large lakes moderate temperatures particularly effectively, since their temperature does not drop below freezing until all the water has changed to ice. At higher temperatures, circulation within the water brings subsurface water to the top. Hence the surface temperature lags behind the air temperature, because the surface water tends to remain at the temperature of the whole lake, or ocean. Thus, the presence of a large body of water tends to maintain a constant, moderate temperature in nearby regions, whether the air is warm or cold.

There is another important way in which a body of water influences the temperature and humidity of the air passing over its surface. Since heat energy is needed to evaporate water, the wind blowing over a lake or ocean is cooled as it participates in the evaporation process. At the same time, it becomes laden with moisture; this is why ocean breezes are cool and moist. But since the land cools more rapidly than the ocean, fog banks often form at night as the warmer ocean air is cooled at the shoreline. A similar process creates the layers of fog that frequently nestle over lakes and rivers and in valleys. In these cases, however, the dense, cool air settles and displaces much of the warmer air from these lowlying areas, which is why fog usually forms first in valleys.

Perhaps the most important oceanic influence on climate results from the flow of the ocean's water in currents which carry it for thousands of miles. These currents are similar to the global air currents. If the continents did not present barriers, the ocean flow pattern would probably be even more like the flow of air above the earth. In fact, much of the energy required for the flow of ocean currents comes from friction forces applied to the ocean's surface by the winds. These long-range ocean currents are shown in Figure 15.18.

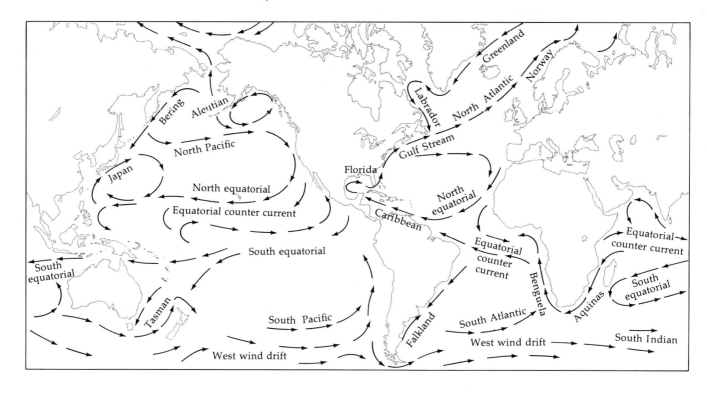

15.18 Persistent ocean currents of the world.

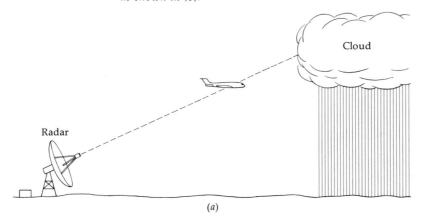
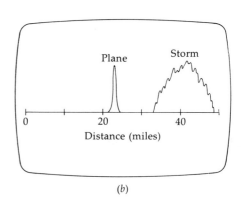

15.19 The radar pulses are reflected from both the plane and the rainstorm, causing echoes that are collected by the bowl-shaped electronic mirror and presented on the radar screen. Each reflecting object has its own characteristic echo, as shown in (b).

15.20 *A radar scan of a hurricane about 50 miles away from the station. (Courtesy of NOAA.)*

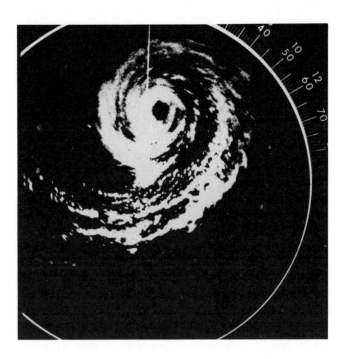

One of the most important of these currents is the Gulf Stream, which carries warm water from the Caribbean up the Atlantic coast of the United States. From there it crosses the Atlantic, flows past England and Scandanavia, and then returns to America by way of Greenland. This fortunate northward flow of warm water greatly moderates the climate along the shores of these otherwise much colder places, but cools New England as it returns from the arctic. Similar currents are found in the other great oceans; their circulation patterns depend markedly upon the general region in which they occur.

Weather Prediction

How often have you listened to a weather forecast and found that the weather was exactly the opposite of what was predicted? We still do not know enough about the weather to predict it precisely. Meteorologists can sometimes make fairly accurate predictions for a day or two ahead, but not for longer periods.

To make short-term predictions, meteorologists look to past experience. Knowing that a particular combination of fronts, clouds, winds, and other conditions have led before to a certain weather pattern, they predict the pattern again when they observe those conditions. They select and store data from the past in computers. Today, weather satellites provide a tremendous amount of information about the cloud patterns of the earth, but even with the help of large computers, meteorologists can analyze and use only a fraction of this information. There is simply too much information to correlate in the limited time available.

One limited but useful technique for locating rain, clouds, and storms is radar. As shown in Figure 15.19a, a pulse of radar waves (very short-wavelength radio waves with $\lambda \approx 10$ cm) is sent out from the radar station. This pulse is reflected strongly by large metal objects, such as an airplane, and by dense clouds of water droplets. By observing the time t required for the reflected pulse to return, and using $s = \bar{v}t$, where $\bar{v} = c$, the speed of light, we can find the distance $s/2$ from the station to the reflecting object. In radar installations, the computation is done electronically, and a screen similar to a TV screen shows the position of the reflecting object. A highly simplified example is shown in Figure 15.19b.

Modern radar screens can show reflecting objects in all directions from the radar station simultaneously. They can monitor the progress of squall lines, thunderstorms, hailstorms, and similar phenomena. A radar scan of a hurricane is shown in Figure 15.20; the eye of the hurricane is approximately 50 mi from the radar station. Radar's ability to provide such instantaneous and continuous weather information is invaluable and cannot be obtained easily by other means.

Pollution of the Atmosphere

In a very real sense, the atmosphere is used as a garbage can by humanity and even by nature. Much of the waste material generated by living organisms as well as machines is dispersed finely enough to float in the atmosphere. These waste products range from the CO_2 we exhale from our lungs to the soot, ash, and grit generated by industrial plants. Although soot, ash, and grit are annoying, their relatively heavy particles soon settle to the ground and so the air is quickly cleansed of them. But close to their source, they are often a serious hazard to both life and property.

Breathing such large pollutant particles can cause respiratory diseases such as emphysema and cancer. This danger is particularly serious for workers in certain occupations. For example, black lung disease has long been an occupational hazard for coal miners, and a recent study has shown that asbestos dust is extraordinarily harmful if inhaled. It is likely that many other such hazards exist that have not yet been identified by systematic studies.

Of course, large pollutant particles are a nuisance to those who live in some industrial areas. Dust, soot, and grit continually cover their cars, their houses, and their bodies. Although this problem is annoying to those who experience it, this is a minor problem compared with other effects of particulate pollution.

Another effect of these large airborne particles is to alter the climate and sunlight intensity of the areas where they abound. For example, until about 1940 most homes were heated by soft-coal furnaces. As a result, each house in a large city belched black smoke all winter, and cities that lay in valleys and did not have strong winds were covered by a blanket of soot. In regions of the world where coal is still used, an entire city in a valley may be invisible to an observer on the nearest hilltop. The quality of life deteriorates in such a situation, and the absence of sunlight endangers health. As you may know, sunlight provides vitamin D. Without sufficient sunlight or vitamin D in the diet, children often develop rickets in such smoke-darkened cities.

Although this example is now growing uncommon, every large city suffers to a certain extent from particulate matter above it. Such pollution is

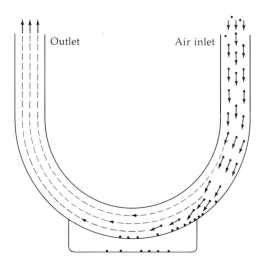

15.21   The basic principle of a cyclone-type dust remover. Since the inertia of the dust particles is high, they cannot follow the swiftly moving gas around the curve and are trapped there by a water spray or a collecting pan.

easy to recognize and remove from the emissions of a factory. Three methods of removal are widely used. Large particles may be washed out of the air by passing the exhaust gases through a fine water spray or mist, which cleans them just as rain cleans the air through which it falls. A second method is to whirl the particles out of the air, as illustrated in Figure 15.21. Exhaust gases are blown at high speed along a circular path. The gas is pulled out of the normal straight-line motion so quickly that air friction cannot hold the particles in the curved path. Instead, the particles continue in a straight line and are thrown out of the gas onto the sides of the confining tube, where they are captured and removed.

A highly efficient method which removes even much smaller particles is the electrostatic precipitator, shown in Figure 15.22. In this device, a strong electric field is produced near thin, charged wires or sharp needle points. A simplified version of this apparatus can be constructed by suspending a thin wire along the axis of a metal tube (or a cardboard mailing tube wrapped in aluminum foil). The center wire is attached to the negative terminal of a high-voltage source, while the outer tube is

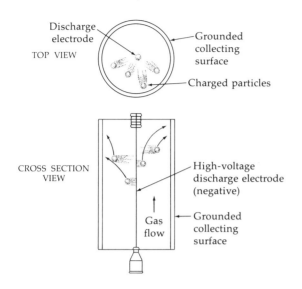

15.22 A simplified (but workable) version of an electrostatic precipitator for removing smoke particles.

connected to the positive terminal. (A spark coil, a readily available high-voltage device, is ideal for this purpose.) As the smoke-filled gas rises in the tube, charges from the center wire fly to the smoke particles and give them a negative charge. They are then attracted to the positive outer tube and, after colliding with it, they settle as dust along the wall. In commercial precipitators, many parallel tubes strip the gas of particulate matter.

Although particulate pollutants are serious, gaseous pollutants are probably more dangerous. These fumes, most of them noxious, emanate chiefly from industrial processes and automobile exhausts. Typical gaseous pollutants are carbon dioxide from the combustion of organic fuels; carbon monoxide from incomplete combustion of fuel, primarily in automobiles; sulfur oxides and other sulfur compounds from coal and oil that contain sulfur as an impurity; oxides of nitrogen formed in the combustion process in automobile engines; and ozone, aldehydes, mercaptans, and lead compounds from industrial processes and automobile exhausts. Although some of these gases can be detected by their odor, most first become apparent when they cause drowsiness (carbon monoxide) or irritation of the eyes and throat. Unlike the particulate pollutants, they cannot be seen and are, therefore, hard to detect and control.

Carbon dioxide is, as far as we know, harmless to living things. This is fortunate, because it is generated in large quantities whenever the products of photosynthesis are consumed. However, the CO_2 in the air is responsible for the greenhouse effect, as discussed on p. 334.

As you know, carbon monoxide, CO, is lethal, particularly in the exhaust of a car engine running in a closed garage. We even have data to show that CO levels on crowded city streets at rush hour are higher than is safe for health. The internal combustion engines of modern automobiles are the main source of CO.

Low-grade coal and oil contain an appreciable amount of sulfur, which is released in oxide form as the fuel is burned. Hence, sulfur oxides are a by-product of nearly all industrial and heat-producing processes. These oxides are not easily removed from the waste gases, nor do they settle out of the air. Therefore, they persist for rather long times. In the air, sulfur dioxide (SO_2) combines with water vapor to produce sulfuric acid (H_2SO_4). Both corrode living tissue and are responsible for many of the eye, throat, and lung irritations we associate with smog. We believe that oxides of sulfur are the principal culprits in smog-induced deaths.

Gaseous pollutants are most dangerous when wind conditions trap them in stagnant air over a city. Many cities suffer severely when the atmosphere above the city contains an *inversion layer*. To understand the meaning of this term, we must recognize that because warm air rises, air will flow upward and away from a city if the temperature is highest at ground level and decreases with altitude. As the warm air over the city rises, it carries pollutants away with it. Sometimes, however, the temperature *increases* with height for the first several hundred feet and only then begins to decrease.

(a)

(b)

15.23 These pictures of Los Angeles were taken in 1956. Notice the top of the inversion layer in (b). A layer much like this, although not necessarily as smog-filled, exists approximately 320 days of the year there. (Courtesy Los Angeles County Air Pollution Control District.)

When the temperature at the surface of the earth is *inverted* this way, with the coldest air next to the ground, the bottom layer of air does not rise. Instead, this cold inversion layer remains stationary over the city and prevents the escape of the airborne pollutants. Inversion layers are particularly bad for a city in a valley or basin. The resulting smog is a mass of highly polluted, usually moist air. Figure 15.23 shows an inversion layer and smog in Los Angeles.

Many cities have had to restrict the production of pollutants when an inversion layer occurs, by such measures as closing down incinerators and smoky factories. If they had not, the pollution would have caused a drastic rise in deaths. (Some cities which did not institute controls soon enough have experienced near catastrophes.) The other possible solution, still an untested dream for the future, is to force the air to rise above the city and break through the inversion layer. Grandiose schemes for accomplishing this have been suggested; even if they prove feasible, they will be expensive.

As we see from this brief review of air pollution, we cannot continue to throw our wastes into the atmosphere in ever-increasing quantities. The atmosphere is large, but it is not infinite. If we are not careful, we may perish in a sea of air filled with our own refuse.

BIBLIOGRAPHY

1 MILLER, A.: *Meteorology* (paperback), Merrill, Columbus, Ohio, 1966. A nonmathematical, readable introduction to the subject. Highly recommended for additional reading.

2 BATTAN, L.: *The Nature of Violent Storms* (paperback), Anchor Science, Doubleday, New York, 1961.

    BATTAN, L.: *Radar Observes the Weather* (paperback), Anchor Science, Doubleday, New York, 1962.

    BATTAN, L.: *Cloud Physics and Cloud Seeding* (paperback), Anchor Science, Doubleday, New York, 1962. This set of three books is directed to the layman and should interest even the most casual reader.

3 MCINTOSH, D. H., THOM, A., and SAUNDERS, V.: *Essentials of Meteorology* (paperback), Wykeham Science, Springer-Verlag, New York, 1969. A more quantitative treatment than reference 1; written for use in British high schools.

4 MEETHAM, A.: *Atmospheric Pollution*, 3d ed., Macmillan, New York, 1964. A nonmathematical but complete coverage of the subject; a good source if you are really interested.

5 ESPOSITO, J.: *Vanishing Air* (paperback), Grossman, New York, 1970. This is the Nader study group report on air pollution.

SUMMARY

The earth's gravitational attraction is insufficient to retain much hydrogen and helium in its atmosphere. Our atmosphere consists of N_2 (78 percent), O_2 (21 percent), and water vapor, with very small amounts of other gases. Only the most energetic molecules can rise to high altitudes; hence the air thins with increasing altitude and has dropped to 10^{-6} of its original density at a height of 60 mi. The temperature of the atmosphere drops with increasing altitude for the first several miles above the earth. The thin atmosphere at high altitudes is heated to high temperatures by the direct rays of the sun.

We divide the atmosphere into layers. The exosphere extends outward from a height of about 300 mi. From 50 to 300 mi is the thermosphere, named for its high temperature. The air in this region is thin enough to be considered a vacuum for most purposes. Below the thermosphere are the mesophere (30 to 50 mi) and the stratosphere (8 to 30 mi). The air in these regions is clear and essentially moisture-free. At the lower boundary, the air is very cold and its density has increased to an appreciable value. Nearest the earth is the troposphere, in which most weather phenomena occur.

Ultraviolet and x-radiation from the sun is largely absorbed in the upper atmosphere, particularly by ozone. When this high-energy radiation is absorbed, it tears molecules and atoms into ions and so produces an ion-filled region called the ionosphere. The ionosphere reflects radio waves, thus enabling them to circle the earth.

Air flows in definite large-scale patterns across the earth. The warm air at the equator rises, while cold air at the poles falls. As air rushes across the earth to compensate for these vertical flows, it is deflected because of the rotation of the earth. This effect, called the Coriolis effect, leads to the prevailing winds observed throughout the world. In the continental United States, the prevailing winds flow from west to east.

Although most of the short- and very long-wavelength radiation from the sun is absorbed by the upper atmosphere, the visible portion of the spectrum reaches the surface of the earth. It is absorbed there and reradiated as infrared or heat radiation. Mainly because of the CO_2 in the atmosphere, this heat is trapped close to the earth by the greenhouse effect.

The relative humidity is the ratio of the quantity of water vapor in the air to the quantity the air is capable of holding at that temperature. Since cold air cannot hold as much water vapor as warm air, warm air becomes saturated when cooled far enough. At this temperature, the dew point, the air holds as much water as it can. If cooled further, the water precipitates as rain, fog, or snow.

A cold front is the boundary between a cold air mass and the warm air mass which it is displacing. A warm front is the boundary where a warm air mass is displacing a cold air mass. Both fronts usually bring stormy weather.

Cyclone motion is not to be confused with the layman's term for a tornado. A cyclone motion is any spiraling air motion; it can easily extend over hundreds of miles. It often originates when a wave in the boundary between oppositely moving warm and cold air masses breaks off and forms a cold and warm front, both of which circle a common point. The center of the circle can be either a low (with bad weather) or a high (good weather), depending on the direction of rotation. Counterclockwise rotation means a low in the northern hemisphere and a high in the southern hemisphere.

Hurricanes are tropical storms resulting from the rapid rise of warm, moist air. They have a cyclonic motion and a low-pressure area called the eye at the center. Their motion is guided by the prevailing wind patterns. Tornadoes are miniature hurricanes although they may not originate in the same way. Their center, corresponding to the eye, is a region of reduced pressure. Much of the damage they do comes from the sudden drop in pressure as the tornado funnel passes by.

The seasons and their associated weather are the result of the earth's yearly movement around the sun and the orientation of the earth's axis to the ecliptic plane. Ocean currents and the thermal inertia of large bodies of water moderate the seasonal temperature changes in certain land areas.

Because of the large number of variables involved, weather predicting is a difficult scientific problem. From weather maps and previous experience, the progress of fronts and other weather details can be charted and predicted with some accuracy. Radar is a valuable tool for locating and tracking storms. Weather satellites are used to map the cloud patterns of the entire earth.

Pollution of the atmosphere results from the particulate and gaseous waste products of combustion and industrial processes. Autos are responsible for a large fraction of this pollution. Carbon monoxide and the oxides of sulfur are among the most serious pollutants.

TOPICAL CHECKLIST

1   What could have furnished the heat needed to melt the earth partially at earliest times?

2   Why is there very little helium on the earth? What is the source of the helium now found here?

3   Discuss the variation of temperature and pres-

sure of the atmosphere as a function of altitude. Explain why the major variations occur.

4   What are the various layers of the atmosphere? What are their distinctive properties?

5   How does the ionosphere originate? What is its function?

6   Of what is the atmosphere composed?

7   Define the terms relative humidity, saturated, and dew point.

8   What is the greenhouse effect? Why is it important?

9   Discuss the major prevailing wind patterns and how they arise.

10 Describe a cold front; a warm front. How are they represented on a weather map? Why do storms often accompany a front?

11 Discuss the origin of cyclonic motion. How does it lead to lows and highs? What weather properties does each have?

12 What is a hurricane? Describe its chief features and interpret them.

13 What is a tornado? Describe and interpret its chief features.

14 Why do we experience seasonal changes in weather? How are they moderated by ocean currents and large bodies of water?

15 Why is it difficult to predict weather? What factors must be considered in formulating a weather prediction? What weather features can be read from a weather map?

16 Describe how radar works. Why is it useful?

17 Discuss the effects of particulate pollutants in the atmosphere. How can smoke and dust particles be removed from industrial gases?

18 What are the major gaseous pollutants? Discuss the sources and deleterious effects of CO_2, CO, and the oxides of sulfur.

19 What is an inversion layer? How does it give rise to smog problems?

QUESTIONS

1   What is the minimum height at which an earth satellite can orbit for prolonged lengths of time?

2   Modern cities generate large amounts of smoke, dust, and molecular air pollutants. Trace the history of a typical smog particle.

3   Compare the appearance of a sunset (or sunrise) on earth to that on the moon.

4   Why are clear nights usually colder than cloudy nights, all other things being comparable?

5   Fruit trees in low areas are more often damaged by frost than those on higher ground. Why? Often smoke pots are burned in the low places to prevent frost. Why are they effective?

6   Where does frost form first in an area with which you are familiar? When does it usually form? Explain.

7   If early primitive man had moved from continent to continent by drifting with the aid of a small sail across the seas, what would his most likely paths of motion have been?

8   In an average 24-hr period, why does the temperature vary much more widely in the Sahara desert than it does in Puerto Rico, even though both are about the same distance from the equator?

9   Discuss the problems one might encounter with radio communication between two points on the moon.

10 Some people claim they cannot get a suntan in the city even though they tan easily when at a lake. Why might this be true?

11 Discuss some of the practical methods and difficulties of eliminating the following pollution problems: automobile exhaust fumes, fumes from a coal-burning electric generating plant, cigarette smoke, solvent fumes in a chemical factory, odors from a city's waste treatment plant.

AMERICAN MUSEUM OF NATURAL HISTORY

Chapter 16 THE EARTH

We have discussed many features of the earth and its oceans in past chapters; this chapter will treat these topics in a more systematic fashion. Geology, the study of the history and structure of the earth, has inspired a great deal of research over a period of many years. In spite of this, our understanding of the earth is far from complete. Oceanography, the study of the oceans, is a younger and less-developed science. The areas of the earth covered by water are vast reservoirs of food, minerals, and energy; they will be very important for civilizations of the future.

Materials of the Earth's Surface

The portion of the solid earth known to us from our everyday experience is far from typical of the earth as a whole. Most of us are only familiar with an extremely thin layer that covers the earth's sphere. This layer is characterized by the presence of living material and the decay products of past centuries of life. For the most part, plant and animal life exist only where earth and air meet, because plants need sunlight and carbon dioxide for cell growth and animals depend on air for oxygen. Although living things can exist elsewhere, they flourish only close to the earth's surface. This layer, which is of overwhelming importance to us as living organisms, is so thin that we cannot even illustrate it in a scale drawing of the earth.

This surface layer is the only portion of the earth that is readily accessible for study; it consists of rocks covered by a thin layer of soil or other loose material such as sand or organic matter. Close inspection of the common rocks found at the surface reveal that they are a very diverse group of objects. Many are clearly not of uniform composition, but are made up of a number of smaller rock or organic fragments or of large and small grains which are somehow cemented together. Others appear uniform to the unaided eye but study through a microscope reveals an aggregate of grains similar to those of the coarser specimens, differing only in their smaller size. Only a small percentage of rocks, those that are structurally similar to glass, contain no grains at all. The grains, the fundamental units of which rocks are composed, are called _minerals_.

A mineral is any naturally occurring inorganic crystalline material of definite chemical composition. Each mineral, therefore, has a chemical formula. For example, calcite is $CaCO_3$ and potassium feldspar is potassium aluminum silicate, $KAlSi_3O_8$. This definition is broad enough so that materials of slightly different composition are often considered to be the same mineral. We can look upon the classification of crustal materials in terms of minerals as a chemical classification scheme.

Like other chemical compounds, each mineral has a characteristic arrangement of atoms, giving it a unique crystal structure. Many physical properties of minerals reflect this unique structure and therefore help us to distinguish one mineral from another. For example, each mineral develops a characteristic crystal form if allowed to grow freely. Pictures of several large mineral crystals are given in Figure 16.1. The hardness, cleavage (the way minerals break along certain planes in the crystal), density (mass per unit of volume), color, and other distinctive characteristics are all related to the crystal structure. Using x-rays, we can figure out in detail the arrangement of the atoms in the structure.

While minerals are classified by their atomic structure and chemical composition, the rocks which they compose are grouped according to their origin and past history. As we have seen, the interior of the earth is much warmer than its surface. Even though the surface is now cool enough to sustain life, deep bore holes indicate that the temperature rises about 1 °C for each 100 feet of depth. Although it is risky to extrapolate from such data, this observed temperature rise implies a temperature of about 1500 °C at a depth of 30 miles. We believe that radioactivity within the earth releases enough energy to be responsible for these enormous temperatures. Temperatures just below the crust are high enough to melt rocks but, because of the high pressures there, the layer beneath the crust is still solid. However, when a defect in the crust allows the release of this pressure, the very hot material becomes molten. This molten rock is called *magma*. When magma cools and solidifies, it forms rocks that constitute

16.1 *The most important rock-forming minerals are feldspars (a), quartz (b), calcite (c), and the ferromagnesian family (d). The feldspars are aluminum silicates with potassium, sodium, or calcium. Usually they are white, light gray, or pink, and glossy in luster. They cleave in two directions at about 90°.*

Quartz is silicon dioxide and looks like white, grey, tinted, or transparent glass. It shatters into sharp-edged, irregular pieces as glass does, since it has no pattern of cleavage. It often forms slender hexagonal prisms (rod-like crystals) that end in sharp, pyramidal points. Quartz is considerably harder than feldspar and will cut deeply into glass. Since it does not react with the atmosphere, it is very durable. As the feldspars in granite rock react with moisture and crumble into soft clay, the quartz grains present drop out and form the deposit we call common sand.

The mineral calcite ($CaCO_3$) varies from white to dark and may be colored or transparent. Limestone and marble are formed of calcite and impurities.

The dark minerals common in granite and similar rocks are usually of the ferromagnesian family, containing iron and magnesium in silicate form. They are normally black, dark green, or dark brown in color and glassy in appearance. Weathering causes most of these minerals to form a rusty coating just as iron does. The rusty reds, yellows, and browns of rock formations in the Grand Canyon and elsewhere are stains from the oxidation of ferromagnesian minerals. Biotite, shown here, is a typical ferromagnesian mineral. [Photos courtesy of U.S. Geological Survey; (a), (b), and (c) by W. T. Schaller, (d) by W. G. Overstreet.]

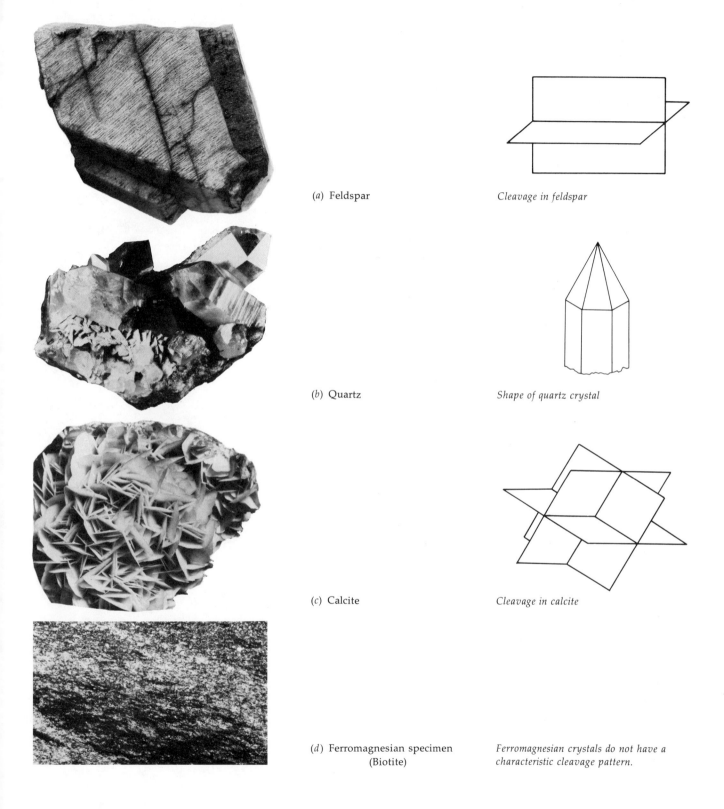

(a) Feldspar — Cleavage in feldspar

(b) Quartz — Shape of quartz crystal

(c) Calcite — Cleavage in calcite

(d) Ferromagnesian specimen (Biotite) — Ferromagnesian crystals do not have a characteristic cleavage pattern.

| IGNEOUS | SEDIMENTARY | METAMORPHIC |

(a) Obsidian

(f) Conglomerate

(j) Gneiss

(b) Rhyolite

(g) Sandstone

(k) Schist

(c) Basalt

(h) Shale

(l) Slate

(d) Granite

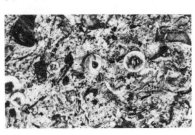
(i) Coquinoid limestone

(m) Quarzite

(e) Gabbro

(n) Marble

16.2 *The rocks in the left-hand column are all igneous rocks, which means that they are formed directly by solidification of magma. The extrusive rocks shown in (a), (b), and (c) form when molten magma is extruded onto the earth's surface. Obsidian is a black volcanic glass; rhyolite (blonde lava rock) forms from molten light-colored rock such as granite, while basalt forms from molten dark-colored rocks. Basalt is the most common extrusive igneous rock.*

Intrusive rocks are formed when magma solidifies below the earth's surface; (d) and (e) show the common intrusive rocks granite and gabbro. Granite is more than 75 percent light-colored feldspars and quartz, with the remainder made up of dark ferromagnesian minerals.

The center column shows sedimentary rocks, which are formed when beds of clay, sand, gravel, or shells become compressed and cemented together to form new rocks. You can see the rounded pebbles cemented together in the conglomerate rock shown in (f). As its name implies, the sandstone in (g) consists of grains of sand cemented together. Layers of clay cement together to form the shale in (h). Limestone is formed from precipitated calcite; when coarse sea shells are cemented together, the result is coquinoid limestone (i). Chalk is formed in this same way from very tiny shells.

The right-hand column shows the third major class of rocks, metamorphic rocks, which are formed by intense heat and pressure far below the earth's surface. Gneiss (j), a coarsely crystalline rock, shows the bands left by various colored minerals as they flowed. Layered rocks showing compressed crystal plates or needles are called schist (k). Slate (l) is formed by high pressure acting on shale. Sand grains in sandstone weld together to form quartzite (m). Limestone, intruded into by magma, is heated and recrystallized to form marble (n). (Courtesy of U.S. Geological Survey.)

the first of the three basic rock types. Rocks formed directly by solidification of magma are called _igneous_ rocks.

The subcrustal material that produces magma contains considerable water. Although the temperature of the water is far above its boiling point, the water is kept from escaping by the high pressures to which it is subjected. The situation is like that which exists in a bottle of soda pop. Carbon dioxide is dissolved in the beverage by pressurizing the liquid; when the bottle is uncapped, the dissolved gas begins to escape and the liquid bubbles or "fizzes."

The water trapped in the magma behaves similarly whenever a volcano or a fissure in the earth's crust allows the magma to escape to the surface; the water vaporizes and causes the magma to bubble. This bubbling, liquid rock which flows out from active volcanoes and fissures is called _lava_. The solidified lava takes different forms depending upon how quickly it is cooled. One form is pumice, which you may know. The tiny bubbles are clearly visible in it.

Igneous rocks like these, which were formed as the magma extruded onto the earth's surface, are called _extrusive_ igneous rocks. Since they cooled so rapidly, the minerals did not have time to form sizable crystals. As a result, the extrusive rocks are usually composed of crystals so tiny that they can be seen clearly only with a microscope. Often, the magma solidifies so rapidly that the solid contains no crystals at all, but is glassy. An example is obsidian, a volcanic glass, shown in Figure 16.2a. Other extrusive rocks are shown in (b) and (c). Most often, the magma does not reach the surface of the earth. Instead, it spreads through fissures or between layers and other imperfections well below the surface. In Figure 16.3, for instance, the magma has risen to an underground reservoir known as a magma chamber. These reservoirs occur at depths of about 30 to 35 miles. From the reservoir the magma moves upward and occasionally reaches the surface, as in the volcano shown at the top. Over the years, the lava outflow has left

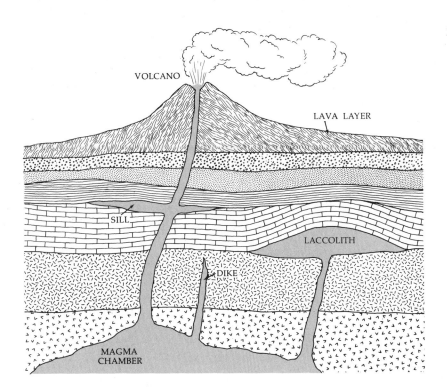

16.3 *The lava flow from the volcano has built up the crust as shown. Rising magma has also formed a dike, a sill, a laccolith, and a magma chamber.*

a volcano cone and built up a lava layer on the surface of the earth near it.

Much more frequently, the rising magma is unable to reach the surface and forms an intrusive region between other layers or within a fissure. When a thin, flat, vertical fissure has filled with solidified magma, we call this a _dike_. A thin inclusion between two horizontal layers is called a _sill_. A thick, lens-shaped sill of small area is called a _laccolith_. A large body of igneous rock, deep beneath the surface and usually covering more than 50 sq mi, is a _batholith_. Unlike dikes and sills, batholiths are rarely associated with volcanoes. Apparently the magma that cools to form batholiths does not approach the surface of the earth before it hardens.

The magma trapped under the surface cools slowly over a period of years. This allows plenty of time for various minerals to form large crystals when their solidification temperatures are reached. As a result, the rocks that form from the magma in these so-called "intrusions" show fairly well-developed crystals throughout. These are called _intrusive_ igneous rocks, and two typical forms are shown in Figure 16.2d and e. You are probably familiar with granite, since it is the brownish-pink speckled stone widely used in monuments.

Since intrusive bodies such as sills and dikes are formed separately from the surrounding rocks, they usually differ in composition and therefore behave differently from their neighboring areas under the forces of erosion. This is reflected in the appearance of terrain exposed by many centuries of weathering. Often the intrusions are more resistant to rain and wind than their surroundings and so they eventually stand as pillars rising high above the nearby terrain. Typical examples are shown in Figures 16.4 and 16.5.

During the long history of the earth, the igneous rocks have undergone _weathering_ and _erosion_. Whenever the rocks are exposed at the surface, wind, water and ice cause many of them to powder, pull apart, or dissolve chemically. In time, many of these fragments and dissolved substances are transported by rivers, glaciers, and wind far from their site of origin and are deposited in a new environment. At first they form loose deposits of clay, sand, gravel, and so on. Eventually, a cementing material such as calcite ($CaCO_3$) or silica (SiO_2), often aided by the weight of material on top of the fragments,

16.4 Erosion of the surrounding terrain has exposed this dike in Arizona. (Courtesy of Tad Nichols.)

16.5 The Devil's Tower in Wyoming, a laccolith that has been exposed to view by the erosion of surrounding material. (Courtesy of John S. Shelton.)

THE EARTH 359

compacts and cements the fragments together. In this way, new rocks are formed; we term this class of rocks *sedimentary rocks*. As the name implies, they are secondary rocks formed from material that has settled from a transporting medium. Several examples are shown in Figure 16.2. Through this process, sand becomes sandstone, clay changes into shale, and gravel forms conglomerate. Organic fragments also undergo this process, producing coal from pieces of plants and coquinoid limestone from bits of shell. Dissolved material may separate out of water and form rocks such as limestone or rock shale.

The third class of rocks is *metamorphic rocks*. These are rocks formed by the action of high pressures and temperatures on already existing rocks. Below the crust of the earth, pressures and temperatures are large enough to cause rocks to flow plastically. Under these conditions new minerals form and produce new rock forms. Examples are schist, slate, gneiss, marble, and quartzite, shown in Figure 16.2*j* through *n*. Since the pressures and temperatures needed for this metamorphism are only available deep below the earth's surface, metamorphic rocks do not appear at the surface unless deep erosion or crustal upheaval has occurred.

To summarize, the three basic rock categories are igneous, sedimentary, and metamorphic. Since each rock class has its own distinctive history, the geologist can use the rocks found in a region to chart the history of the region. Moreover, since the radioactivity within a rock often helps to determine the time of its formation, modern geology can provide a quantitative history of the earth.

Seismic Waves and Earth Structure

A study of the rocks exposed at the surface of the earth or even those found in deep mines gives us information only about the very thin surface layer of this planet. To extend our knowledge, a program to drill through the crust of the earth was initiated in 1961. Unfortunately, technical and financial dif-

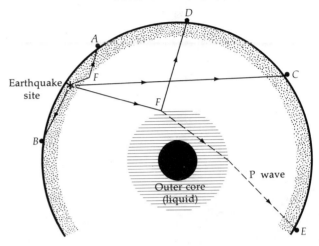

16.6 *Typical shock-wave paths sent out from an earthquake site. Only the P wave is able to reach E through the liquid outer core, but both P and S waves reach A, B, C, and D.*

ficulties prevented the experimenters from reaching rocks beneath the crust.

Most of our knowledge about the internal structure of the earth has been gained more indirectly, much of it through observation of *seismic waves*. These are the jolting disturbances that radiate from the site of an earthquake or underground explosion. An earthquake is the shaking due to sudden fracturing and movement of rocks at depths ranging from just below the surface to 400 miles within the earth. Like the intense sound or shock wave sent through the air by an explosion, disturbance pulses are sent through the earth by a sudden movement beneath its surface. Typical paths of the shock waves emanating from an earthquake (or man-made explosion) are shown in Figure 16.6.

Two entirely different kinds of waves are sent through the earth by such a disturbance. One is a compression wave (a sound wave), called a P wave. The P stands for "primary," since this wave travels faster and therefore reaches a distant point earlier. (It is sometimes convenient to associate the P with "push," because compression waves push back and forth along the direction of propagation.) Since sound waves can travel through water, the P waves travel through both solids and liquids.

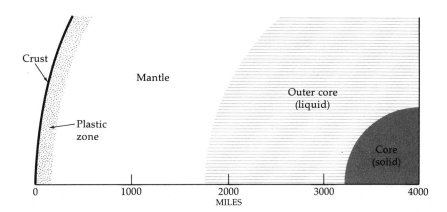

16.7 *Cross section of the earth drawn to scale.*

The second type of wave is called an S wave, for "secondary." This is a transverse wave, like a wave on a string. (You may associate the S with "shake" since you must shake a string up and down to generate a transverse wave.) In an S wave, the material of the earth vibrates back and forth perpendicular to the line of propagation of the wave. This type of wave cannot travel far through a liquid, and so a solid can be distinguished from a liquid by observing whether or not S waves pass through it.

The paths of P and S waves are refracted as they pass from one material to another, because they travel with different speeds in materials of different densities. This is similar to the refraction of light waves passing from air into water. In addition, at the boundary between two materials of different hardness, the waves are reflected as shown at F in Figure 16.6. This is comparable to the way light waves are partly reflected at the air-water boundary.

The shock waves reaching various points on the earth such as A, B, C, D, and E travel different paths and therefore reach the observation points at different times following an earthquake or man-made explosion. From these arrival times, we can infer the path taken by the wave and draw a diagram such as Figure 16.6. Moreover, since only the P wave reaches point E, part of its path must have been through liquid. From such evidence, we can conclude that the portion of the earth's interior through which it passed is molten.

Seismic waves are widely used for geological exploration. Since various rock layers reflect and refract the waves differently, it is possible to study the disposition of rocks beneath the surface and locate sites where oil might accumulate. However, the interpretation of the seismic wave patterns is not easy and so predictions concerning oil and mineral deposits are not completely reliable.

By studying travel paths and travel times of seismic waves we can infer the existence of the areas shown in Figure 16.7. The outermost layer, the crust, varies in thickness from about 3 miles beneath the oceans to a maximum of about 50 miles in the areas covered by the continents. There is a rather sharp boundary at the lower limit of the crust called the Mohorovičić discontinuity, or Moho.

From the minerals and rocks at the surface, we know the constituents of the earth's crust reasonably well; its chemical composition is given in Table 16.1. The crust is composed primarily of silicon and oxygen, with appreciable amounts of aluminum, iron, calcium, sodium, potassium, and magnesium.

Table 16.1   Composition of the Earth's Crust

Element	Percent
Oxygen	49
Silicon	26
Aluminum	7.6
Iron	4.6
Calcium	3.4
Sodium	2.7
Potassium	2.4
Magnesium	1.9
Others	4

The abundance of silicon and oxygen reflects the fact that most common minerals have a silicon-oxygen compound as a basic structural unit. These minerals are called *silicates*. With small variations in the content of other metal atoms, the silicates are the major constituents of most rocks.

Beneath the crust in Figure 16.7 is a region called the plastic zone, which is usually considered part of the mantle. Although it is not molten, it has some of the properties of a very viscous liquid or glass. Over the centuries, it can flow over large distances, so the crust acts as if it is floating on a very viscous liquid.

Since direct measurement is not possible, we can only guess the composition of the mantle. We believe it, too, is composed chiefly of silicates. However, the silicates in this layer must contain much larger amounts of iron, magnesium, and other metal atoms than those in the crust. From seismic waves, we know this layer is solid; pressures at these depths must raise the melting point of the rock above the high temperatures prevailing there.

The next layer, the outer core, has been shown to be liquid. We believe that the earth's magnetic field results from large-scale liquid motion within the outer core. We surmise that both the inner and outer cores are composed chiefly of iron and nickel; this is based on the metallic content of meteorites.

The inner core appears to be solid. The pressure has apparently increased enough to negate the effects of the high temperatures there. We believe these high pressures have compressed the core so that it is about 15 times more dense than water. For comparison, the earth's crust is about 3 times as dense as water, the earth as a whole is about 5.5 times as dense, and iron is 7.9 times as dense. We can only guess the properties of matter under these tremendous pressures and densities, since the necessary pressures are not yet easily reproduced in the laboratory. Certainly, materials under such high pressure will show markedly different properties from those under the conditions we know.

Isostasy

Now that we have an idea of the internal structure of the earth, we can take a closer look at the conformation of the crustal layer. The crust floats upon a very viscous foundation which can change in volume, shape, or position slowly and continuously without breaking. Such continuous flow is described as *plastic*. When a rock body changes volume, shape, or position in the earth's crust, it undergoes *deformation*. Although the plastic zone shown in Figure 16.7 is not liquid, it still can undergo considerable plastic flow over time spans of centuries. This model helps explain many features of the earth's crust.

The crust can be divided into two main parts, the continents and the ocean basins, which are separated from each other by the continental margins. As shown in Figure 16.8, under both the continents and ocean basins is a lower crust composed of basalt, designated as *sima*. (This word comes from the first two letters of its distinctive elements, silicon and magnesium.) Floating on this

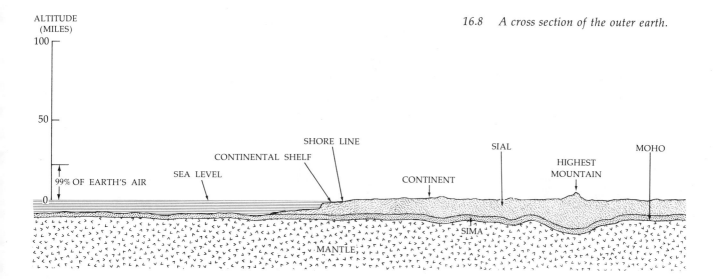

16.8 *A cross section of the outer earth.*

are the continents, composed chiefly of a layer of lighter rock more like granite which is called *sial* (from silicon and aluminum). A thin veneer of sedimentary rocks covers much of the continental and oceanic crust and forms thick wedges along many of the continental margins.

As we see in Figure 16.8, the continent depresses the Moho, the boundary between the crust and the mantle. The density of granite is less than that of basalt, and basalt, in turn, is less dense than the material of the mantle. As a result, the basalt floats like a scum on the more dense mantle while the continent (called a continental block) floats above both of them. In a sense, a continental block is like an iceberg floating on an ocean. The submerged portion of the granitic continental mass is lighter than the material of the mantle which it displaces. Hence a buoyant force is produced, causing the top of the continent to float above the surrounding surface just as a small portion of an iceberg floats above water. Moreover, as erosion occurs on the continent, the various parts float at different levels. The theory that the continents float this way upon the mantle is called *isostasy*.

Even though the mantle is not a liquid, its plastic flow properties give it many features of a liquid as flow continues over thousands of years. Indeed, considerable evidence shows that the North American continent is still readjusting its floating position from the last ice age, about 10,000 years ago. As the glaciers moved back into the arctic regions, the portions of the continent that they had covered lost mass. These regions then began to rise because of the buoyant effect of the mantle below. In fact, the Lake Superior region is still rising slowly (about 16 ft each 1,000 years).

The Continents

On the continents, two main subdivisions can be recognized: relatively stable interior regions known as *shields*, and more mobile regions called *belts* which have been subjected to extensive amounts of disruption through time. Where this disruption has occurred recently, the belts are mountainous regions. Many mountains are the result of volcanoes. Typical are Mt. Rainier in Washington, Mt. Shasta in Cal-

16.9 *The Mexican volcano, Parícutin, in 1944, one year after it first erupted out of a cornfield; it grew over 2,000 ft high before the volcanic activity ceased nine years later. (Courtesy of Tad Nichols.)*

16.10 *(a) Erosion of the folded crust leads to an uplifted range of mountains. (b) This mountain in Montana's Glacier National Park shows the effect of folding on strata that were originally horizontal. (Courtesy of Tad Nichols.)*

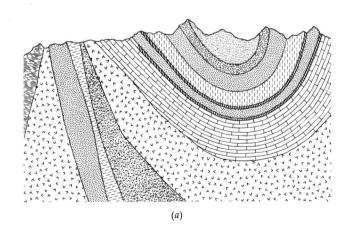

(a)

(b)

16.11 *This view of the Rockies gives some idea of the massive folding and buckling processes that created them. (Courtesy of John S. Shelton.)*

ifornia, and the volcano shown in Figure 16.9. However, most mountain ranges have been built up over many millenia by large-scale motion of the earth's crust. Such motion, which buckles and distorts large regions of the earth's crust, is called *diastrophism*.

If we examine the earth's crust near a mountain range such as the Alps, Andes, Appalachians, or Rockies, we find unmistakable evidence that the earth's crust has been folded. A type of structure often found is shown in Figure 16.10a. The way the crustal layers have been folded is clearly marked by the originally horizontal sedimentary strata (layers). The edges of these folded layers are exposed at the weathered surface and are often clearly visible, as Figure 16.10b shows. Moreover, holes bored through and into the mountain range show the indicated layering. Since various strata erode at different rates, ridges and valleys are formed along the upraised edges of the layers. As erosion continues, isostasy causes the remaining material to be uplifted so as to float near the level that existed before erosion. The majestic appearance of such a mountain range is shown in Figure 16.11. However, erosion levels even mountains. In time, these mountains will be rounded and flattened until finally they are no more than rolling hills. It is apparent from the structures shown in Figure 16.11 that tremendous forces must have existed to compress the earth's crust in this way. Even today, these processes are working in some ranges.

Not all mountain ranges are the result of buckling, however. The Sierra Nevada mountain range in California and the Tetons in Wyoming are the result of a somewhat different mechanism. If the crust of the earth is too brittle or is deformed too sharply, it will fracture rather than buckle. Such a sharp fracture surface is called a *fault*. The basic structure of a normal fault is shown in Figure 16.12a. Notice that the crust has slipped along the surface of the fault, so that there is a sharp discontinuity across the surface of the earth, as shown in (b). In practice, the slippage occurs jerkily as the slip-producing forces rise until they overcome the friction forces at the fault.

You will understand the situation if you have ever tried to slide a heavy box along a rough floor. You push harder and harder until the box suddenly "gives" and moves with a jerk. The sudden movement causes your pushing force to diminish, and the

THE EARTH 365

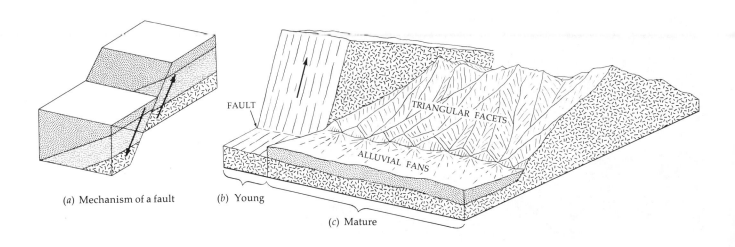

16.12 Stages in the development of a fault block mountain range; the mechanism of a fault is shown in (a).

box stops immediately. Similarly, when the earth displaces suddenly along the fault line, the stress is removed and the motion ceases. Although the causes of the stress vary from one fault to another, they are often associated with isostasy effects or continental drift, a topic we will discuss shortly. No matter what the cause, however, the sudden movement at the fault causes an earthquake in the region nearby; its severity depends upon how large the stress becomes before the slippage occurs. Over the years, many displacements take place along the fault, and the discontinuity increases in height (see Figure 16.12b).

Simultaneously, erosion changes the face of the fault. In extremely arid regions, the face (or escarpment) is eroded slowly. But eventually the sharp fault escarpment deteriorates, as shown in Figure 16.12c. Typical of this and later stages of development is the eastern front of the Sierra Nevada range, shown in Figure 16.13.

Lateral slippage is also possible at a fault. A famous example of this effect is taking place along the coast of California, as the Pacific side of the crust moves northward past the interior of the continent. This is the San Andreas fault; it runs east of Los Angeles and through San Francisco as it parallels the coast for about 400 mi. This fault line shows clearly in Figure 16.14.

Here too, friction retards motion along the fracture surface, and so movement occurs in sharp jerks as the forces build up and exceed the friction force. The San Francisco earthquake in 1906 was caused by a sudden slip along the San Andreas fault. The motions involved were both horizontal and vertical, and the coastal side of the fault moved northward. At one point some 30 mi from San Francisco, the relative displacement was 21 ft. It is estimated that the average motion along the fault line is about 2 in. per year. Figure 16.14 shows the cumulative effect of this motion, since the terrain is mismatched along the fault.

Earthquakes occur frequently but most are too slight to be noticed except with delicate equipment. Fortunately, most slippages occur long before the forces have built up to destructive magnitude. It is hoped that the San Andreas fault will continue to release strains periodically without building up strains as large as those that caused the 1906 earthquake. However, only now are we acquiring the technological ability to monitor strains in the

16.13 The east front of the Sierra Nevada Range as seen from Owens Valley, California, showing a typical portion of the great escarpment which is nearly 2 mi in height. (Courtesy of W. G. Mendenhall, U.S. Geological Survey.)

16.14 The San Andreas fault in California's Carrizo Plain. Notice that the terrain on the left does not match that on the right. (Courtesy of John S. Shelton.)

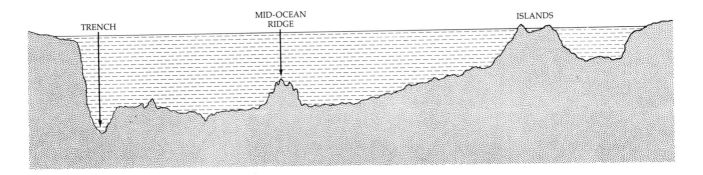

16.15 Vertical distances in this schematic cross section of an ocean between two continents are grossly exaggerated. Even so, the intricate canyons and mountains can only be suggested. The ocean might be about 15,000 ft (about 3 mi) deep, while its width would be more than 1,000 mi.

earth's crust with enough exactness to contemplate the prediction of earthquakes.

When an earthquake or other disturbance occurs in the ocean floor, it may produce a <u>tsunami</u>, a wave perhaps 3 ft high and hundreds of miles long. (A tsunami is often erroneously called a tidal wave.) When a tsunami is generated in the sea, it moves across it at a speed close to 500 mi/hr. While far at sea, it is too low to be seen easily, but when it enters shallower water near the coast, it grows much higher as it slows in speed. Just before it arrives, the water on the shore recedes, leaving fish and boats stranded on the beach. Then the tsunami strikes violently, throwing every obstruction out of its path. Tsunamis occasionally strike islands in the central Pacific, such as Hawaii, with destructive force.

The inner areas of the continents, the shield areas, are more stable. They are characterized by complex patterns of igneous and metamorphic rocks of considerable age. Studies of the younger and more deformed regions (our mountain ranges) suggest that the shield areas are the remnants of similar but much older mountain belts which have been worn down by erosion. The shield areas seem to form the nuclei of the continents, fringed by the deformed regions.

The Ocean Basins

Little was known about the ocean basins until recently. Detailed mapping of the topography of the ocean floor had to await the development of accurate instruments, such as <u>sonar</u>, to measure water depths. From a ship on the surface, a sonar set sends a sound pulse to the ocean bottom and records it when it is reflected back to the surface. From the speed of sound in water and the time taken for the pulse to go to the bottom and back, it is possible to compute the depth of the ocean at that point. In addition to sonar, sophisticated navigational aids are required to give the precise locations at which depth observations are made. Satellites have helped considerably in this task.

With the help of sonar mapping, we now know that the ocean floor has widely varied terrain. Much of it is a deep, flat expanse known as the abyssal plain. Like the land masses, it has high mountains, deep trenches, and winding canyons; but these are distributed much differently from those on the continents. A cross section of the ocean between two continents (say the coast of Africa and South America) might appear as shown in Figure 16.15. Many islands, particularly those in the

16.16 *The edge of the ocean (not to scale).*

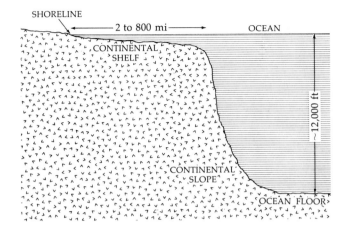

Pacific Ocean, are actually the tops of volcanoes rising from the ocean floor; the Hawaiian Islands are familiar examples. There are many more volcanic mountains dotting the ocean, deep beneath the surface. Far more impressive than the volcanic mountains, however, is the mid-oceanic ridge. This is a continuous mountain system which wends its way through the centers of the major oceans. The mid-oceanic ridge is characterized by many earthquakes and considerable volcanic activity.

Where the oceans meet the continents, the crustal surface has the general features shown in Figure 16.16. As we see, the shore line does not mark the edge of the continents. There is more than enough water to fill the regions between the continental blocks, and the excess covers the low portions of the continents bordering the oceans. These *continental shelf* areas vary in width from 800 mi off Alaska to 2 or 3 mi off the coast of Chile in South America. The average width is close to 40 mi, and this distance has been proposed as the territorial limit for nations bordering oceans. The question of fishing rights in deep oceans within the 40-mile limit has caused many disputes between fishermen and governments.

Around the Pacific Ocean, between the ocean floor and the edges of the continents, arc-shaped island chains and trenches replace the stable continental shelf. For example, the Philippines Trench, over 6 mi deep, curves east of the Philippine island chain. Such arcs may be the sites of new mountain belts forming on the outer edges of the continents. Eventually they will become part of the continent, just as the older mountain belts were added at the margins of the continental shields.

Since seacoast cities often use the ocean as dumping ground for their sewage and other refuse, the shallowness of the waters on the continental shelf is important. New York, for example, is troubled by pollution of this entire shallow area since the tides tend to carry refuse inland from the ocean. In the near future, strict limits will almost certainly be placed on all dumping within the continental-shelf areas of the United States.

The continental shelf has recently proved to be rich in oil and minerals. Offshore oil wells and gas fields are now common in many areas, and some mining operations have begun there. It is still too early to predict the economic usefulness of minerals beneath the floor of the deeper portions of the oceans. Since more than half the surface of the earth is ocean, we can clearly expect increased efforts to exploit these possibilities.

Perhaps the most puzzling feature of the shelf and slope regions is the existence of deep canyons, some of which are as large as the Grand Canyon. They are obviously the result of erosion, and some are simply underwater extensions of continental rivers. Many, however, seem to have no relation to the rivers on land and the cause of their erosion is still not known. Although the limits of the oceans have changed markedly over the centuries, it does not seem possible that the bottoms of these canyons, thousands of feet deep, were ever above water level. These submarine canyons are probably eroded by occasional flows of very muddy seawater, which may be started by earthquakes. The suspension of mud in water forms a very dense mixture which behaves like a muddy stream in the ocean and flows rapidly down the continental slope. As it flows along the ocean floor, it can erode a canyon in the relatively soft sediments which make up the floor.

In addition to the recent mapping of ocean floor topography, we have accumulated other information about this relatively mysterious part of the crust. Magnetic measurements of the ocean floor have shown a striking pattern of stripes which differ in their magnetic characteristics. Apparently the magnetic field of the earth has reversed its polarity numerous times in the past. When this happens, the north and south magnetic poles exchange positions; why it occurs, no one knows. When a rock forms containing magnetic minerals (magnetite, Fe_3O_4, is the most common), the magnetized minerals will line up like compass needles in the earth's magnetic field. Measurements of the alignment of magnetic minerals in the basalts of the ocean floor reveal stripes in which the mineral compasses show alternately normal magnetization (pointing north) and then reversed (pointing south). These magnetic stripes are parallel to the mid-oceanic ridge system, and are symmetrical on either side of that ridge.

The deep-sea drilling program has recently revealed another surprising pattern. In the ocean basins, the youngest rocks are along the mid-oceanic ridges. As you move away from the ridges toward the continents, the age of the bottom rocks increases. This contrasts with the continents, where the older rocks of the shields form a nucleus surrounded by younger belts of deformed rocks.

Continental Drift

In the last few years, a theory has been developing that explains the features of the continents and the oceans in a way no other has before. According to this theory, until about 200 million years ago the continental blocks we know today were combined

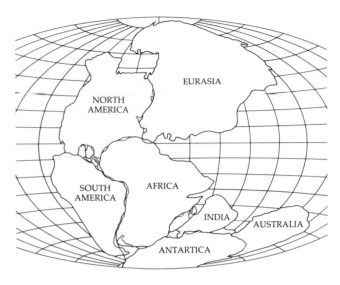

16.17 *The probable conformation of the supercontinent Pangaea.*

into one huge land mass. The developers of the theory refer to the single mass as *Pangaea*; their idea of how our continents were once joined together is illustrated in Figure 16.17.

Why Pangaea began to break up is not known with certainty. Possibly temperature differences caused fluidlike circulatory motion within the subcrustal regions, and this movement pushed the present continents apart. We do know that this process is still continuing today; for example, the Atlantic is slowly widening at a rate of a few inches per year. This may seem a hopelessly small movement, but in 100 million years it would amount to a distance of thousands of miles. Hence, even such a slow motion could result in the distances we now observe between the continents.

We do not yet know what driving force is responsible for the separation motion, although several theories have been suggested. However, it is agreed that the separation involves the growth of new ocean floor; in most oceans, this growth is taking place along the mid-ocean ridge. To understand what seems to be happening, consider the crustal behavior that is thought to cause the recession of one continent from another—South America and Africa, for example.

Figure 16.18 shows the two continental blocks

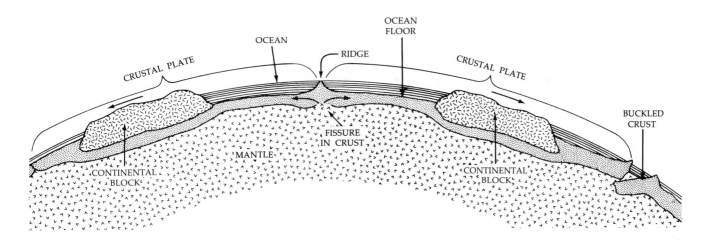

16.18 *As the continents separate, molten rock flows in from the mantle to fill the cracks in the mid-ocean floor (not to scale).*

and a portion of the ocean floor, all floating on the plastic zone of the mantle. Each continental block floats on a separate region of the crustal layer. These raftlike pieces of the crust are called *crustal plates*. At the former contact line of the two plates, a weak region in the crust must have appeared as the plates separated. Even though the mechanism for separation is unknown, let us say for simplicity that a stretching force in the earth's crust pulls the plates away from each other. The fracture line between the plates would have lower than normal pressure, since stresses would be released by the separation. As a result, pressure would be released from the portion of the mantle directly below. Since the rocks of the mantle melt at a lower temperature when the pressure is lowered, the mantle in this region changes to magma, flows into the fissure in the crust, and forms new ocean-bottom material of basalt.

As the crustal plates move apart, the fissure in the crust is filled from below. If this theory is true, the rocks of the ocean floor will be youngest along the dividing line between the crustal plates. This proves to be the case. Moreover, the basaltic nature of the ocean floor tends to support such a picture of its formation.

Since this process has occurred over millions of years, the gradually growing ocean floor will preserve a record of past events, such as the magnetic reversals discussed earlier. The new crust forming at the mid-oceanic ridge becomes magnetized parallel to the magnetic field of the earth at the time of its formation. When the field reverses, the solidified crust has its polarity frozen in. But new crust is continually being produced, spreading outward from the ridge. The new crust will show the reversed magnetization. Thus each time the magnetic field reverses, it is recorded in the new crust and a new magnetic stripe starts to form. Since the crust spreads away from the ridge on both sides, the stripes must be symmetrical on either side of the ridge.

Elsewhere, the edges of crustal plates must be pushing together, as shown at the far right in Figure 16.18. Along these lines the earth's crust is compressed and buckled. The edge of one plate may be pushed under the edge of the other, or both may buckle and pile up crustal material. In these regions, deep trenches, volcanic island arcs, and folded mountain belts would be formed by the compressional forces. This is what appears to be taking place in the western Pacific Ocean.

In spite of its plausibility, this mode of separation is far from proved. We only know with certainty that a drift process has occurred.

THE EARTH 371

16.19 *The gullies shown here were cut in a sandy slope in a period of only a few years. (Courtesy of John S. Shelton.)*

16.20 *Water spilling from canyons onto the plains deposits its sediments in alluvial fans. (Courtesy of Tad Nichols.)*

16.21 *The Grand Canyon in Arizona has been cut by the Colorado River. Notice that the sharp walls are being eroded away by tributary streams. (Courtesy of John S. Shelton.)*

Erosion

One reason the topography of the continents is more complex and varied than that of the ocean floor is that the features of land are drastically shaped by _erosion_. Erosion is the physical and chemical breaking down and transport of rock or earth from one place to another. There are four major sources of erosion: water, wind, ice, and the action of frost. Let us examine each in turn.

Except in the extremely arid or cold portions of the earth, water is the primary source of erosion. It may begin in an innocuous way, like the infant erosion pattern shown in Figure 16.19. Left to itself, that gully might develop into a deep ravine, depending upon such variables as the steepness of the slope and the nature of the earth strata. The erosion in this particular case became serious only after local vegetation had decreased until it no longer held water in place during heavy rains.

Erosion by a stream or river depends on many variables, since anything that influences the volume or speed of water flow will affect the way a stream cuts across the surface of the earth. Moreover, the nature of the terrain itself is important, because rock that can be dissolved or pulverized easily will erode first. Though the water will carry away much of the eroded material, this material, even if it is dissolved in the water, must eventually collect somewhere. Most of the dissolved minerals eventually reach the sea or lake into which the river deposits its water. The particulate matter, however, drops from the water wherever the flow rate becomes slow enough. The large delta regions at the mouth of the Mississippi and Nile rivers are examples of such action. Similar but smaller deposits called _alluvial fans_, fan-shaped bodies of sediment, form at the mouths of river canyons as shown in Figure 16.20.

Rivers follow a general pattern of development. We have seen in Figure 16.19 how an infant stream is formed in a nearly level plane. Gullies such as this tend to cut deeply and sharply at first. Under certain conditions, streams will retain steep, narrow valleys for many years. The Grand Canyon, shown in Figure 16.21, is an awesome example of the cutting action of a stream in an arid region accompanied by the characteristic uplifting of isostasy. Usually, however, the walls of the stream are more fragile than this. For example, the gullies in Figure 16.19 will rapidly widen as the walls collapse.

16.22 *A mature stream winds across a flood plain. (Courtesy of Allan Forsyth.)*

Typically, a stream bed widens as erosion proceeds. Moreover, sediments moved into it by landslides and other forms of erosion cause the valley walls to slope more gently.

The mature stream, then, loses its sharp edges; it no longer runs as fast (since the steep slopes and rapids have been eroded away), and it now winds across a flat valley floor or flood plain. An example of a mature stream in this stage is shown in Figure 16.22. Later still, the river valley may continue to widen as tributaries erode away the surrounding upland areas. In this elderly stage, a river often flows very slowly as it meanders back and forth through its now very wide, gently sloping valley.

This age cycle is rarely true in detail. So many factors influence the process, such as the original topography, vegetation, climate, and rock layer structure, that no two streams behave identically. For example, Figure 16.23 shows a stream that has reverted to the youthful stage because of uplift of the terrain and is cutting a narrow gully through its own mature bed. But despite many such exceptions, the general pattern of development is a helpful guide provided due caution is observed.

In very arid climates, the wind also causes erosion even though water is still the more important agent. The wind erodes the surroundings by blowing across terrain that contains small, loose particles, transporting the particles from place to place. A typical example is a sand or dust storm. The sand dunes shown in Figure 16.24 are evidence of this kind of erosion. Often the wind-driven sand itself causes erosion as it abrades the objects it strikes. This form of wind abrasion can be seen in wind-sculptured rocks and abraded wooden telephone or electric poles. During the highly damaging dust storms of the 1930s, wind erosion stripped off much of the topsoil of the central United States, turning once-fertile plains into a "dust bowl."

A third type of erosion occurs in freezing, moist climates; it is the damaging effect of alternate freezing and thawing of water. You already know that water expands as it freezes to form ice. This is visible wherever asphalt and concrete roads buckle and deteriorate rapidly as subsurface water alternately expands and contracts with freezing and thawing. A similar process occurs in fissures and other water-retaining areas of rock strata. The formation of ice in a water-filled crevice exerts tremendous expansion forces upon the rock. Sooner

16.23 This river in California's Hungry Valley has cut a new canyon in its mature bed. (Courtesy of John S. Shelton.)

16.24 Sand dunes are one result of wind erosion. (Courtesy of Tad Nichols.)

16.25 The Teton Mountains of Wyoming; frost wedging makes the peaks sharp and steep. (Courtesy of L. L. Huff, U.S. Geological Survey.)

16.26 A glaciated valley in Fjordland, New Zealand. (Courtesy of Tad Nichols.)

or later, the rock layers are split apart and the rocks fragment. The fragments are then often small enough so that rain and wind can quickly erode them further. At the base of cliffs in the Rocky Moutains, the fragments of this *frost-wedging* process are always evident. Frost wedging is responsible for the sharp, spirelike peaks of the Teton Mountains shown in Figure 16.25.

More important than frost wedging is the erosive power of frozen water in the form of glaciers. During the last ice age, which ended only 10,000 years ago, vast ice sheets covered and planed off most of central and eastern Canada and the northern portions of the United States. The ice cap on Greenland is a small remnant of this huge ice mass. In high, mountainous areas, where snow does not melt in the summer, the snow piles up and turns into glacier ice. The glaciers slide down the mountains, scouring the valley sides and floors, and producing the characteristic U-shaped valleys of the Tetons, the northern Rockies, and the Sierra Nevada. Figure 16.26 shows a glaciated valley in another part of the world.

Geologic Time Determination

In the preceding sections, we have referred to events in the past. Let us see how geologists trace the course of the earth's history by using two basic methods to determine the age of objects found on and within the earth. You already know one of these, radioactive-isotope age determination. You will recall that it is based on the measurement of radioactivity in rocks and wood. The age of the object is obtained in terms of the half-life of the radioactive isotope. When an object is one half-life old, the radioactivity has decreased by half; when it is two half-lives old, the radioactivity is only one-fourth as large as it was originally, and so on. Knowing the half-life of the isotope concerned, a geologist can determine the age of the object directly in years. Using this method, we have determined that the earth itself is about 4.5 billion years old.

Long before radioactivity was discovered, however, geologists were charting the history of the earth. Their techniques are based upon the layered structure of the earth's crust. Each layer was formed from the debris of things that existed on the earth's surface at a particular time in the past. The successive layers contain a continuous record of the history of the earth. In the past, it was difficult to assign a scale of years to this recorded history, and geologists had to estimate roughly when each stratum of rock was laid down. Today, radioactive dating techniques give the dates of these layers with good accuracy.

The major divisions of geologic time are shown in Table 16.2. Notice that the history of living organisms begins at the edge of the Precambrian era. We have found no evidence that complex life forms existed prior to that time. Thus, more than 80 percent of the history of the earth had passed before complex life forms appeared.

In many places on the earth, the whole history of millions of years lies exposed in the sharp cliffs standing in arid areas. Layer after layer of historical documentation can be seen on the walls of the Grand Canyon, for example. The fossils and other materials present in these layers provide us with a vivid picture of life in that area throughout its history.

Development of Life on Earth

As we examine the successive layers of historical debris, we find evidence of a progression of life forms. Much of this evidence is in the form of *fossils*, which are traces of ancient organisms—skeletons, footprints, leaves, and so on—that were imbedded and preserved in geological layers. Al-

Table 16.2 Major Divisions of Geologic Time

Millions of years ago	Era	Period	Explanation of name
0	Cenozoic era (age of recent life)	Quaternary period	The several geologic eras were originally named Primary, Secondary, Tertiary, and Quaternary. The first two names are no longer used; Tertiary and Quaternary have been retained, but are used as period designations.
		Tertiary period	
65			
		Cretaceous period	Derived from the Latin word for chalk (*creta*) and first applied to extensive deposits that form white cliffs along the English Channel.
135	Mesozoic era (age of middle life)	Jurassic period	Named for the Jura Mountains, located between France and Switzerland, where rocks of this age were first studied.
190			
		Triassic period	Taken from Latin *trias*, three, in recognition of the threefold character of these rocks in Europe.
225			
		Permian period	Named after the province of Perm, U.S.S.R., where rocks of this period were first examined.
280			
		Pennsylvania period	Named for the State of Pennsylvania, where these rocks have produced much coal.
320	Paleozoic era (age of ancient life)	Mississippi period	Named for the Mississippi River Valley, where these rocks are well-exposed.
345			
		Devonian period	Named after Devonshire County, England, where these rocks were first studied.
395			
		Silurian period	Named after Celtic tribes, the Silures and the Ordovices, that once lived in Wales, where rocks of these periods were discovered.
440		Ordovician period	
500			
		Cambrian period	Taken from Roman name for Wales (Cambria), where rocks containing the earliest evidence of complex forms of life were first studied.
600			
	Pre-Cambrian era	----------	The time between the birth of the planet and the appearance of complex forms of life. More than 80 percent of earth's estimated 4½ billion years falls within this era.

Source: "Geologic Time," U.S. Geological Survey, Washington, D.C., 1970.

16.27 *Algae beds found in pre-Cambrian rocks of Glacier National Park, Montana. (Courtesy of R. Rezhak, U.S. Geological Survey.)*

though conditions for preservation of fossils were good only at certain locations, we can still piece together a clear history of life on the earth. Fossilized remains of complex life forms have been found in layers dating back to the end of the Precambrian era about 600 million years ago. But life during the previous four billion years of the earth's existence is still largely a mystery.

1 *Pre-Cambrian era* (before 600 million years ago). For the period prior to about three billion years ago, no fossil forms have been found. The oldest fossils known are the outlines of tiny, one-celled microorganisms found in one form of quartz (chert). The Pre-Cambrian rocks shown in Figure 16.27 contain fossilized algae, a one-celled primitive form of plant life related to the algae that form green slime in fishponds today. Evidence of worm holes and other simple life forms have also been found in a few localities. Two factors explain the scarcity of fossils in late Pre-Cambrian rocks even after life probably appeared. First, many simple forms of life have no hard parts to be preserved. Second, the various destructive forces at work over such a long period of time would effectively obliterate most evidence of former life.

2 *Paleozoic era* (225 to 600 million years ago). About 600 million years ago, life forms began to leave an abundant record. This date marks the beginning of Paleozoic time. At first, the chief forms of life were marine invertebrates; many of these were relatives of present day crabs, lobsters, and crayfish. Figure 16.28 shows typical life forms from this period. Then the first fish appeared, heavily armored creatures quite different from modern fish. The amphibians came next, marking animals' first attempts to live on land, at least on a part-time basis. Some of the early amphibians resembled enlarged modern salamanders. About the same time, plants characteristic of swampland appeared. By the end of Paleozoic time, these plants became so abundant that they furnished the source material for coal in many areas of the earth, including the Appalachians. The Pennsylvania period of this era was so named because of the vast deposits of coal formed then in what is now the state of Pennsylvania.

3 *Mesozoic era* (65 to 225 million years ago). At the close of the Paleozoic era, the first small reptiles and modern cone-bearing plants (conifers) appeared. There was also glacial activity. By the start of the Mesozoic era, reptiles reigned on land, in the

16.28 *Typical forms of marine life early in the Paleozoic era (225 to 600 million years ago). (Courtesy of the American Museum of Natural History.)*

16.29 *A typical scene in the American West about 180 million years ago. (Courtesy of the American Museum of Natural History.)*

16.30 *Two flying creatures that evolved during the mid-Mesozoic era. The reptile form (right) became extinct, while the bird form (left) developed into the species we know today. (Courtesy of the American Museum of Natural History.)*

sea, and even in the air. Figure 16.29 shows a typical landscape of about 180 million years ago, when dinosaurs and a wide variety of other strange animals roamed the earth. Some of the largest were 120 ft long from nose to tail and weighed more than 20 tons. These huge animals disappeared toward the end of the Mesozoic era. The reason for their extinction is not certain. Overspecialization, colder climate, food shortage, epidemic, or perhaps some small animal that destroyed their eggs are a few possibilities.

Also living in this era were great soaring reptiles (see Figure 16.30) with wingspreads ranging to 25 ft. In addition, many varieties of marine reptiles swam in the earth's waters. But at the close of the Mesozoic, most of these strange reptiles had vanished. Only the simple ones, such as snakes and lizards and the deep sea turtle, survived.

Well along in the Mesozoic era the first birds appeared. The archeornis in Figure 16.30 differed from the modern bird in three ways: he had a reptilian head with teeth (instead of a beak), claws at the midpoint bend in his wings, and a fairly long, fleshy tail as a support for his tailfeathers (instead of the short flap of flesh found today). In every other respect he looked like a modern bird. Small mammals and flowering plants also began to appear at the close of this era.

4 *Cenozoic era* (began 65 million years ago). At the start of this era the ancestors of most modern living forms had come into being. The most significant new events were the increase in and diversification of mammals and flowering plants, both of which were now of prime importance.

Relatively late in this era we begin to find traces of man. About 20 million years ago there were creatures that had some features much like those of modern man. For example, apelike fossils possessing manlike jaw and tooth distinctions provide evidence that such creatures existed in Kenya, Africa at this time. However, the first truly manlike creature seems to have developed much later.

Man

There is no clear-cut date that marks the beginning of man on earth. Long before man appeared, apelike creatures were already living on the continents.

Since they have so many features in common with man, it is difficult to draw a line between early human forms and variants of apes. Even in the best instances, only a portion of a skeleton remains for study. Moreover, the behavior patterns of primitive man were evidently apelike, and so a clear distinction is practically impossible. The best we can do is classify these ancient creatures according to their skeletal similarities and differences compared to modern humans.

Animals with manlike features lived in central Africa as early as the beginning of the Quaternary period, about 3 million years ago. This was the time of the Great Ice Age, a period of nearly 2 million years during which glaciers covered the United States as far south as the Ohio River valley. Four times the glaciers moved southward and then retreated again. The last such glacial advance began about 30,000 years ago and retreated only 10,000 years ago. In spite of these difficult conditions, animal life flourished. This was the time of the huge mastodons, fur-covered rhinoceroses, and mammoths. Some of their carcasses have been found perfectly preserved, frozen in the tundra of Alaska and Siberia.

But the precursors of man were too intelligent and mobile to be caught easily in the ice or the swamp areas which could freeze or fossilize their bodies. One of the rare finds from this epoch is the creature *Australopithecus* (southern ape), which inhabited Kenya about 5.5 million years ago. This immediate precursor of man walked erect, lived on the ground, and made crude tools. More important, its teeth were manlike and its 4-ft-high body had a skull and brain size approaching that of modern humans.

At about the same time, a more advanced creature appeared. He is called *Homo habilis* ("capable man"), because the tools found with him indicate that he was more intelligent than his predecessors and contemporaries. Unlike *Australopithecus*, which developed rapidly at first but then became extinct, *Homo habilis* progressed to an even more manlike form, called *Homo erectus* because of its upright stance. This creature, also known as Java man and Peking man after the places of its earliest discovery, existed about 500,000 years ago. Although he had an apelike brow, his rather small brain had distinctively human contours and arrangement, as did his teeth. We believe he was a man, not an ape; he was an adept toolmaker and had learned the use of fire. He also appears to have been a cannibal.

How man's development progressed is not completely clear. One might expect the progression from Peking and Java man to be straightforward, culminating in modern man. However, about 150,000 years ago, Neanderthal man occupied France, Germany, Italy, the Middle East, and Northern Africa as well as other parts of the world. He was a brutish-looking creature, as Figure 16.31 shows, but recent discoveries indicate that he was more gentle than he looked. Up to about 35,000 years ago, he continued to develop. But, strangely, he appears to have become more apelike and less like modern man as time went by. This regression may have been an adaptation forced on Neanderthal man by the difficulties of his environment or his inability to compete with more aggressive manlike creatures.

About 35,000 years ago, Neanderthal man began to disappear and was succeeded, in Europe, by Cro-Magnon man of the species *Homo sapiens* ("thinking man"). *Homo sapiens* is the human predecessor of the modern European races. (Little is known about the predecessors of the other races.) In appearance, he was not unlike modern man. He progressed rapidly from his originally primitive way of life, and we know much about him from the magnificent paintings with which he decorated the walls of his caves. By 5,000 B.C. (7,000 years ago) and probably even earlier, he had begun to live in well-established villages and to cultivate the soil. In the stability and continuity of village life, civilization began. Although some sections of the world moved ahead faster than others, civilization progressed at an ever-accelerating pace.

16.31 *Neanderthal man occupied much of Europe, the Middle East, and Northern Africa about 150,000 years ago. (Courtesy of the American Museum of Natural History.)*

Looking over the history of mankind from our present vantage point, we cannot help but be impressed by the way *Homo sapiens* has subjugated both inanimate and animate things of the earth. He can manipulate at will nearly everything from atoms and cellular life to large aircraft and even the clouds above which they fly. In spite of this tremendous power, he has thus far failed in the most important task that faces him, the ability to live peacefully and with due regard for his fellow creatures. His superior intelligence has made it possible to build marvelous machines and to utilize the great resources of the earth. But, in so doing, he has often taken steps that could eventually destroy him. The fantastic power which he now holds can erase all living things from the earth in an almost instantaneous blast of light and radiation. Or, it can lead him to design a slow, painful, lingering death, literally suffocated in his own refuse. But we do not believe that *Homo sapiens* will extinguish himself. Although it is late, we will come to realize that self-destruction can and must be avoided. This is a task that requires self-sacrifice, intelligent analysis, and, above all, more than a little love and consideration for one another.

BIBLIOGRAPHY

1 SHELTON, J.: *Geology Illustrated*, Freeman, San Francisco, 1966. As its title suggests, this is primarily a picture book with supporting text. Over 400 pages of beautiful photographs makes this delightful reading for the layman.

2 VIORST, J.: *The Changing Earth* (paperback), Bantam, New York, 1967. Clear, concise explanations of geologic topics written as an encyclopedia would treat them.

3 FOSTER, R.: *Geology* (paperback), Merrill, Columbus, Ohio, 1966. A 130-page summary of the most important topics of geology, written for the beginner.

4 DIETZ R., and HOLDEN, J.: "The Breakup of Pangaea," *Scientific American*, October, 1970. A discussion of the continental-drift theory written by its best-known expositors. Authoritative but non-mathematical.

5 *National Geographic Magazine*, vol. 138, no. 3, August, 1970. A beautifully illustrated study of the ocean floor.

6 *Scientific American*, September, 1969. A complete issue devoted to the oceans.

7 *Scientific American*, September, 1970. A complete issue devoted to the biosphere.

8 HURLEY, P.: *How old is the Earth?* (paperback), Anchor Science, Doubleday, 1959. An interesting discussion of methods of dating the age of the earth and its crustal layers.

9 STRAHLER, A.: *Physical Geography*, 3d ed., Wiley, New York, 1965. A well-written standard geology text which should be consulted for an in-depth treatment of the material in this chapter.

10 LONGWELL, C., and FLINT, R.: *Introduction to Physical Geology*, 2d ed., Wiley, New York, 1964. An elementary geology text, clearly written and well illustrated.

11 ANDERSON, D.: "The San Andreas Fault," *Scientific American*, November, 1971 (page 53). A readable discussion of the fault system along the California coast.

SUMMARY

A mineral is any naturally occurring, inorganic crystalline material of definite chemical composition. Most have well-defined cleavage planes. Rocks are aggregates of various minerals. There are three basic rock classes: igneous rocks, formed directly from the cooling and solidification of magma; sedimentary rocks, which are compacted sediments; and metamorphic rocks, which are formed by high pressures and temperatures acting on igneous and sedimentary rocks and on sediments.

Igneous rocks can be divided into two groups. Extrusive rocks were formed when magma spewed out onto the earth's surface and cooled rapidly; intrusive rocks are found in dikes, sills, laccoliths, and batholiths, where the magma solidified slowly while still beneath the surface.

Most of our knowledge about the internal structure of the earth has come to us through observation of seismic waves which travel through the earth. The P wave is compressional and the S wave is transverse. The S wave cannot travel through liquid and thus helps distinguish between solid and liquid.

The crust of the earth is largely composed of silicates. Its atomic composition is 49 percent oxygen, 26 percent silicon, and 8 percent aluminum, with smaller amounts of other atoms, especially iron, calcium, sodium, potassium, and magnesium.

The earth has several layers. The crust is 3 to 10 miles thick; it has a sharp lower boundary called the Moho. The plastic zone is a very viscous part of the mantle. Below the mantle is the liquid outer core and then the solid inner core. Temperatures and pressures rise rapidly with distance deeper into the earth. The high pressures prevent the plastic zone and mantle from being molten.

The crust of the earth floats on the mantle. It is depressed under the continents, since the less-dense portions of the continents float like icebergs. This is the theory of isostasy.

Mountains are formed in three ways. Volcanic mountains are formed when magma spews out onto the surface of the earth through deep fissures. Mountain ranges such as the Alps have been formed by buckling of the earth followed by erosion of the softened strata. Fault-block mountains, for example the Sierra Nevada range, result from a vertical slip along a fault line. Lateral slippage can also occur along a fault, as it does on the San Andreas fault. Slippage along a fault produces an earthquake. When an earthquake occurs in the ocean floor (which has mountain ridges too), it can produce a huge wave called a tsunami.

Our knowledge of the ocean floor comes largely from sonar, the use of reflections of sound. The ocean floor has a varied terrain of mountains, trenches, and canyons. Volcanoes rising from the ocean bottom sometimes stand above the ocean surface as islands. Extending from the shoreline is the continental shelf, an underwater portion of the continent where the water is still shallow. Further out, the continental slope drops quickly to the bottom of the ocean proper. The continental shelves are already providing oil and gas; eventually the seas will provide another source of minerals and food. Magnetic measurements of the ocean floor have shown that the earth has reversed its magnetic field periodically. This gives us a way to trace the history of the ocean floor.

We believe that about 200 million years ago the continents were combined in one huge land mass called Pangaea, and that they drifted apart due to forces we do not yet understand. The continents are still drifting today, according to the theory, and the ocean floor is enlarged by magma produced at the site of the mid-ocean ridges.

The features of the continents are much more complex than those of the ocean floor because of erosion. Water erosion is responsible for the formation of our rivers. Wind erodes surfaces by sand and dust storms and by the abrasive action of the materials it carries. Frost wedging breaks up large rocks during freeze-thaw conditions. Glaciers are also responsible for erosion.

Geologists trace the history of the earth by means of radioactivity and by studying the layer formations of the earth's crust. They divide the earth's development into large time sequences called eras, which are subdivided into periods.

The Precambrian era ended about 600 million years ago; only the most elementary life forms appear to have existed then. The Paleozoic era extends from 225 to 600 million years ago; in this era plants of the swamp developed; the animal forms of life were the marine invertebrates, fish, and then amphibians. Late in the Paleozoic era the first small reptiles and conifers appeared. The Mesozoic era (65 to 225 million years ago) was the age of giant reptiles on land, in the water, and in the air. Later in the era, birds and then small mammals and flowering plants appeared. The Cenozoic era, which started about 65 million years ago, saw the development of life forms now on the earth.

Man has appeared only recently. Although *Australopithecus*, an "ape-man," lived 5.5 million years ago, *Homo erectus*, typified by Peking and Java man, apparently did not appear until about 500,000 years ago. Neanderthal man existed in Europe, the Middle East, and Northern Africa about 150,000 years ago. He was succeeded by Cro-Magnon man, the direct predecessor of the modern European races. By 5,000 B.C., man was beginning to form communities and cultivate the soil.

TOPICAL CHECKLIST

1 Distinguish between the following sets of terms: (*a*) mineral and rock; (*b*) granite and basalt; (*c*) igneous, metamorphic, and sedimentary rocks; (*d*) extrusive and intrusive rocks.

2 Distinguish between extrusive and intrusive activity. Apply the terms in describing the origin of volcanoes, dikes, sills, laccoliths, and batholiths.

3 (*a*) Discuss the origin and motion of the magma observed in volcanoes and sills. (*b*) How do seismic waves tell us the outer core region is molten? What is its composition thought to be? What are P waves? S waves?

4 About how long ago was the earth formed?

5 Why does the earth's crust differ in chemical composition from its interior? Discuss the composition of the various earth layers.

6 About how thick is the crustal layer? What is the Moho? What are the chief elements and molecular compounds in the crust?

7 How high is the estimated temperature 30 mi below the earth's surface?

8 What is one mechanism by which heat energy is continually produced within the earth?

9 Discuss the composition of the earth's mantle. What is the plastic zone?

10 Why is the mantle solid even though ordinary rocks melt at its temperature?

11 How does the density of the inner core compare to ordinary densities of the same material?

12 Define the principle of isostasy; continental block; continental plate.

13 Define diastrophism. Give examples to illustrate your definition.

14 Describe how folding can produce mountain ranges. What part does erosion play in the process?

15 What is a fault? Discuss how a fault gives rise to earthquake activity.

16 What is the San Andreas fault and what is its importance to Californians?

17 What are tsunamis and how do they originate?

18 Discuss the mechanism of formation of the three basic mountain types.

19 What is the mid-ocean ridge? What mechanism may have caused it?

20 Discuss the origin of the Hawaiian Islands.

21 Define the following: continental shelf, ocean trenches.

22 Discuss the continental-drift theory. What is meant by Pangaea?

23 Describe the stages in development of a river and relate it to water erosion.

24 Describe the effects of wind erosion and of frost wedging.

25 How does the geologist establish a time scale for the history of the earth?

26 What are the geological eras and what are their time limits?

27 Discuss the basic life forms existing during each era.

28 Trace the historical development of man. In doing so, refer to *Australopithecus*, *Homo habilis*, *Homo erectus*, Peking man, Neanderthal man, *Homo sapiens*, Cro-magnon man.

QUESTIONS

1 The earth becomes warmer as we go deeper below its surface. If the source of the heat is radioactivity, does this mean the earth is more radioactive at these depths than it is at the surface?

2 Attempts are being made to probe into the earth by drilling deep holes. Discuss the possible danger of initiating a new volcano accidentally in the process.

3 Small earth tremors occur frequently along the San Andreas fault. Suppose such tremors stopped for a period of a few years. Is this likely to be a bad or a good omen? Do you see any way man could control earthquakes along such a fault?

4 Sometimes a seashell is found imbedded in a layer of mountain rock. Does this mean that the mountain was covered by a sea at some previous stage of history?

5 Sometimes erosion levels the terrain and at other times it makes it more jagged. Why doesn't erosion always lead to leveling? Over the millions of years, why haven't the continents been eroded to flat planes?

6 The fossil shown in Figure 16.32 was found in a quarry in Wyoming. Try to estimate the earliest time that this fish might have lived. What data would you need to determine its true age?

7 Suppose we built a gigantic reservoir by erecting 5-mi-high walls enclosing Iowa, and then filled the reservoir with seawater. What changes would this be likely to cause on the face of the earth?

8 The magnetic poles of the earth have reversed several times during the existence of the earth. Along the ocean floor it is found that the orientation of magnetic rocks reverses in bands parallel to the mid-ocean ridge. Can you connect these two facts?

16.32 *A fossil found in a Wyoming quarry. (Courtesy of W. T. Lee, U.S. Geological Survey.)*

9   Suppose the gravitational force constant suddenly became ten times larger than it is now. Discuss the evolutionary changes you might expect as a result. In so doing, consider such topics as erosion, plant and animal evolutionary selection, climate, and pollution.

10 The earth has experienced great ice ages when glaciers have grown and extended into the midsection of the United States. Discuss possible reasons why this may have happened. We do not know for sure what caused the Ice Age. How might such an ice age come once again to the earth?

APPENDIXES

Appendix I MATHEMATICS

We review in this Appendix the mathematical operations necessary to solve the problems in the text. Although you will not have to know most of these operations if your teacher deemphasizes the mathematics, none of the material here is beyond your abilities.

Addition

The order in which quantities are added is not important. That is, $2 + 4 = 6$ and $4 + 2 = 6$. Similarly, $a + b = b + a$, where a and b are real numbers.

Subtraction

You know that $10 - 3 = 7$. It is also true that $-10 + 3 = -7$. Further, to subtract a negative number, simply change its sign and add. Thus $10 - (-3) = 10 + 3 = 13$ and $a - (-b) = a + b$.

Multiplication

The following rules involving signs are self-explanatory:

$(6) \times (3) = (3) \times (6) = 18$ or $a \times b = ab = ba$
$(-3) \times (6) = -18$ or $(-a) \times b = -ab$
$(-3) \times (-6) = +18$ or $(-a) \times (-b) = +ab$

Division

Here, too, the given rules are self-explanatory:

$3 \div 6 = \dfrac{3}{6} = 0.50$ or $a \div b = \dfrac{a}{b}$

$-3 \div 6 = -\dfrac{3}{6} = -0.50$ or $-a \div b = -\dfrac{a}{b}$

$-3 \div (-6) = \dfrac{-3}{-6} = +0.50$ or $-a \div -b = \dfrac{-a}{-b} = +\dfrac{a}{b}$

Fractions

In a fraction such as $\dfrac{a}{b}$, a is the numerator and b the denominator. For most purposes we shall replace fractions in computations by their decimal equivalent. Thus, $\tfrac{1}{2} = 0.50$, $\tfrac{1}{3} = 0.33$, $\tfrac{3}{10} = 0.30$, and $\tfrac{5}{20} = 5 \div 20 = 0.25$.

When the denominator of a fraction is itself a fraction, $2/\frac{3}{4}$ for example, the fraction in the denominator can be inverted and multiplied as follows:

$$2/\tfrac{3}{4} = 2 \times \tfrac{4}{3} = 2 \times 1.33 = 2.66$$

Another frequently used method is:

$$2/\tfrac{3}{4} = \frac{2}{0.75} = 2 \div 0.75 = 2.66$$

We can also multiply the numerator and denominator of a fraction by the same number. Hence

$$\frac{a}{b} = \frac{3 \cdot a}{3 \cdot b} = \frac{3a}{3b}$$

Similarly, we can divide the numerator and denominator by the same number. For example,

$$\frac{8a + 4b}{2a + 8b} \quad \text{becomes} \quad \frac{(8a + 4b) \div 2}{(2a + 8b) \div 2}$$

which works out to be $\dfrac{4a + 2b}{a + 4b}$, which is equivalent to the original fraction. We often use this rule when we cancel a quantity from both numerator and denominator. For example, $\dfrac{8a}{8b}$ is simply $\dfrac{a}{b}$ since both the numerator and denominator can be divided by 8. We say we have *canceled* 8 from both numerator and denominator.

Equations

Usually we want to *solve an equation*. That is, given an equation such as

$$2x + 16 = 32x - 4$$

we wish to find what x is. The following operations involving equations help us to find x.

a Quantities equal to the same quantity are equal to each other. Thus, given

$$a + b = c \quad \text{and} \quad x + y + z = c$$

we know that since both are equal to c, they are equal to each other; hence

$$a + b = x + y + z$$

b Both sides of an equation can be multiplied by the same quantity without invalidating the equation. Therefore, if

$$\tfrac{1}{7}c = a + b$$

then

$$7 \times \tfrac{1}{7}c = 7 \times (a + b)$$

or

$$c = 7a + 7b$$

c Both sides of an equation can be divided by the same quantity. For example, if

$$16c = 3a + 2b$$

then

$$\frac{16c}{16} = \frac{3a + 2b}{16}$$

or

$$c = \frac{3a}{16} + \frac{2b}{16}$$

which can be written as

$$c = \tfrac{3}{16}a + \tfrac{1}{8}b$$

(How did we change $\tfrac{2}{16}$ to $\tfrac{1}{8}$?)

d An equation of the general form

$$\frac{a}{b} = \frac{c}{d}$$

can be cross-multiplied. That is,

$$\frac{a}{b} \times \frac{c}{d} \quad \text{gives} \quad ad = bc$$

The same result is achieved by multiplying the numerators of both sides of the equation by bd, giving

$$\frac{abd}{b} = \frac{cbd}{d}$$

which gives, after canceling,

$$ad = cb$$

as we found by cross-multiplying.

e We can square both sides of an equation. Hence

$$3x = 4 \quad \text{is equivalent to} \quad 9x^2 = 16$$

That is, we can use (b) to multiply one side of the equation by $3x$ and the other by 4 since $3x = 4$.

f The reverse of (e) is also permitted; that is, we can take the square root of both sides of an equation. Then

$$9x^2 = 16 \quad \text{becomes} \quad 3x = 4$$

(See below for an explanation of the square root.)

g An isolated quantity can be moved from one side of an equation to the other, or *transposed*, provided we change its sign. Thus

$$3 + x = 16$$

can be written

$$x = 16 - 3 \quad \text{or} \quad x = 13$$

This rule simply involves subtracting the same quantity (3 in this case) from both sides of the equation.

Square roots

We recall that the square root of 9 is 3. By this we mean that when 3 is squared (multiplied by itself once), the result is 9. In symbols, we write $\sqrt{9} = 3$. Similarly,

$$\sqrt{16} = 4 \quad \text{and} \quad \sqrt{169} = 13$$

The latter result is correct since

$$13 \times 13 = 13^2 = 169$$

A situation we will sometimes encounter is $\sqrt{1 - (0.6)^2}$, which presents no problem if we take it step by step. First, square 0.6. Then, since $(0.6)^2 = 0.36$, we have

$$\sqrt{1 - (0.6)^2} = \sqrt{1 - 0.36}$$

But $1 - 0.36$ is simply 0.64 and so we find

$$\sqrt{1 - (0.6)^2} = \sqrt{0.64}$$

In other words, we are looking for the number which, when squared, is 0.64. Since $0.8 \times 0.8 = 0.64$, we see that

$$\sqrt{1 - (0.6)^2} = 0.8$$

Exponents

You know that "3^2" means "3 squared," that is, "3 multiplied by itself once." Hence $3^2 = 3 \times 3 = 9$. We say that 3 is raised to the exponent 2, or to the second power. Similarly, when we write $(16)^5$ we mean $16 \times 16 \times 16 \times 16 \times 16$. As another example, we have

$$10^5 = 10 \times 10 \times 10 \times 10 \times 10 = 100,000$$

Using symbols rather than numbers, we have

$$a^3 = a \times a \times a$$

A similar notation is used for negative exponents. The symbolism 3^{-2} is equivalent to 1 divided by 3^2, or $\frac{1}{3 \times 3}$. For example,

$$18 \times 3^{-2} = \frac{18}{1} \times \frac{1}{3^2} = \frac{18}{3^2} = \frac{18}{3 \times 3} = \frac{18}{9} = 2$$

Similarly,

$$10^{-3} = \frac{1}{10^3} = \frac{1}{1,000} = 0.001$$

One way to use this symbolism is shown in the following example.

$$\frac{3^1}{3^2} = \frac{3}{3^2} = \frac{3}{9} = \frac{1}{3}$$

But $\frac{1}{3^2}$ can also be represented by 3^{-2}, where the $-$ sign tells us that 3 should be in the denominator, and hence we also have

$$\frac{3^1}{3^2} = 3^1 \times 3^{-2} = 3^{-1} = \frac{1}{3}$$

Notice that we have added the exponents to obtain our result. The rule we used to accomplish this is stated as follows: When one quantity raised to an exponent is multiplied by the *same quantity* raised to another exponent, the exponents add. As an example,

$$10^5 \times 10^3 = 10^8 \quad \text{or} \quad 10^6 \times 10^{-4} = 10^2$$

We recall that $a^{-3} = \frac{1}{a^3}$. Likewise,

$$a^3 = \frac{1}{a^{-3}}$$

Therefore

$$\frac{a^7}{a^{-2}} = a^7 \times a^2 = a^9$$

For example, using the above rules, we have

$$\frac{2 \times 10^{-6} \times 10^2 \times 3}{4 \times 10^9 \times 9 \times 10^{-6}} = \frac{6 \times 10^{-4}}{36 \times 10^3}$$

$$= \frac{6 \times 10^{-4} \times 10^{-3}}{36}$$

$$= \frac{6 \times 10^{-7}}{36}$$

$$= \frac{6}{36} \times 10^{-7}$$

$$= 0.167 \times 10^{-7}$$

$$= 0.0000000167$$

since 10^{-7} means to divide by 10 seven times.

We can summarize these rules for exponents in the following way:

$$a^n \cdot a^m = a^{n+m}$$

$$a^n a^{-m} = a^{n-m}$$

$$\frac{a^n}{a^m} = a^n \cdot a^{-m} = a^{n-m}$$

$$\frac{a^n}{a^{-m}} = a^n \cdot a^m = a^{n+m}$$

One special case should be noted. We have

$$\frac{a^2}{a^2} = a^{2-2} = a^0$$

Now, since the numerator of the original fraction is the same as the denominator, the fraction must be unity; that is, $a^0 = 1$. In other words, *any quantity raised to the zeroth power is 1.*

Appendix II

TEMPERATURE SCALES

You will recall that the freezing point of water is 32 °F or 0°C and the boiling point is 212°F or 100°C. Clearly, it takes more Fahrenheit degrees than centigrade degrees to cover the same temperature range. The magic numbers in this regard are $\frac{5}{9}$ and $\frac{9}{5}$. For example, 90 Fahrenheit degrees covers the same range that $90 \times \frac{5}{9}$ or 50 centigrade degrees covers. Stated the opposite way, 50 centigrade degrees covers the same range that $50 \times \frac{9}{5}$ or 90 Fahrenheit degrees covers. To convert a temperature on one scale to the equivalent temperature on the other scale, we simply remember these numbers, $\frac{9}{5}$ or $\frac{5}{9}$, and ask how far away the temperature is from 0°C or 32°F. Let us illustrate with some examples.

When room temperature is 30°C, what will a Fahrenheit thermometer read? We note that 30 centigrade degrees covers the same range as $30 \times \frac{9}{5} = 54$ Fahrenheit degrees. But when it is 0°C, it is 32°F. So we must add 32 to the quantity obtained by multiplying the centigrade temperature by $\frac{9}{5}$. Hence,

Fahrenheit temperature = $\frac{9}{5}(30) + 32 = 54 + 32 = 86$

If the temperature of a bowl of soup is 162°F, what would a centigrade thermometer read when dipped in it? We noted that when it is 32°F it is 0° centigrade. Hence, before we can multiply the Fahrenheit temperature, 162, by $\frac{5}{9}$ to obtain the centigrade temperature, we must subtract 32. Thus we have

Centigrade temperature = $(162 - 32)\frac{5}{9} = 130(\frac{5}{9}) = 72.2$

Hence the temperature of the soup is about 72°C. Symbolically, we have

$$°F = \tfrac{9}{5}°C + 32$$

and

$$°C = \tfrac{5}{9}(°F - 32)$$

Mercury freezes at a temperature of −39°C. What temperature is this on the Fahrenheit scale? The temperature is 39 centigrade degrees *below* freezing. Or, it is $\frac{9}{5} \times 39 = 70$ Fahrenheit degrees below freezing. Hence the temperature is $32 - 70 = -38°F$. Notice that the numbers on the two thermometers nearly agree near this temperature.

If a winter temperature dips to −18°F, what will a centigrade thermometer read? Since −18°F is $32 - (-18) = 50$ Fahrenheit degrees below freezing, the temperature is $50 \times \frac{5}{9} = 27.8$ centigrade degrees below freezing. Therefore, the temperature is $0 - 27.8 = -27.8°C$.

Appendix III — LAWS AND FORMULAS

Page 5   Law of conservation of angular momentum —part I
A freely rotating system will rotate more rapidly if the mass of the system is brought close to the axis of rotation.

Page 23   Law of conservation of angular momentum —part II
In the absence of externally applied twisting forces, the axis of a spinning object will retain its orientation relative to the distant stars forever.

Page 38   Kepler's laws
1. All planets move in elliptical orbits about the sun. The sun is at the focus point of the ellipse. This is called the *elliptical orbit law*.
2. The line joining the sun and the planet sweeps out equal areas in equal times. This is called the *equal area law*.
3. The square of the time taken to make one trip around the orbit (i.e., the square of the period of motion) is directly proportional to the cube of the average orbit radius. This is called the *harmonic law*.

Page 46
$$\bar{v} = \frac{s}{t} \qquad (3.1)$$

Page 47   Acceleration
The acceleration of an object is equal to the amount its velocity changes in a unit of time, usually a second. The direction of the acceleration is the same as the direction of the velocity change.

Page 47
$$a = \frac{v - v_0}{t} \qquad (3.2)$$

Page 49   Acceleration due to gravity
In the absence of all forces except gravity, every object on the earth accelerates toward the earth with the acceleration $g = 32$ ft/sec².

Page 51   Newton's first law of motion
1. A body at rest will remain at rest.
2. A body in motion will remain in motion in the same direction *unless* an unbalanced force is applied to the object to cause it to change its state of rest or motion.

Page 52   Newton's third law of motion
For every force exerted on one object (the *action* force), an equal but opposite force (*reaction* force) is always exerted on some other object.

Page 54
$$F = ma \qquad (3.4)$$

Page 54 Newton's second law of motion
1. When a given mass is accelerated, the acceleration is proportional to the applied force.
2. The force required to provide a given acceleration is proportional to the mass of the object being accelerated.

Page 55 Weight
The weight of an object is the force of gravitation pulling on it.

Page 55 $\qquad W = mg \qquad$ (3.5)

Page 60 $\qquad mv - mv_0 = Ft$

Page 60 Linear momentum $= mv$
The change in momentum of an object, caused by a force F acting on it for a time t, is Ft.

Page 61 Law of conservation of linear momentum
In the absence of unbalanced external forces, the linear momentum of a system will remain constant.

Page 70 $\qquad$ Work $= Fs \qquad$ (4.1)

Page 72
$\qquad$ Work done during acceleration $= \tfrac{1}{2}mv^2 - \tfrac{1}{2}mv_0^2 \qquad$ (4.2)

Page 73 Kinetic energy $= \tfrac{1}{2}mv^2$
The work required to accelerate an object of mass m from speed v_0 to v is equal to the change in kinetic energy of the object.

Page 73 Gravitational potential energy
An object of weight $W = mg$ at a height h has a gravitational potential energy of Wh, or mgh.

Page 75 Law of conservation of energy and work-energy relation
1. The total energy of an object, kinetic plus potential, is constant provided that no unbalanced force except gravity acts on the object.
2. The original energy of an object plus the work done by accelerating forces minus the work done by stopping forces equals the final energy of the object.

Page 83 $\qquad F = G\dfrac{m_1 m_2}{r^2} \qquad$ (4.3)

Page 95 $\qquad PV = NkT \qquad$ (5.1)

Page 98 $\qquad PV = \tfrac{1}{3}Nmv^2 \qquad$ (5.2)

Page 98 $\qquad T = \tfrac{2}{3}k(\tfrac{1}{2})(mv^2) \qquad$ (5.3)

Page 98
Kelvin (or absolute) temperature is a direct measure of molecular kinetic energy.

Page 114
Like charges repel each other whereas unlike charges attract each other.

Page 115 $\qquad F = 9 \times 10^9 \dfrac{q_1 q_2}{r^2} \qquad$ (6.1)

Page 121 $\qquad I = \dfrac{q}{t} \qquad$ (6.2)

Page 124 Potential difference
The potential difference between two points (or plates) is the work done in carrying a $+1$ coulomb charge from one of the points to the other. The unit of potential difference is the volt.

Page 124
The work done in carrying a charge q through a potential difference V is qV. When a charge q falls through a potential difference V, it acquires an energy qV.

Page 127 Ohm's law
$\qquad I = \dfrac{V}{R} \quad$ or $\quad V = IR \quad$ Ohm's law $\qquad$ (6.3)

Page 128 $\qquad$ Power $= VI \qquad$ (6.4)

Page 128 Power
The power lost as a current I flows downhill through a potential difference V is VI. It is measured in watts.

Page 141 $\qquad B = \dfrac{2 \times 10^{-7} I}{r} \quad$ for a long wire $\qquad$ (7.1)

Page 142 $\qquad B$ inside solenoid $= 4\pi \times 10^{-7} \dfrac{NI}{L} \qquad$ (7.2)

Page 144 Right-hand rule
If you hold your opened right hand with fingers

pointing in the direction of the field and thumb pointing along the wire in the direction of current flow, then the palm of your hand will push in the direction of the force.

Page 148
$$r = \frac{mv}{Bq} \quad (7.3)$$

Page 150 Faraday's law
The induced emf in a coil is proportional to both the *rate of change* of flux through the coil and the number of loops of wire on the coil.

Page 151 Lenz's law
The induced emf is always in such a direction as to keep the flux through the coil from changing.

Page 161
$$\tau = \frac{1}{f} = \frac{1}{\nu} \quad (8.1)$$

Page 162
$$\lambda = v\tau = \frac{v}{\nu} \quad (8.2)$$

Page 196 Huygen's principle
Each point on the crest of a wave acts as a new source for wave crests.

Page 209 Speed of light in vacuum
The speed of light (and of all electromagnetic radiation) is always the same, c, regardless of motion of the radiation source or of the observer.

Page 210
Only relative velocities can be measured.

Page 213 Einstein's conclusions
When an object or timing mechanism moves with speed v past an observer, the stationary observer notices that, where c is the speed of light, 2.988×10^8 m/sec,
1. The moving timing mechanism is ticking out time too slowly, slower by a factor of $\sqrt{1 - (v/c)^2}$.
2. All moving objects appear to shrink along the line of motion; lengths in the direction of motion decrease by a factor of $\sqrt{1 - (v/c)^2}$.
3. No material object can be accelerated to speeds greater than the speed of light in vacuum c.

Page 214 Mass-energy interconversion
An object which has a mass m_0 when at rest (its rest mass) will appear to have a mass
$$m = \frac{m_0}{\sqrt{1 - (v/c)^2}}$$
when moving with speed v.

Page 215
When the energy of an object is changed by an amount ΔE, its mass will change by an amount Δm such that
$$\Delta E = \Delta m c^2$$
where c is the speed of light.

Page 220 Photons and light
A beam of light with wavelength λ (or frequency $\nu = c/\lambda$) consists of a stream of energy quanta (called photons), and the energy of each photon is $h\nu$ (that is, hc/λ).

Page 226 Bohr's first assumption
Bohr assumed that the atom has certain special orbits in which the electron can revolve without radiating energy, and these are called Bohr orbits.

Page 228 Bohr's second assumption
When an electron falls from orbit p to a smaller orbit n, the atom emits energy in the form of a photon; the photon's energy is given by $h\nu = E_p - E_n$.

Page 228 Bohr's third assumption
The stable electron orbits are those orbits for which the angular momentum of the electron is an integer multiple of $h/2\pi$.

Page 229 De Broglie Wavelength
$$\lambda = \frac{h}{mv} \quad (10.1)$$

Page 266
$$E_n = -\frac{13.6}{n^2} \text{ electron volts} = \text{Bohr orbit energies} \quad (12.1)$$

Page 273 Pauli exclusion principle
Only one electron can exist in each state of an atom or molecule. Two electrons cannot occupy the same state.

Appendix IV TABLE OF THE ELEMENTS

The values listed are based on $_6C^{12} = 12u$. A value in parentheses is the mass number of the most stable (long-lived) of the known isotopes.

ELEMENT	SYMBOL	ATOMIC NUMBER Z	AVERAGE ATOMIC WEIGHT	ELEMENT	SYMBOL	ATOMIC NUMBER Z	AVERAGE ATOMIC WEIGHT
Actinium	Ac	89	(227)	Erbium	Er	68	167.26
Aluminum	Al	13	26.98	Europium	Eu	63	152.0
Americium	Am	95	(243)	Fermium	Fm	100	(253)
Antimony	Sb	51	121.75	Fluorine	F	9	19.00
Argon	A	18	39.948	Francium	Fr	87	(223)
Arsenic	As	33	74.92	Gadolinium	Gd	64	157.25
Astatine	At	85	(210)	Gallium	Ga	31	69.72
Barium	Ba	56	137.34	Germanium	Ge	32	72.59
Berkelium	Bk	97	(247)	Gold	Au	79	196.97
Beryllium	Be	4	9.012	Hafnium	Hf	72	178.49
Bismuth	Bi	83	208.98	Helium	He	2	4.003
Boron	B	5	10.81	Holmium	Ho	67	164.93
Bromine	Br	35	79.909	Hydrogen	H	1	1.00797
Cadmium	Cd	48	112.40	Indium	In	49	114.82
Calcium	Ca	20	40.08	Iodine	I	53	126.90
Californium	Cf	98	(251)	Iridium	Ir	77	192.2
Carbon	C	6	12.011	Iron	Fe	26	55.85
Cerium	Ce	58	140.12	Krypton	Kr	36	83.80
Cesium	Cs	55	132.905	Lanthanum	La	57	138.91
Chlorine	Cl	17	35.453	Lawrencium	Lw	103	(257)
Chromium	Cr	24	52.00	Lead	Pb	82	207.19
Cobalt	Co	27	58.93	Lithium	Li	3	6.939
Copper	Cu	29	63.54	Lutetium	Lu	71	174.97
Curium	Cm	96	(247)	Magnesium	Mg	12	24.31
Dysprosium	Dy	66	162.50	Manganese	Mn	25	54.94
Einsteinium	E	99	(254)	Mendeleevium	Mv	101	(256)

ELEMENT	SYMBOL	ATOMIC NUMBER Z	AVERAGE ATOMIC WEIGHT	ELEMENT	SYMBOL	ATOMIC NUMBER Z	AVERAGE ATOMIC WEIGHT
Mercury	Hg	80	200.59	Samarium	Sm	62	150.35
Molybdenum	Mo	42	95.94	Scandium	Sc	21	44.96
Neodymium	Nd	60	144.24	Selenium	Se	34	78.96
Neon	Ne	10	20.183	Silicon	Si	14	28.09
Neptunium	Np	93	(237)	Silver	Ag	47	107.870
Nickel	Ni	28	58.71	Sodium	Na	11	22.990
Niobium	Nb	41	92.91	Strontium	Sr	38	87.62
Nitrogen	N	7	14.007	Sulfur	S	16	32.064
Nobelium	No	102	(253)	Tantalum	Ta	73	180.95
Osmium	Os	76	190.2	Technetium	Tc	43	(97)
Oxygen	O	8	15.9994	Tellurium	Te	52	127.60
Palladium	Pd	46	106.4	Terbium	Tb	65	158.92
Phosphorus	P	15	30.974	Thallium	Tl	81	204.37
Platinum	Pt	78	195.09	Thorium	Th	90	232.04
Plutonium	Pu	94	(242)	Thulium	Tm	69	168.93
Polonium	Po	84	(210)	Tin	Sn	50	118.69
Potassium	K	19	39.102	Titanium	Ti	22	47.90
Praseodymium	Pr	59	140.91	Tungsten	W	74	183.85
Promethium	Pm	61	(147)	Uranium	U	92	238.03
Protactinium	Pa	91	(231)	Vanadium	V	23	50.94
Radium	Ra	88	(226)	Xenon	Xe	54	131.30
Radon	Rn	86	222	Ytterbium	Yb	70	173.04
Rhenium	Re	75	186.2	Yttrium	Y	39	88.905
Rhodium	Rh	45	102.905	Zinc	Zn	30	65.37
Rubidium	Rb	37	85.47	Zirconium	Zr	40	91.22
Ruthenium	Ru	44	101.1				

Appendix V

SOLUTIONS TO THE ODD-NUMBERED PROBLEMS

CHAPTER 3

1. Using $s = \bar{v}t$ with $s = 1{,}800$ mi and $t = 4$ hr gives $\bar{v} = s/t = 450$ mi/hr.

3. Since $a = (v - v_0)/t$, with $v_0 = 0$, $v = 40$ in./sec, and $t = 3$ sec, we have $a = (40 \text{ in./sec})/(3 \text{ sec})$, which is $a = 13.3$ in./sec². We can find $\bar{v}$ since it is $(v_0 + v)/2$, so $\bar{v} = 20$ in./sec. Then $s = \bar{v}t$ gives us $s = 60$ in. To find its speed 2 sec after release we use $a = (v - v_0)/t$, which is $v = v_0 + at$. Placing in the values, we have $v = 0 + (13.3 \text{ in./sec}^2)(2 \text{ sec}) = 26.6$ in./sec.

5. The acceleration of a freely falling object is g, so, in this case, $a = 32$ ft/sec². Since $v_0 = 0$ and $t = 2.5$ sec, we find from $v = v_0 + at$ that $v = 0 + (32 \text{ ft/sec}^2)(2.5 \text{ sec}) = 80$ ft/sec. Its average speed is $\frac{1}{2}(v_0 + v) = 40$ ft/sec, so $s = \bar{v}t$ gives $s = (40 \text{ ft/sec})(2.5 \text{ sec}) = 100$ ft.

7. The easiest approach is to note that a ball dropped from a height of 20 m will have the same speed at the bottom as the ball must have if thrown up to that height. Therefore let us work the falling-ball problem. For it, $v_0 = 0$, $s = 20$ m, and $a = 9.8$ m/sec². (Notice we use g in the metric system.) Then, since $s = v_0 t + \frac{1}{2}at^2$, we have $20 = 0 + 4.9 t^2$ from which $t = \sqrt{4.08}$ sec. But $\sqrt{4} = 2$, and so t is about 2 sec. Using $v = v_0 + at$, we find $v = 0 + 9.8(2) \approx 20$ m/sec.

9. To find $a = (v - v_0)/t$ we have $a = (400 \text{ m/sec} - 0)/(0.002 \text{ sec})$ or $a = 200{,}000$ m/sec². Then, using $F = ma$ with $m = 0.010$ kg, we find $F = 2{,}000$ newtons.

11. We find $\bar{v} = \frac{1}{2}(v + v_0) = \frac{1}{2}(0 + 400 \text{ m/sec}) = 200$ m/sec. Using $s = \bar{v}t$, we have $t = s/\bar{v} = (0.02 \text{ m})/(200 \text{ m/sec})$ or $t = 0.00010$ sec. The deceleration is $-a = -(v - v_0)/t = 4{,}000{,}000$ m/sec². Then $F = ma$ gives $F = 40{,}000$ newtons, since $m = 0.010$ kg.

13. Before collision, one of the balls has a momentum $m(10 \text{ ft/sec})$, while the momentum of the other is $-m(10 \text{ ft/sec})$. The minus sign arises since momentum has direction, and, since the motions are opposite in direction, one ball must have negative momentum. The law of conservation of linear momentum tells us:

Momentum before collision
$$m(10 \text{ ft/sec}) - m(10 \text{ ft/sec}) = mv + mv$$

We use the same v for both after collision because they are stuck together. Dividing each side of the equation by m, we have $10 - 10 = 2v$, from which $v = 0$.

CHAPTER 4

1. Since the force needed to hold the child is directed *upward* and the motion is *sideward*, there is no motion in the direction of F. In the second case, both $F = 40$ lb and $s = 15$ ft are upward, so $W = Fs = 600$ ft-lb.

3. The work done equals the gain in kinetic energy. Thus
$$W = \tfrac{1}{2}mv^2 - \tfrac{1}{2}mv_0^2$$
or
$$W = \tfrac{1}{2}(2{,}400/32)(50)^2 - 0 = 94{,}000 \text{ ft-lb}$$

5. . . . .

7. Work done by the pushing force at the wheels equals the gain in kinetic energy of the car.
$$Fs = \tfrac{1}{2}mv^2 - \tfrac{1}{2}mv_0^2$$
or
$$(500 \text{ lb})(s) = \frac{1}{2}\left(\frac{3{,}200 \text{ lb}}{32 \text{ ft/sec}^2}\right)(60 \text{ ft/sec})^2 - 0$$
from which $s = 360$ ft.

9. The stone will rise until all its kinetic energy is changed to potential energy. Then
$$\text{KE lost by stone} = \text{PE gained by stone}$$
$$\tfrac{1}{2}mv_0^2 - \tfrac{1}{2}mv^2 = mgh$$

Canceling m from each term and remembering that $v = 0$ at the very top, we find $\tfrac{1}{2}v_0^2 = gh$, or $\tfrac{1}{2}(50 \text{ ft/sec})^2 = (32 \text{ ft/sec}^2)h$. Solving for h, we get $h = 39$ ft.

11. As the ball falls through the 0.5 ft to its lowest position, its potential energy is changed to kinetic energy.
$$\text{PE lost by ball} = \text{KE gained by it}$$
$$mgh = \tfrac{1}{2}mv^2 - \tfrac{1}{2}mv_0^2$$
Canceling through by m and placing in the values, we get
$$(32 \text{ ft/sec}^2)(0.5 \text{ ft}) = \tfrac{1}{2}v^2 - 0$$
from which
$$v = \sqrt{32} \approx 5.7 \text{ ft/sec}$$

13. Now some of the lost potential energy, as well as the kinetic energy of the car, goes into doing friction work:
$$\text{PE lost} = \text{KE gained} + \text{friction work done}$$
$$mgh = \tfrac{1}{2}mv^2 - \tfrac{1}{2}mv_0^2 + Fs$$
Notice m does not cancel out in this case. It is $m = (2{,}400 \text{ lb})/(32 \text{ ft/sec}^2)$, while $h = 20$ ft, $v = 10$ ft/sec, $v_0 = 0$, and $s = 300$ ft. Placing in these values and solving for F yields $F = 147$ lb.

15. The earth's force of attraction for the object is 20 lb. Since the sun is about 3×10^5 times as massive as the earth, the sun attracts it 3×10^5 times more strongly because of this factor. However, it is 20,000 times as far from the sun's center and since the gravitation force varies inversely as the square of the distance, the attraction is $(1/20{,}000)^2$ as large. Combining both effects, the 20-lb object feels an

attraction force to the sun of $(20 \text{ lb}) \times (3 \times 10^5) \times (1/20,000)^2 = 0.015$ lb. At night, the object is on the far side of the earth and so this force adds to the weight. During daytime, it subtracts. Hence the weight variation is about 0.030 lb.

17 ....

CHAPTER 5

1. 72°F is 40 Fahrenheit degrees above freezing. Since this is $40(\frac{5}{9}) = 22.2$ centigrade degrees above freezing, the centigrade temperature is $0 + 22.2$ or 22.2°C. The same procedure is followed for (b), (c), and (d). However, (c) is 22 F° below freezing while −13°F is 45 F° below freezing. The results are: (b) 35°C, (c) −12.2°C, and (d) −25°C.

3. Since $P_1V_1 = NkT_1$ and $P_2V_2 = NkT_2$, with $V_1 = V_2$, the ratio of the two equations gives

$$\frac{P_2}{P_1} = \frac{T_2}{T_1}$$

with $T_1 = 273$°K and $T_2 = 373$°K. Therefore

$$\frac{P_2}{P_1} = \frac{373}{273} = 1.37$$

5. Since $P_1V_1 = NkT_1$ and $P_2V_2 = NkT_2$, with $T_1 = T_2$, the ratio of the two equations gives

$$\frac{P_1V_1}{P_2V_2} = 1 \quad \text{or} \quad P_1V_1 = P_2V_2$$

In our case $P_1 = 100$ psi, $P_2 = 15$ psi, and $V_1 = 1$ liter. Placing these values in the equation yields $V_2 \cong 6.7$ liters.

7. We saw in this chapter that $T = (2/3k)(\frac{1}{2}mv^2)$. In this case

$$v = 3 \times 10^8 \text{ m/sec}$$

$$m = 20 \times 10^{-28} \text{ kg}$$

$$k = 1.38 \times 10^{-23} \text{ joule/°K}$$

Then

$$T = \frac{2}{4.14 \times 10^{-23}} \frac{20 \times 10^{-27} \times 3 \times 10^8 \times 3 \times 10^8}{2}$$

$$= \frac{180 \times 10^{-27} \times 10^{16}}{4.14 \times 10^{-23}}$$

$$= 43.5 \times 10^{-27} \times 10^{16} \times 10^{23}$$

$$= 43.5 \times 10^{-27} \times 10^{39}$$

$$= 43.5 \times 10^{12} \text{ °K}$$

9. The work done by a 50-kg person climbing a 3-m-high stairs is mgh or $50(9.8)(3) = 1,470$ joules. Now 2,000,000 calories is equivalent to 8,400,000 joules (multiply 2×10^6 by 4.2). Therefore the number of times N the person could climb the stairs is

$$N = \frac{8,400,000}{1,470} = 5,710$$

CHAPTER 6

1. According to Coulomb's law, $F = 9 \times 10^9 q_1q_2/r^2$. In this case $F = 0.010 \times 9.8$ newtons, $r = 2$ m, and $q_1 = q_2 = q$. Placing in these values gives

$$9.8 \times 10^{-2} = \frac{9 \times 10^9 q^2}{4}$$

After cross-multiplying and solving for q^2, one finds

$$q^2 = \frac{39.2 \times 10^{-2}}{9 \times 10^9} = \frac{39.2}{9} \times 10^{-2} \times 10^{-9}$$

$$= 4.4 \times 10^{-11} = 44 \times 10^{-12} \text{ coul}^2$$

But $10^{-6} \times 10^{-6} = 10^{-12}$ and $\sqrt{44} \approx 6.6$, so $q = 6.6 \times 10^{-6}$ coul.

3 From Coulomb's law,

$$F = 9 \times 10^9 \frac{160(160)}{(2)^2} = 576 \times 10^{11} \text{ newtons}$$

Since, from Chapter 3, one newton = 0.225 lb, this is equivalent to a force of 130×10^{11} lb. This is an extremely large force, and so we must conclude that a penny could not easily be given a charge this large.

5 As an alpha particle falls through a potential difference V it acquires a kinetic energy Vq, where $q = 3.2 \times 10^{-19}$ coul. Hence, $Vq = \frac{1}{2}mv^2$ gives

$$V(3.2 \times 10^{-19}) = \frac{1}{2}(6.68 \times 10^{-27})(4 \times 10^{14})$$

from which

$$V = \frac{13.36 \times 10^{-27} \times 10^{14}}{3.2 \times 10^{-19}}$$

$$= 4.2 \times 10^{-27} \times 10^{14} \times 10^{19}$$

$$= 4.2 \times 10^{-27} \times 10^{33}$$

$$= 4.2 \times 10^{6}$$

$$= 4{,}200{,}000 \text{ volts}$$

If it started with this speed initially, it would have to rise through the same potential difference in order to be stopped.

7 Since V is the work done in carrying unit charge through a potential V, the work done in carrying a charge q through a potential difference V is qV. In the present instance $q = 2 \times 10^{-6}$ coul, while $V = 500$ volts, so

Work $= (2 \times 10^{-6})(500) = 1{,}000 \times 10^{-6} = 0.0010$ joules

9 (a) The battery is connected so that the bottom wire in the figure is at the high potential end of the battery. Since current flows from high to low potential, it flows from $B \rightarrow A$ in the 6-ohm resistor. (b) The 2-ohm resistor offers only $\frac{1}{3}$ as much resistance as does the 6-ohm resistor. Therefore 3 times as much current will flow through it compared to the 6 ohm, since both are connected across the same potential difference. Thus the current is 9 amp. (c) Since $9 + 3 = 12$ amp flows from the bottom wire to the top wire, the battery must furnish a current of 12 amp. (d) Ohm's law gives the potential across the 6-ohm resistor to be $V = IR = 3(6) = 18$ volts. This is the potential from B to A and is therefore the potential difference of the battery.

11 Since current is charge per second flowing through a wire or device, a current of $\frac{1}{2}$ amp means that a charge of 0.50 coul per second flow through it. The charge on an electron is 1.6×10^{-19} coul, so the number of electrons flowing through it each second is (0.50 coul/sec)/$(1.6 \times 10^{-19}$ coul$) = 0.50/1.6 \times 10^{19} = 0.31 \times 10^{19}$ electrons each second.

13 (a) Since $V = 120$ volts and power $= VI$, in this case $60 = 120I$ and so $I = 0.50$ amp. (b) Since $V = 120$ volts and $V = IR$, we have $I = V/R = 120/20 = 6$ amp. (c) The current through both flows past A, so the current there is $I_b + I_t$. Only the current through the toaster flows past point B, so the current there is I_b.

15 ....

CHAPTER 7

1 Using the right-hand rule, you can see that the fields due to the two wires will be in opposite directions at the midpoint between the wires if the currents are in the same direction. Since the field due to each will have the same magni-

tude, the two fields will cancel at the point, $B = 0$. If the currents are in opposite directions, the two fields will be in the same direction at the midpoint. From Eq. (7.1), the field due to one wire is

$$B_{one} = \frac{2 \times 10^{-7} I}{r} = \frac{(2 \times 10^{-7})(30)}{0.05}$$

$$= (2 \times 10^{-7})(30)(20)$$

$$= 1,200 \times 10^{-7} \text{ tesla}$$

For the two wires together, $B = 24 \times 10^{-5}$ tesla.

3   According to Eq. (7.2), the value of B inside a solenoid is $4\pi \times 10^{-7} NI/L$. Placing in our values gives

$$B = 4(3.14) \times 10^{-7} \times \frac{1,000(0.5)}{0.40}$$

$$= 15.7(10^3)(10^{-7})$$

$$= 15.7 \times 10^{-4} \text{ tesla}$$

Notice that meters, not centimeters, must be used since our formulas are in the *mks* system.

5   When the proton falls through a potential difference V it acquires a kinetic energy Vq. Hence $Vq = \frac{1}{2}mv^2$, or

$$2,000(1.6 \times 10^{-19}) = \tfrac{1}{2}(1.67 \times 10^{-27})v^2$$

which gives

$$v^2 = 3,800 \times 10^8$$

from which

$$v \cong 62 \times 10^4 \text{ m/sec}$$

From Eq. (7.3) we have $r = mv/Bq$, where B must be in teslas. Since 10,000 gauss = 1 tesla, we have

$$r = \frac{(1.67 \times 10^{-27})(62 \times 10^4)}{0.20(1.6 \times 10^{-19})}$$

$$= 320 \times 10^{-4}$$

$$= 0.032 \text{ m}$$

CHAPTER 8

1   From Eq. (8.2) we have $\lambda = v\tau$ or $\lambda = v/\nu$. Solving for v and placing in the values gives

$$v = \lambda \nu = (2 \text{ ft})(40 \text{ Hz}) = 80 \text{ ft/sec}$$

3   The frequency of the eardrum is the same as the sound wave frequency, 264 Hz. Using Eq. (8.1), we have $\tau = 1/\nu = 0.0038$ sec.

5   The wavelength to which the string will now resonate is only half as large as previously. But since $\lambda \sim 1/\nu$, to decrease λ by a factor of 2 we must increase ν by that same factor. Hence the resonance frequency is $2 \times 198 = 396$ Hz.

7   A small coke bottle is about 0.20 m long. Consider the bottle to be a tube closed at one end; the open end will be an antinode and the other end will be a node. Hence the length of the bottle will be equivalent to about $\lambda/4$, and so we have $\lambda \cong 0.80$ m. Using Eq. (8.2), $\lambda = v/\nu$, and solving for ν gives $\nu = v/\lambda = (330 \text{ m/sec})(0.80 \text{ m}) \cong 400$ Hz.

9   For the tube to resonate, the wavelength must be just long enough to fit the tube properly. Since Eq. (8.2) tells us $\lambda = v/\nu$, we see that any change in v must be accompanied by a like change in ν if the wavelength (and resonance) is to be maintained. In our case, the speed v is changed by a factor $1,270/330 = 3.85$ as we go from air to hydrogen. To maintain

resonance, the frequency of the sound must also be increased by this same factor. Therefore the new resonance frequency will be $3.85 \times 200 = 770$ Hz. This accounts for the fact that a person who talks when his lungs are filled with hydrogen will have a very high-pitched voice.

11 Since the ends are free, the rod should vibrate with antinodes at the two ends and a node at the center. Hence $L = \lambda/2 = 0.60$ m. Therefore $\lambda = 1.20$ m. Using $\lambda = v/\nu$ to find v, we have $v = \lambda\nu = 1.20(4,000) = 4,800$ m/sec.

CHAPTER 9

1 Since $\lambda = v\tau = v/\nu$, we have $\nu = v/\lambda$, or

$$\nu = (3 \times 10^8 \text{ m/sec})/(0.03 \text{ m}) = 100 \times 10^8 \text{ Hz}$$

3 The flat plate displaces the beam slightly but does not alter its direction.

5 Compute the time taken for the pulse of light to travel 1,000 m. We have $s = \bar{v}t$, which gives

$$t = (1{,}000 \text{ m})/(3 \times 10^8 \text{ m/sec}) = 333 \times 10^{-8} \text{ sec}$$

This is far too short a time for him to be able to measure.

7 Using Eq. (9.1), we have

$$\frac{x}{D} = \frac{\lambda}{d}$$

with $D = 100$ cm, $x = 5.0$ cm, and $\lambda = 3 \times 10^{-8}$ cm. After cross-multiplying, we have $(5)d = 100(3 \times 10^{-8})$, giving $d = 60 \times 10^{-8}$ cm. This means $d = 60$ Å, and so the slits could only be a few atoms apart—too small to be practical.

CHAPTER 10

1 From Table 10.1, when $v/c = 0.9$ the relativistic factor is 0.44. Therefore the man on the spaceship will age only 0.44 as fast, and so in 1 year on earth he will age only 0.44 year.

3 Since the astronaut is in an inertial reference frame, he will obtain the same result as if he were at rest on the earth, namely, m_0. To the man on earth, it will be $m_0/\sqrt{1 - (v/c)^2}$. From Table 10.1, at $v/c = 0.90$ the relativistic factor is 0.44. The earth observer will see the mass to be $m_0/0.44 = 2.27\, m_0$.

5 Since $\Delta E = \Delta mc^2$ and since we change the object's energy by mgh, we have $mgh = \Delta mc^2$, or, dividing by mc^2, $\Delta m/m = gh/c^2$. But $g = 9.8$ m/sec², $h = 2{,}000$ m, and $c^2 = 9 \times 10^{16}$, from which $\Delta m/m = 2 \times 10^{-13}$, a completely negligible change.

7 The mass changed to energy is $\Delta m = 2(9 \times 10^{-31})$ kg and this must be equivalent to an energy $\Delta E = \Delta mc^2$. Hence $\Delta E = (18 \times 10^{-31}) \times (9 \times 10^{16}) = 162 \times 10^{-15}$ joule.

9 The energy of a photon is $h\nu = hc/\lambda$. But this must equal mc^2, where m is the relativistic mass of the photon. Equating $mc^2 = hc/\lambda$ and solving for m gives

$$m = \frac{h}{c\lambda} = \frac{6.6 \times 10^{-34}}{(3 \times 10^8)(6.6 \times 10^{-7})} = 0.33 \times 10^{-35} \text{ kg}$$

11 This wavelength photon is emitted as the electron falls from the third to the second orbit. Since the photon energy is hc/λ, we have

$$E_3 - E_2 = \frac{hc}{\lambda} = \frac{(6.6 \times 10^{-34})(3 \times 10^8)}{6.6 \times 10^{-7}} = 3 \times 10^{-19} \text{ joule}$$

13 A ping-pong ball has a radius of about 0.01 m. As stated in Section 10.8 the atom radius is about 10^{-10} m, while the nuclear radius (corresponding to the ping-pong ball) has a radius of about 10^{-15} m. Hence the atom is 10^5 or 100,000 times larger than the nucleus. The electron should therefore be at a distance of about 0.01(100,000) m from the center of the ping-pong ball, that is, at a distance of about 1,000 m, which is close to 0.6 mi.

CHAPTER 11

1 The mass increase is $18.01263 - 18.01142 = 0.00121u$. Since $1u$ is equivalent to an energy of 931 MeV (see Section 11.4), by proportion

$$\frac{0.00121}{1} = \frac{x}{931}$$

which gives $x = 1.13$ MeV as the minimum energy required.

3 From Fig. 11.4 we see that the mass defect per nucleon for Fe^{56} is about $0.0093u$. Since there are 56 nucleons in this nucleus, the total mass one would need to create to tear it apart is $56(0.0093) = 0.52u$.

5 The energy acquired by a charge q when it falls through a potential difference V is qV joules or qV/e electron volts (eV). For both an electron and proton, $q = e$, and so both of them will have energies 2×10^6 eV or 2 MeV. The alpha particle has a charge $2e$, and so its energy will be twice as large, 4 MeV.

7 From the graph, the time taken until only $\frac{1}{4}$ remains is about 2.0 half-lives. Therefore the time taken would be about $2(1,620) = 3,240$ years.

9 Since 1 MeV is $(1 \times 10^6)/\frac{1}{40} = 40 \times 10^6$ times as large as $\frac{1}{40}$ eV, the absolute temperature will have to be increased by that factor. But room temperature is about $27°C = 300°K$, and so the required temperature is $300(40 \times 10^6) = 12$ billion degrees absolute.

CHAPTER 12

1 Since $E_n = -13.6/n^2$ eV, the photon energy will be $13.6/36 - 13.6/49 = 0.10$ eV. A wavelength of 12,400 Å corresponds to 1 eV, and so this photon will have a wavelength ten times as large, that is, 124,000 Å.

3 The electrons will have energies of 100 eV. They can, at best, give rise to photons of this energy. Since 1 eV corresponds to a wavelength of 12,400 Å, these photons will have wavelengths no shorter than 124 Å.

5 Since 5798 Å corresponds to an energy of $12,400/5,798 = 2.14$ eV, the atom must have a vacant level 2.14 eV above one of its filled levels.

CHAPTER 13

1 Consider the case of a sodium atom isolated from a chlorine atom. Picture the following process done in steps:
   (a) Tear electron loose from Na (uses 5.1 eV).
   (b) Give the electron to Cl to form Cl^- (gives 3.6 eV).
   (c) Let Na^+ and Cl^- come together (gives 6.0 eV).
   We had to furnish 5.1 eV of energy, but the other steps gave us back 9.6 eV. Hence we gain $9.6 - 5.1 = 4.5$ eV of energy in forming NaCl from Na and Cl.

3 A molecule of nitrogen gas has the formula N_2. It consists of two atoms. From the periodic chart we see that the atomic weight of nitrogen is 14. Hence, from the statement of Problem 2, 28 g of nitrogen *molecules* contains 6×10^{23}

molecules. Thus, the mass of a nitrogen molecule is

$$\frac{28 \text{ g}}{6 \times 10^{23} \text{ molecules}} = 4.7 \times 10^{-23} \text{ g/molecule}$$

5   As in Problems 2 and 3, the mass of a carbon atom is $12/(6 \times 10^{23}) = 2 \times 10^{-23}$ g. If there are 1×10^{23} atoms in a cube 1 cm on a side, then the mass for this cube (mass/cm³) will be

$$(1 \times 10^{23} \text{ atoms/cm}^3)(2 \times 10^{-23} \text{ g/atom}) = 2 \text{ g/cm}^3$$

7   According to the reaction, four H atoms combine with two O atoms to form two molecules of H_2O. Since the molecular weight of H_2O is 18, we know 18 g of H_2O contains 6×10^{23} molecules. We wish to form 3 g of H_2O, so we will have only $\frac{1}{6}(6 \times 10^{23})$ or 1×10^{23} molecules of H_2O formed. This will require the same number of O atoms since the number of H_2O molecules formed equals the number of O atoms used. The mass of 1×10^{23} atoms of oxygen is $16(1 \times 10^{23}/6 \times 10^{23})$ or 2.67 g since 16 g of oxygen contains 6×10^{23} atoms. Similarly, since twice as many H atoms are needed to form this many water molecules, we will need 2×10^{23} atoms of hydrogen. Since each atom has a mass $1/(6 \times 10^{23})$ g, we need $\frac{1}{3}$ g of hydrogen. Therefore we need 0.33 g H and 2.67 g O.

9   In order to break the bond, a photon must have an energy of at least 4.7 eV. Remember that 1 eV corresponds to 12,400 Å and that the relationship is an inverse proportion. Therefore, the photon wavelength would be 12,400 Å ÷ 4.7 = 2640 Å. Hence any wavelength less than 2640 Å will be capable of breaking the bond.

Appendix VI ANSWERS TO THE PROBLEMS

CHAPTER 3

1. 450 mi/hr
2. 11.1 yd/sec
3. $a = 13.3$ in./sec^2; 60 in.; 26.6 in./sec
4. 40 ft/sec; 240 ft; 13.3 ft/sec^2
5. 80 ft/sec; 100 ft
6. 1.56 sec; 39 ft
7. 20 m/sec
8. −800 lb
9. 2,000 newtons
10. 5 ft/sec; 6 sec; $\frac{5}{3}$ ft/sec^2; $\frac{10}{3}$ lb
11. 200 m/sec; 0.00010 sec; 4,000,000 m/sec^2; 40,000 newtons
12. . . . .
13. $v = 0$
14. It bounces ahead at 4 ft/sec.

CHAPTER 4

1. None; 600 ft-lb
2. 19.6×10^5 joules
3. 94,000 ft-lb
4. 300,000 joules
5. . . . .
6. 180 ft
7. 360 ft
8. 710×10^5 ft-lb/sec; 1.3×10^5 hp
9. 39 ft
10. 36 ft/sec
11. 5.7 ft/sec
12. . . . .
13. 147 lb
14. . . . .
15. 0.030 lb
16. 2,000; 0.002; 0.016; 5.00
17. . . . .

CHAPTER 5

1. 22.2°C; 35°C; −12.2°C; −25°C
2. 86°F; 194°F; −40°F

3   1.37
4   225°K
5   6.7 liters
6   76×10^{-6} cm Hg
7   43.5×10^{12} °K
8   157 m/sec
9   5,710
10   19.5°C

CHAPTER 6

1   6.6×10^{-6} coul
2    $r = 1.52 \times 10^5$ m
3   130×10^{11} lb
4   2,500 volts
5   4,200,000 volts
6   -1.5×10^6 newtons/coul
7   0.0010 joules
8   6 amp; 5 ohms; 40 watts
9    $B \rightarrow A$; 9 amp; 12 amp; 18 volts
10   0.50 amp
11   0.50 coul; 0.31×10^{19} electrons/sec
12   40 ohms; 96 watts; 5.8 amp; at point 1
13   0.50 amp; 6 amp; $I_b + I_t$; I_b
14   0.77 amp
15   . . . .

CHAPTER 7

1    $B = 0$; 24×10^{-5} tesla
2    $B = 5 \times 10^{-5}$ tesla; $B = 1 \times 10^{-5}$ tesla
3   15.7×10^{-4} tesla
4    $v = 1.9 \times 10^4$ m/sec; $v = 3.5 \times 10^7$ m/sec
5   62×10^4 m/sec; 0.032 m
6   2.65×10^7 m/sec; 0.00076 m

CHAPTER 8

1   80 ft/sec
2   0.50 m
3   264 Hz; 0.0038 sec
4   198 Hz, 2×198, 3×198, etc.
5   396 Hz
6   550 Hz
7   400 Hz
8   25 cm, 75 cm, 125 cm, etc
9   770 Hz
10   12 ft; 27.5 Hz
11   4,800 m/sec
12   . . . .

CHAPTER 9

1   100×10^8 Hz
2   1,200 kHz; 250 m
3   The flat plate displaces the beam slightly but does not alter its direction.
4   2.53 sec
5   333×10^{-8} sec
6   . . . .
7   60 Å
8   12 m

CHAPTER 10

1   0.44 year
2   0.995
3   m_0; $2.27\, m_0$
4    $v/c = 0.995$; 7.4×10^{-13} joules
5   2×10^{-13}
6    $n = 9 \times 10^{14}$
7   162×10^{-15} joules

8 The pendulum could swing 5 cm high, 10 cm, etc., but no distances in between.
9 0.33×10^{-35} kg
10 ....
11 3.4×10^{-19} joule
12 16.3×10^{-19} joule
13 1,000 m
14 ....

CHAPTER 11

1 1.13 MeV
2 93.1 MeV
3 $0.52u$
4 480 MeV
5 2 MeV; 2 MeV; 4 MeV
6 2×10^7 m/sec
7 3,240 years
8 $\frac{1}{2}; \frac{1}{4}; \frac{1}{32}$
9 12 billion degrees absolute
10 ....

CHAPTER 12

1 124,000 Å
2 0.0095 Å
3 124 Å
4 Ionization energy of these atoms is 21.7 eV.
5 The atom must have a vacant level 2.14 eV above one of its filled levels.

CHAPTER 13

1 4.5 eV
2 1.67×10^{-24} g/atom
3 4.7×10^{-23} g/molecule
4 1.25×10^{23} atoms
5 2 g/cm³
6 71 g; 73 g
7 0.33 g H and 2.67 g O
8 1.13 g
9 Any wavelength less than 2640 Å

GLOSSARY

ABSOLUTE SCALE (Kelvin scale): The temperature scale on which zero is taken to be absolute zero. On it, water freezes at 273°K and boils at 373°K.

ABSOLUTE ZERO: The temperature at which the pressure of an ideal gas would become zero; $0°K = -273°C$.

ABSORBTION SPECTRUM: The wavelengths of light which are absorbed as the light passes through atoms of a gas, liquid, or solid.

ACCELERATION: The amount velocity changes in unit time. It is a vector whose direction is the same as that of the velocity change.

ACID: Most frequently, a hydrogen-containing molecule which produces hydrogen ions (protons) when dissolved in water. Weak acids ionize little, whereas strong acids readily ionize.

ALCOHOL: Molecules characterized by the group: CH_2OH. For example, ethyl alcohol or ethanol:

$$CH_3CH_2OH$$

ALPHA PARTICLE: A helium atom nucleus consisting of two protons and two neutrons. It is emitted by many heavy radioactive nuclei.

ALTERNATING EMF: The emf obtained by rotating a coil in a fixed magnetic field, as in the common ac generator of a power-generating station. It reverses its direction in a cyclic manner. Normally, its frequency is 60 cycles/sec or Hz.

AMINO ACID: A molecule, one end of which is an organic acid and the other an amine group. Amino acids are the "building blocks" or units from which protein molecules are built.

AMPERE: The unit of electric current equivalent to one coulomb per second.

ANGSTROM UNIT: A length unit equivalent to 10^{-10} m. It is useful since atoms are about this large in radius. Abbreviation: Å.

ANTINODES: The points of maximum vibration in a standing wave.

APOGEE: Farthest point from the earth for an earth orbiting satellite; also used in other similar orbiting situations (opposite of perigee).

APPARENT WEIGHT: The force which an object exerts upon its support. It is not equal to the object's weight if the object is accelerating.

ASTEROID: A planetlike celestial body orbiting the sun. Asteroids are at most only a few hundred miles in diameter, and their orbits lie chiefly between Mars and Jupiter. A planetoid.

BAROMETER: A device used for measuring atmospheric pressure.

ATOMIC NUMBER: The number of protons in the nucleus of an atom; equal to the number of electrons in the neutral atom.

BASALT: A heavy, dark gray or black, fine-grained igneous rock.

BASE: Most frequently, a molecule which produces OH^- ions when dissolved in water. Strong bases produce many such ions while weak bases produce only a few.

BATHOLITH: A vast (many miles) subsurface region of solid rock that is formed when a large pocket of magma (molten rock) solidifies.

BATTERY: A device for providing electric energy. Most common batteries change chemical energy to electrical energy.

BETA PARTICLES: Fast-moving electrons (usually those emitted in radioactive decay processes).

BINARY STARS: Two close stars which orbit each other.

BINDING ENERGY: The energy required to tear an atomic particle from its position in an atom; the energy needed to tear a nucleus into its constituent nucleons.

BLACK HOLE: A star so dense that light cannot escape from it.

BOILING POINT: The temperature at which the vapor pressure of a liquid equals the atmospheric pressure acting on it. At this temperature bubbles form and grow within the liquid.

BOLTZMANN'S CONSTANT: A constant of nature, Boltzmann's constant $k = 1.38 \times 10^{-23}$ joule/°K.

BROWNIAN MOTION: The zigzag motion observed for very small particles in a gas or liquid resulting from collisions between the particles and the surrounding molecules.

CARBOHYDRATES: Carbohydrate molecules are composed of carbon, hydrogen, and oxygen. They have the general formula $C_n(H_2O)_x$, and are typified by sugars, starches, and cellulose.

CELSIUS SCALE: A temperature scale on which the freezing and boiling points of water are 0 and 100°C respectively; also commonly referred to as centigrade.

CENTIGRADE SCALE: The Celsius scale (see definition above).

CHLOROPHYLL: The green pigment found in plants which absorbs sunlight and initiates the process of photosynthesis.

COHERENT WAVES: Waves which are identical in form and frequency.

COLD FRONT: The region of contact between an advancing cold mass of air and a receding warmer mass.

COMET: A celestial body composed of a large aggregate of meteors, ice, dust and gas. Some comets revolve about the sun in well-defined orbits of high eccentricity.

CONDENSATION OF A GAS: The coalescing of gas molecules into droplets; i.e., water vapor condenses to form fog and rain.

CONDUCTOR OF ELECTRICITY: A material which contains many charges which are free to move. Metals are the best such conductors.

CONSTELLATIONS: Groupings of visible stars seen in the night sky at various times of the year. Constellations are named after various mythological characters, inanimate objects, and animals.

CONTINENTAL-DRIFT THEORY: Theory that the continents have drifted apart, floating like blocks on the plastic zone, and are still in motion today.

CONTINENTAL SHELF: Those portions of the continents near their edges, below the ocean surface, extending to cliffs which descend steeply to the ocean floor.

CORIOLIS EFFECT (meteorology): An effect due to the rotation of the earth which causes air currents to deflect to the right in the Northern Hemisphere and to the left in the Southern Hemisphere.

COULOMB: The mks unit of electrical charge equal to the charge transferred in one second by a steady current of one ampere.

COVALENT BOND: A sharing of electrons between two atoms to effectively complete the outer shell of each. When one pair of electrons is shared, a single covalent bond results. A double covalent bond consists of two pairs of shared electrons.

COVALENT CRYSTALS: Solids in which the atoms are held in a well-defined lattice structure by covalent bonds.

CRAB NEBULA: The remnant of the supernova of 1054.

CRITICAL MASS: The mass of fissionable material just large enough so that a chain reaction can be maintained at a steady rate.

CRUST OF THE EARTH: The outer layer of the earth's sphere; varies from about 3 to 50 miles thick.

CRYSTAL: A solid in which the atoms are arranged in an ordered pattern.

CURRENT, ELECTRIC: The quantity of charge passing any given point in a wire each second. Measured in amperes.

CYCLONE MOTION: A rotating motion of the atmosphere often extending over hundreds of miles. Cyclones rotate clockwise in the southern and counterclockwise in the northern hemisphere.

DEUTERIUM: The isotope of hydrogen whose nucleus consists of one proton and one neutron; heavy hydrogen.

DEW POINT: That temperature below which, if air is cooled, water vapor will begin to condense out.

DIASTROPHISM: The formation of the earth's features (continents, mountains, and so on) by large-scale distortion, buckling, and sometimes fracture of the earth's crust.

DIFFRACTION: The bending of light or other waves into the region of shadow behind a barrier.

DIFFRACTION GRATING: A large number of very narrow, closely spaced parallel slits used to disperse light into its various colors.

DIKE: Solidified magma filling a vertical slab-like fissure in the earth's crust.

DOPPLER EFFECT: The phenomenon that sound heard from an approaching source is raised in pitch while sound heard from a receding source is lowered.

ECLIPSE: The partial or complete obscuring, relative to the observer, of one celestial body by another. In an eclipse of the sun, the moon passes between the earth and the sun, blocking an observer's view of it.

ELECTRIC FIELD: A region in which a stationary test charge experiences an electrical force.

ELECTROMAGNET: A magnet formed from a soft iron bar by passing a current through a cylindrical coil of insulated wire wound around the bar.

ELECTROMAGNETIC SPECTRUM: The various electromagnetic waves extending from long to short as follows: radio, radar and microwave, infrared and heat, light, ultraviolet, x-rays, and gamma rays.

ELECTROMAGNETIC WAVES: The oscillating combined electric and magnetic fields sent out by accelerating charges in an antenna, molecule, or atom.

ELECTRON: A fundamental negative particle found in the outer portion of all atoms: $q = -1.6 \times 10^{-19}$ coul; $m = 9.1 \times 10^{-31}$ kg.

ELECTRON SHELLS: In the Bohr picture of the atom, the series of concentric shells about the nucleus in which electrons can exist. Each shell is characterized by a quantum number and energy; complete shells contain definite numbers of electrons.

ELECTRON VOLT (eV): The unit of energy equivalent to the energy acquired by a charge of $+e$ as it falls through a potential difference of one volt; 1 eV $= 1.6 \times 10^{-19}$ joule.

ELEMENT: A substance which cannot be made by chemical union or separated into simpler substances by chemical processes.

EMF: The electrical energy imparted per unit charge from an energy source. For a battery, the "voltage" of the battery.

EMISSION SPECTRUM: The wavelengths of light radiated by an atom or molecule as its electrons fall to lower energy levels.

ENERGY LEVEL: One of the permitted energy values that an atom or molecule can possess.

EQUINOX: The two times during a year when days and nights are of equal length because the earth's axis is perpendicular to its ecliptic plane; about March 21 and September 21.

EROSION: The pulverizing, wearing away, and transport of rock and soil by wind, water, glacier action, and frost wedging.

ESCAPE VELOCITY: Minimum speed at which an object must be thrown from the earth to escape into space.

EVAPORATION: The escape of molecules from the surface of a liquid, whereupon they form vapor.

EVENING STAR: The planet Venus seen in the reflected rays of the sun.

EXOSPHERE: That portion of the atmosphere beyond 300 miles above the earth.

EXTRUSIVE ROCK: Igneous rock formed from magma which has flowed out onto the surface of the earth.

FAHRENHEIT SCALE: The temperature scale on which the freezing and boiling points of water are 32 and 212°F respectively.

FAULT LINE: A fracture line in the earth's crust. Earthquakes occur along a fault line as one fracture surface periodically slips along the other.

FISSION REACTION: The splitting of a heavy nucleus into two or more smaller nuclei with the consequent release of energy. Used as the energy source in nuclear reactors.

FLUX, MAGNETIC: The number of magnetic field lines passing through a given area, usually through a coil.

FREQUENCY OF VIBRATION: The number of vibration cycles (i.e., to-and-fro motions) completed in unit time; frequency = one divided by the period.

FUSION REACTION: The joining together of nuclei of light elements with the consequent release of energy.

GALAXY: A large grouping of stars that is often hundreds of thousands of light years in diameter.

GAMMA RAYS: Electromagnetic waves with wavelength shorter than about an angstrom which originate from transitions within nuclei of atoms (identical to x-rays except for their mode of origin).

GAUSS: A non-mks unit of magnetic field strength. 10,000 gauss = 1 tesla.

GENERATOR, ELECTRIC: An emf source consisting of one or more coils rotating in a magnetic field. Makes use of Faraday's law of induction.

GRANITE: A coarse-grained igneous rock containing clearly visible crystals ranging from pink to gray.

GRAVITATIONAL ACCELERATION: The acceleration of an object in free fall; 32.2 ft/sec^2 or 9.8 m/sec^2 on earth.

GREENHOUSE EFFECT: A result of the atmosphere above the earth (particularly the CO_2 in it) trapping heat radiation energy from the earth, tending to cause the atmosphere to warm.

HALF-LIFE: The time taken for half of an original sample of radioactive atoms to decay to a new form.

HEAT: The energy resident in the random motion of molecules.

HEAVY HYDROGEN: The isotope of hydrogen whose nucleus consists of one proton and one neutron. Deuterium.

HORSEPOWER: A power unit equivalent to 550 ft-lb/sec or 746 watts.

HURRICANE: A violent low-pressure cyclonic disturbance extending over many miles with wind speeds often exceeding 100 mi/hr.

HYDROCARBONS: Molecules composed predominantly of hydrogen and carbon atoms.

HYDROGEN BOND: A bond characterized by the bond between adjacent water molecules where one molecule is held to the next by the electrostatic attractive force between the exposed proton of one water molecule and the oxygen of the other.

HYDROXIDE: A molecule which contains an OH grouping that ionizes to form OH⁻ ions in water.

IGNEOUS ROCKS: Rocks formed when magma solidifies on or near the surface of the earth.

INDEX OF REFRACTION: The ratio of the speed of light in vacuum to the speed of light in the material in question.

INDUCED CHARGE: An excess charge on some portion of a conductor which has been forced there by a nearby excess of charge on another object.

INERTIA: The tendency of an object at rest to remain at rest and of an object in motion to remain in motion. The magnitude of an object's inertia is given by its mass.

INERTIAL REFERENCE FRAME: A coordinate system which is not accelerating. Newton's laws of motion apply in this type of frame.

INFRARED WAVES: Electromagnetic waves with wavelengths longer than visible red light (7000 Å) and shorter than about a millimeter.

INSULATOR: A material containing negligible charges which are free to move. Does not conduct electricity.

INTERFERENCE, COMPLETE DESTRUCTIVE: Interference occurring when two waves acting at the same place exactly cancel each other.

INTERFERENCE, CONSTRUCTIVE: Interference occurring when two waves acting at the same place reinforce each other.

INTRUSIVE ROCK: Igneous rock formed when magma solidifies beneath the earth's crust.

INVERSION LAYER: A condition which results when the air near the surface of the earth is cooler than the air a thousand or so feet higher. The surface air remains stagnant, leading to increasing air-pollution.

ION: An atom which has lost or gained one or more electrons and therefore carries a net charge.

IONIC BOND: The electrostatic attractive forces holding two or more ions together.

IONIC SOLUTION: A solution in which at least some of the dissolved molecules have split apart so as to form ions.

IONIZATION ENERGY (or potential): The energy required to tear an electron away from an atom.

IONOSPHERE: That portion of the atmosphere which is highly ionized by the rays of the sun. Its altitude varies but is in the range of 40 to 150 miles.

ISOSTASY: The theory that portions of the earth's crust rise and fall as erosion and deposition affect the way they float on the plastic layer.

ISOTOPES: Two atoms whose nuclei contain the same number of protons but different numbers of neutrons are called isotopes of the same element.

JOULE: The mks unit of work and energy.

KELVIN SCALE: The absolute temperature scale on which zero is taken to be absolute zero. On the Kelvin Scale, water freezes at 273°K and boils at 373°K.

KINETIC ENERGY: The ability of an object to do work because of its motion. For speeds much less than that of light, it is given by $\frac{1}{2}mv^2$ measured in joules.

KINETIC THEORY: The theory of molecular motion in gases, liquids, and solids first developed by Boltzmann and Maxwell.

LASER: (Light Amplification by Stimulated Emission of Radiation.) A light source in which the atoms are stimulated to emit their waves all in unison; a highly coherent light source.

LATTICE (of a crystal): The geometric pattern formed by the center points of the atoms in a crystalline solid.

LAVA: Solidified magma which has spewed out onto the surface of the earth from a volcano.

LIGHT QUANTUM: A photon. A small packet of light energy having energy of $h\nu$ or hc/λ.

LIGHT YEAR: The distance light travels in one year, 5.9×10^{12} miles.

LINE OF FORCE: A line used to represent the direction of a magnetic or electric field.

MAGMA: Molten rock.

MAGNETIC FIELD: A region in space where a magnet experiences an aligning force. A compass needle lines up along the field lines.

MAGNETIC NORTH POLE: The end of a freely suspended bar magnet which points north on the earth.

MANTLE: The region of the earth below the crust extending down to the molten outer core.

MASS: The quantity intrinsic to an object which is a measure of the object's inertia. In terms of its weight W, $m = W/g$.

MASS DEFECT: The mass difference between the mass of a nucleus and the mass of its isolated constituent particles. When multiplied by c^2, it is equal to the binding energy of the nucleus.

MASS SPECTROGRAPH: A device to measure the masses of atomic-sized charged particles. Makes use of deflection in a magnetic field.

MESOSPHERE: Atmospheric region extending between altitudes of 30 and 50 miles.

METAL CRYSTALS: Solids in which the valence electrons are only very loosely held to the atoms and, upon escaping, roam through the lattice in such a way as to provide the binding energy of the crystal.

METAMORPHIC ROCK: Rock which has undergone plastic flow as a result of high pressures and temperature.

METASTABLE STATE: An excited configuration of an atom which loses energy only after an extended period of time.

METEOR: Any small aggregate object shooting through space.

METEORITES: Remnants of objects which have entered the earth's atmosphere and impacted the earth.

METEOROLOGY: The study of weather and climate.

METRIC SYSTEM: The system of measure in which the unit of length is the meter and the unit of mass is the kilogram.

MID-OCEAN RIDGE: A subsurface mountain system which winds its way through the centers of the earth's oceans. It is there that new ocean floor is being formed.

MILKY WAY: The galaxy in which our solar system is situated.

MINERAL: Any naturally occurring, inorganic crystalline material of definite chemical composition.

MKS UNITS: The metric system of units; based upon the meter, kilogram, and second.

MOHO or MOHOROVIČIĆ DISCONTINUITY: The lower boundary of the earth's crust.

MOLECULAR CRYSTALS: Solids having a lattice made of intact molecules, one molecule in each lattice site. They are usually held together by Van der Waals forces.

MOLECULE: The simplest structural unit that displays the characteristic physical and chemical properties of a compound.

MOMENTUM, LINEAR: The measure of straight-line motion represented by the product of an object's mass and velocity. It is a vector having the same direction as the velocity.

MUON (mu meson): A particle having a rest mass about 210 times larger than the electron mass.

MUTATION: An inheritable alteration of the genes or chromosomes of an organism.

NATURAL FREQUENCY: The frequency with which a system vibrates after it is set into vibration and allowed to vibrate freely.

NEUTRALIZATION REACTION: The chemical reaction between equivalent amounts of an acid and base; the OH^- ions from one and the H^+ ions from the other combine to form water molecules.

NEUTRINO: A particle having no charge and zero rest mass. Not to be confused with the neutron.

NEUTRON: A neutral, fundamental particle having about the same mass as a proton (1.67×10^{-27} kg). It is a major constituent of most atomic nuclei.

NEUTRON STAR: A tiny, dense, hot star which has essentially all its mass in neutrons and energy.

NEWTON: The force unit in the mks system. One newton gives a 1-kg object an acceleration of 1 m/sec² in the direction of the force.

NOBLE GASES: Those unreactive atoms which have filled outer shells: helium, neon, argon, krypton, xenon, and radon.

NODES: The places in a standing wave where no motion occurs.

NOVA: A star which has increased its brightness by a factor of thousands in a period of about a day or less.

NUCLEIC ACIDS: Giant organic chain molecules found in all living cells which direct the course of cell growth and function. The most important nucleic acids are RNA and DNA.

NUCLEON: A constituent of an atomic nucleus; a proton or neutron.

NUCLEUS OF AN ATOM: The tiny region at the center of an atom which contains all its positive charge and most of its mass.

ORGANIC CHEMISTRY: The chemistry of carbon compounds.

PANGAEA: The name given by proponents of one continental-drift theory to the vast land mass into which all the continents were presumably joined about 200 million years ago.

PERIGEE: Closest point to the earth for an orbiting satellite; also used in other similar orbiting situations (opposite of apogee).

PERIOD OF VIBRATION: The time taken for one complete vibration cycle.

PHOTOELECTRIC EFFECT: The ejection of electrons from a substance by light incident on the substance. First explained by Einstein in terms of photons.

PHOTON: A small packet of light energy (a light quantum) having energy $h\nu$ or hc/λ.

PHOTOSYNTHESIS: The process by which plants convert light energy into chemical energy, using energy from the sun to synthesize CO_2 and H_2O into plant cellular material and O_2.

PION (pi meson): The particle associated with the nuclear force field.

PLANCK'S CONSTANT: A constant of nature which appears in many relations that are concerned with the quantized nature of energy and the wave nature of particles.

PLANE WAVE: A wave for which the wave crests lie along straight lines or flat planes. Characteristic of waves far from their source.

PLANET: A large nonstellar celestial body circling a sun.

PLASTIC ZONE: The layer of the earth directly below the crust. It is solid but undergoes flow over a period of many centuries.

POLARIS: The North or Pole Star which is seen directly overhead by a person at the North Pole.

POLE STAR: See Polaris.

POLYMER: A very long molecule formed by adding many small molecules to yield a long chain of atoms.

POTENTIAL DIFFERENCE, ELECTRICAL: The energy required to carry a unit positive test charge from one of two points being considered to the other. Measured in volts.

POTENTIAL ENERGY, GRAVITATIONAL: The ability of an object to do work by virtue of its position. Given by mgh and measured in joules.

POTENTIAL ENERGY, SPRING: The ability to do work residing in a distorted spring.

POWER: Rate at which work is done; given by the formula, power = work/time taken. Measured in joules/sec, or watts.

PRESSURE: Force per unit area.

PRIMARY COIL: The coil of a transformer which is connected to the power source.

PRIMEVAL FIREBALL: According to the "big bang" theory, the extremely dense, hot ball which was our original universe.

PRISM: A device used to disperse the colors of light. Usually a flat triangle of glass.

PROTEINS: The giant chain molecules that make up the bulk of most animal cells. Composed of amino acids attached end to end.

PROTON: A fundamental positive particle found in the nuclei of all atoms. It is the hydrogen atom nucleus. $q = 1.6 \times 10^{-19}$ coul; $m = 1.67 \times 10^{-27}$ kg.

PULSAR: Astronomical radar signal sources which pulsate in energy with a frequency of a few seconds or less.

P WAVE: A compressive (like sound) seismic wave. It can travel through a liquid.

QUANTUM OF CHARGE: The smallest charge found in nature, $\pm 1.60 \times 10^{-19}$ coul. Equal in magnitude to charge of the electron and of the proton.

QUANTUM NUMBERS: A set of integers (and integers $+ \frac{1}{2}$) which are used to characterize a given structure for an atom.

QUASAR: Quasi-stellar radio source. A strong source of radio waves showing large red shifts.

RADAR WAVES: Very short-wavelength radio waves (a few centimeters in wavelength).

RADIAL FORCE: A force directed toward or away from the center of a circle.

RADIOACTIVE ATOM: An atom whose nucleus is unstable and which changes its nature by emitting a high-energy particle.

RADIOACTIVE DATING: Measuring the age of a material by observing how much of its original radioactivity still remains.

RADIOACTIVE TRACER: The use of a radioactive material to trace the path followed by a substance.

REACTOR, NUCLEAR: A power-generating device which uses the fission reaction as an energy source.

RED GIANT: A large, cool star of high luminosity, one of the later stages in the life of a star.

RED SHIFT: The color change from the blue to the red end of the spectrum undergone by light reaching us from a swiftly receding source.

REFERENCE FRAME: The coordinates or other reference system with respect to which we measure the position of objects.

REFRACTION: The bending of a beam of light as it enters one material from another as a result of different light speeds in the two materials.

RELATIVE HUMIDITY: The ratio of the actual amount of water vapor in the air to the maximum capacity the air could hold at that temperature (often multiplied by 100 and called percent humidity).

RESISTANCE, ELECTRICAL: Defined by Ohm's law, $R = V/I$, as the potential difference required to produce a current of one ampere through the element in question. Measured in ohms.

RESONANCE: The situation which occurs when the source of a wave exactly reinforces the vibration or reflected waves sent out previously. A maximum vibration occurs since the energy source continuously reinforces the vibration.

REST MASS: The mass of a particle (or object) as measured by an observer at rest relative to the particle.

RIGHT-HAND RULE: A convenient rule for remembering the direction of forces on currents in a magnetic field as well as for finding the direction of the field about current-carrying wires.

ROCHE'S LIMIT: The closest distance at which a moon can orbit a planet and still not be destroyed by tidal action.

ROCK: Any solid aggregate of inorganic material found naturally and in quantity in the earth's crust.

SATURATED AIR: Air which contains as much water vapor as it can hold. (relative humidity is 100 percent).

SCALAR: A quantity, such as mass or length, having magnitude but not direction. Compare a scalar quantity, such as speed, with a vector quantity, such as velocity.

SEDIMENTARY ROCK: Rocks formed from erosion products and similar sediments by pressure and cementing action.

SEISMIC WAVES: Disturbances sent through the earth from the sites of earthquakes or explosions.

SEMICONDUCTORS: Any of various substances, such as germanium or silicon, having electrical conductivity intermediate between insulators and conductors.

SHIELD REGION: The oldest portion of a continent; characterized by its rocks, usually the roots of an ancient mountain system.

SIMPLE HARMONIC MOTION: The back-and-forth cyclical motion characteristic of vibrating systems. Sinusoidal motion.

SINUSOIDAL MOTION: See simple harmonic motion.

SMOG: Air containing appreciable amounts of smoke, other pollutants, and fog, often occurring in conjunction with an inversion layer.

SOLAR SYSTEM: The sun, together with the nine planets and other celestial bodies captured by the sun's gravitational pull.

SOLENOID: A coil of wire wound on a cylindrical form.

SOLSTICE: The date in June when the sun is highest in the sky and in December when it is lowest; the longest and shortest days of the year, respectively. (Reversed in Southern Hemisphere.)

SONIC BOOM: The crash of sound and its destructive energy when the huge sound-wave crest built up by a supersonic aircraft passes by an observer.

SPECTRAL LINES: The individual wavelengths as they appear in a spectrometer when a line-spectrum light source is being used.

SPECTROMETER: A device for observing and measuring the wavelengths present in a beam of light. Either a prism or a diffraction grating can be used as the dispersing element.

SPEED: Distance moved divided by the time taken. This is the average speed.

SPIN: The characteristic of fundamental particles which measures their rotational and magnetic properties.

SPIRAL GALAXY: A large grouping of stars which forms a flat, spiral shape, probably as a result of its rotation.

STANDING WAVE: A stationary pattern of nodes and antinodes formed by the interference of two or more waves. It exists when the system is resonating.

STAR: A large, self luminous celestial body composed of gas in which the temperatures are high enough to sustain thermonuclear reactions.

STRATOSPHERE: The portion of the atmosphere extending between altitudes of 8 and 30 miles.

SUPERHEATED LIQUID: A liquid heated above its boiling point in which boiling has not yet begun.

SUPERNOVA: A star whose brilliance suddenly increases by tens of millions.

S WAVE: A transverse seismic wave, incapable of passing through fluid.

TERMINAL SPEED: The constant speed with which an object falls when it is falling fast enough that the friction forces equal the gravitational pull.

TERMINATOR: The dividing line between dark and bright at the edge of a shadow.

TESLA: The unit of magnetic field strength. The earth's magnetic field is about 10^{-4} tesla.

THERMOSPHERE: The region of the atmosphere between altitudes of from 50 to 300 miles.

TIDES: The twice-daily rise and fall of the ocean level at a given point on the earth. Results primarily from the attraction of the moon.

TORNADO: A restricted low-pressure region a mile or less in radius with high-velocity cyclonic winds near its center.

TRANSVERSE WAVE: A wave in which the vibration to-and-fro motion is perpendicular to the direction in which the wave is moving. (String, water, electromagnetic waves.)

TRIPLE ALPHA PROCESS: A fusion reaction in which three helium nuclei fuse to form carbon. It is the source of much of the sun's energy.

TROPOSPHERE: The lowest region of the atmosphere existing up to altitudes of about 5 miles at the poles and 10 miles at the equator.

TSUNAMI: A destructive ocean-wave system generated by an earthquake beneath the ocean.

TWIN PARADOX: The result of relativity theory which claims that if one twin remains on earth while the other travels through space at very high speed, the twin remaining on earth will be older than the twin in space after a specified period of "earth" time.

ULTRAVIOLET WAVES: Electromagnetic waves with wavelengths shorter than light (4000 Å) and longer than a few hundred angstroms.

UNSATURATED MOLECULE: A molecule which contains double or triple covalent bonds.

VALENCE: A measure indicating the number of other atoms an atom may bond with; equal to the charge (in units of e) which ions of that atom display in chemical reactions.

VALENCE ELECTRONS: The outermost electrons of an atom; i.e., the electrons in the outer incomplete shell of the atom.

VAN DER WAALS CRYSTALS: Solids in which the atoms and molecules are held in a well-defined lattice by van der Waals bonds.

VAN DER WAALS FORCE: A very weak attractive force always present between nearby molecules; the principle binding force in most very low boiling point liquids.

VAPOR PRESSURE: The pressure of the molecules which have evaporated from a liquid when they are confined in an enclosure.

VECTOR: A quantity which has both magnitude and direction.

VELOCITY: A vector quantity, the magnitude of which is a body's speed and the direction of which is the body's direction of motion.

VISCOSITY: The measure of a liquid's resistance to flow. High-viscosity liquids do not flow readily.

VOLT: The unit of potential difference equivalent to one joule per coulomb.

WARM FRONT: The region of contact between an approaching warm air mass and a receding colder mass of air.

WATT: The unit for power in the mks system; one joule per second.

WAVELENGTH: The distance between two adjacent crests of a wave. Related to wavespeed v and frequency ν by $\lambda = v/\nu$.

WEIGHT: The force with which an object is attracted by the earth (or, on the moon, by the moon).

WHITE DWARF: A very dense, hot star formed by the collapse of a red giant. Smaller than our sun.

WORK: The product of the force acting on an object and the distance that the object is moved in the direction of the force. Measured in joules.

X-RAYS: Electromagnetic waves with wavelengths shorter than a few hundred angstroms. They originate from electron collisions and transitions within atoms.

INDEX

Absolute temperature scale, 95
Absolute zero, 94
Absorption of light, 268
Abyssal plane, 368
Acceleration
   definition, 47
   gravitational on earth, 49–50, 57
   gravitational on moon, 50
   negative, 58
Accelerators, 254–257
Acetic acid, 300, 314
Acids, 299
Action-reaction law, 52
Affinity energy, 289
Age of earth, 377
Age determination by radioactivity, 260
Air molecules, speed of, 99
Air pollution, 345–349
Alcohols, 313
Algae beds, 379
Aliphatic hydrocarbons, 309
Alkali metals, 293
Alluvial fan, 372
Alpha Centauri, distance to, 14
Alpha particle, 252
Alternating current, 152
Alternating voltage, 152
Amides, 316
Amines, 316

Amino acids, 319
Ammeter, 127, 146
Ammonia, 316
Ammonium hydroxide, 300
Ampere, unit, 121
Ampere's theory of magnetism, 143
Amplitude of vibration, 162
Andromeda galaxy, 7
Angstrom unit, 191
Angular momentum conservation law, 5, 123
Angular momentum quantized, 228
Antenna, radio, 185–189
Antimatter, 258
Antinode definition, 165
Apogee, of moon, 23
Apparent weight, 86
Aristotle, and falling objects, 44
Aromatic hydrocarbons, 314–316
Asteroids, 37
Astronaut in orbit, 86
Astronomy, 19–41
Atmosphere
   composition, 329
   density vs height, 329
   layers of, 328–332
   pollution of, 345–349
   pressure of, 101
   temperature vs height, 330

Atoms and the periodic table, 265–282
Atom size, 106
Atomic bomb production, 249–251
Atomic energy levels, 266
Atomic magnets, 143
Atomic mass unit, 224, 241–242
Atomic number, 222, 241
Atomic state, 273
Australopithecus, 382
Avogadro's number, 307

Balmer formula, 236
Balmer series, 225
Barometer, 102
Barringer meteorite, 26, 37
Basalt, 356, 362, 363, 371
Bases, 300
Batholith, 358
Batteries, 121
Bead on a wire analogy, 81
Beats, in sound, 183
Benzene, 314
Beta Particle, 252
Bicarbonate of soda, 306
Big bang theory, 1
Binary star, 10
Binding energy of nuclei, 244–247
Biotite, 355

Bird forms, early, 381
Black body radiation, 216
Black hole, 17
Blue norther, 336
Bohr, Niels, 227
Bohr model of hydrogen atom, 226–228
Bohr orbits, 227
Boiling, 105–106
Boltzmann, Ludwig, 101
Boltzmann's constant, 95
Bond
  covalent, 287
    double, 288, 315
    ionic, 289, 293
Bonding in solids and liquids, 291
Bond representations, 288
British system of units, 56
Brownian motion, 99
Bumping of liquids, 105
Burning of hydrocarbons, 310
Butane, 309–312

Calcite, 355
Calcite crystal, 107–108
Calcium bicarbonate, 301
Calcium fluoride, 294
Caloric, supposed fluid, 91
Calorie and work unit, 90
Canyons, in ocean floor, 369
Carbohydrates, 323
Carbon dating, 261
Carbonic acid, 301
Carboxyl group, 314
Cell damage, 322
Celsius scale, 93, 95
Cenozoic era, 381
Centigrade scale, 93
Cesium chloride crystal, 107
Chain reaction, 251
Chemical formula, 288
Chemical names, summary, 301
Chemical reaction, definition, 287
Chemistry
  inorganic, 285–308
  organic, 309–326
Chlorine gas molecule, 287
Chlorophyll, 323
Charge
  acceleration of, 124

forces between, 115
force in magnetic field, 146
induced, 117
kinds of, 114
quantum of, 113
types, 114
Circuit breaker, 130
Circuits, 126–135
Circular motion in magnetic field, 148
Climate, effect of oceans, 343
Clocks and relativity, 213
Cobalt-60, 260
Coherent waves, light, 196, 279
Coil in magnetic field, 144
Cold front, 336–339
Collisions, 60–63
Color of light, 194–196
Color and wavelength, 190
Comets, 37–39
Compass, 139
Components of vectors, 63
Composition of earth's crust, 362
Compound, definition, 285
Compressional waves, 169
Compton effect, 238
Condensation of galactic dust, 3
Condensation of gases, 104
Condensation reaction, 318
Conduction of heat, 103
Conductor, 116
Conglomerate, 356
Conservation
  of linear momentum, 61
  of energy, 75
Constellations, 20–22
Continental block, 363
Continental drift, 370
Continental shelf, 369
Continents, 363–368
Continuous spectrum, 195
Contrails, 331
Controlled experiment, 44
Conversion of temperature scales, 93
Copernicus, 38
Coquinoid limestone, 356
Core of earth, 362
Coriolis effect, meteorology, 332–333, 340
Coulomb's law, 115
Coulomb, unit, 114
Covalent bond, 287

Covalent crystals, 294
Crab nebula, 11
Crest of wave, 162
Critical mass, 251
Crust of earth, 361
Crustal plates, 371
Crystalline solids, 106–108, 293
Current
  eddy, 159
  electric, 120
Cyclohexane, 314
Cyclone circulation, 338–342
Cyclone type dust remover, 346

Dating, by radioactivity, 260
de Broglie, Louis, 230
de Broglie waves, 228
de Broglie wavelength, 229
Decay curve, 253
Deceleration, 58
Definite proportions law, 241
Deflection in magnetic field, 146
Deimos, Mars moon, 32
Delineator, 22
Delocalized bond, 294, 315
Density
  of atmosphere, 329
  of earth, 362
Deuterium, 245
Deuteron, 248
Devil's tower, 359
Dew point, 335
Diamond lattice, 295
Diastrophism, 365
Dielectric constant
  alcohols, 313
  and solubility, 298
Diesel engine, 110
Diffraction, 201–204
  and detail, 203
Diffraction grating, 200
Dipoles, molecular, 297
Direct voltage, 129
DNA, 320
Doppler effect, 175–177
Double bond, delocalized, 294, 315
Double helix, proteins, 320
Double slit experiment, 200
Dyes, 316

Ear damage, 178, 182
Ear response, 178, 182
Earth
  age, 260, 377
  atmosphere, 328–332
  composition, 353–362
  crust, composition, 362
  density, 362
  history, 327
  magnetic field, 140
  structure, 360–362
  temperature of interior, 354
Earthquakes, 366
Eclipse
  of sun, 28
  of moon, 26
Ecliptic plane, 22
Eddy currents, 159
Einstein, Albert, 211
Einstein letter to Roosevelt, 250
Electric current, 120
Electric field, 118
Electric field strength, 119
Electric generator, 151
Electric potential difference, 123
Electric power, 128
Electrical safety, 130
Electrical shock, 130
Electricity, 113–159
Electromagnets, 143
Electromagnetic spectrum, 189–191
Electromagnetic waves, 184–208
Electromagnetism, 138–155
Electromotive force
  induced, 149
  meaning of, 121
Electron affinity energy, 289
Electron charge, 119
Electron spin, 270
Electron in hydrogen atom, 265
Electron volt unit, 246
Electronegativity
  definition, 291
  table, 277
Electrostatic precipitator, 346
Elements, table of, 276
Elevator, weight in, 86
Elliptical galaxy, 7–8
Emission of light, 268
Emission spectra, 278

emf, induced, 149
Energy conservation, KE and PE, 75
Energy, electric, 122
Energy level diagram, atomic, 267
Equinox, 22
Erosion, 358, 365, 372–377
Escape velocity, 2
Esters, 314
Evaporation, 104–106
Evening star, 31
Exclusion principle, 271–274
Exosphere, 330
Exothermic reaction, 326
Expanding universe, 2, 16
Extrusive rocks, 357
Eye of hurricane, 341
Eye sensitivity, 190

Fahrenheit scale, 93
Fallout, radioactive, 264
Faraday, Michael, 150
Faraday's law, 150
Fault, 365–368
Fault block mountains, 365–366
Feldspar, 355
Fermi, Enrico, 250
Fibers, synthetic, 318
Field
  electric, 118
  gravitational, 118
  magnetic, 139
Field lines, magnetic, 139–141
Fireball, primeval, 1
Fish, fossilized, 387
Fission reaction, 249
Flood plain, 374
Flux, magnetic, 150
Focal point
  of lens, 194
  of mirror, 193
Folded mountains, 364–365
Foot pound unit, 69
Force
  charge in electric field, 115
  charge in magnetic field, 146
  coil in magnetic field, 144
  inverse square law, 83
  and motion, 51–61
  nuclear, 244, 258

  on wire in magnetic field, 144
  ratio, electric to gravitational, 115
  van der Waals, 291
Forces at a distance, 118
Formic acid, 299, 314
Formula, chemical, 288
Fossils, 377, 387
Frames of reference, 210
Free electrons, 116
Free-fall acceleration, 49
Freely falling bodies, 49
Frequency
  definition, 161
  relation to period, 161
Fronts, weather, 336–339
Fuse, 130
Fusion reaction, 247
  in stars, 8–12

Gabbro, 356
Galaxies, 3–8
  formation of, 3
Galileo
  life, 45
  and free fall, 45, 51
Gamma rays, 253
  wavelength range, 190
Gas law, 95
Gas, molecular speeds, 99
Gas, natural, 309
Gas pressure, molecular model, 96–100
Gas thermometer, 94
Gasoline, 309
Gauss, unit, 141
Geiger and Marsden experiment, 223
Generator, electric, 151
Genetic accidents, 322
Geology, 353–387
Geologic time, 377
Germanium 295
Girl, charged, 134
Glacier, 376–377
Glass
  as viscous liquid, 106
  volcanic, 357
Glycerol, 313
Gneiss, 356
Grand Canyon, 373, 377
Granite, 356

Graphite, 107, 294
Grating, diffraction, 200
Gravitation force vs height, 84
Gravitation
    law of, 83
    on moon, 83
Gravitational field, 118
Gravitational potential energy, 73–74
Greenhouse effect, 334–335
Ground wire, 129
Group, of periodic table, 275
Gulf stream, 345
Gun, recoil, 61

Half life, 252
Halley's comet, 37–39
Halogens, 293
Hawaiian Islands, 369
Head rest, use, 52, 62
Heat conduction, 103
Heat energy
    in liquids and solids, 103
    and molecular motion, 91–109
Heisenberg, Werner, 233
Heisenberg uncertainty principle, 231
Helix form of proteins, 319
Helium, 281
    atom, 270
    origin on earth, 327
Heredity, molecular basis, 321
Hertz unit, 161
High side of battery, 123
High, weather, 340
Homo erectus, 382
Homo habilis, 382
Homo sapiens, 382
Horsepower unit, 78
House circuits, 129
Humidity, 335–336
Hurricane, 340
Huygen's principle, 196
Hydrocarbon chain molecules, 309
Hydrochloric acid, 299
Hydrogen atom, 224–230, 265–269
    wave picture, 265
Hydrogen bomb, 248
Hydrogen bond, 297
    in alcohols, 313

Hydrogen energy levels, 266
Hydrogen molecule, 285–287
Hydrogen spectrum, 225
Hydrogen star, 9
Hydrogen sulfide, 290
Hydronium ion, 299
Hydroxides, 300

Ice age, 382
Ice crystal, 108
Ice, melting behavior, 297
Ice wedging, 374
Ideal gas law, 95
Igneous rock, 356, 357
Impulse, 60
Index of refraction, 193
Induced charge, 117
Induced emf, 149
Induced magnetization, 138–139
Inertia, 51–55
Inertial reference frame, 210
Infra red light, 190
Inorganic chemistry, 285–304
Instantaneous speed, 47
Instantaneous velocity, 47
Insulator, 116
Interchange of PE and KE, 79–82
Interference of light, 196–200
    of particle waves, 228–231
    of sound, 174
    of waves on string, 163
Intrusive rock, 358
Inverse square law force, 83, 115
Inversion layer, 347–349
Ion, definition, 289
Ionic bond, 289, 293
Ionic crystals, 293
Ionic solutions, 298
Ionization in water, 298–299
Ionization energy, definition, 268
    table, 274
Ionosphere, 331
Iron
    effect of, 143
    soft and hard, 138
Isomer, 312
Isostasy, 362
Isotope, 243

Java man, 382
Jet propulsion, 62
Joule, James, 100
Joule unit, 70
Jupiter, 13, 31–35

Kelvin scale, 95
Kepler, Johannes, 38, 40
Kepler's laws, 38–40
Kilogram, definition, 57
Kinetic energy
    definition, 72
    and temperature, 98
    and work, 72
Kinetic theory of gases, 99–105
Krakatoa, effect on weather, 335

Laccolith, 358
Lake Superior uplift, 363
Laser, 279
Lattice, 293
Lava, 357
Law
    of definite proportions, 241
    of energy conservation, 75
    of gravitation, 83
Length contraction, relativity, 213
Lenz's law, 151
Life cycle for energy, 323
Life
    development on earth, 377–383
    in other solar systems, 14
Light
    diffraction of, 201–204
    emission and absorption, 268
    from hydrogen, 225
    interference of, 196–200
    quanta of, 219–221
    reflection of, 191–193
    refraction of, 193–196
    speed of, 193
    speed and relativity, 209
Light year, 4
Lighting circuits, 129
Limestone, 356
Limiting speed, relativity, 214
Line spectrum, 196
Linear accelerator, 256

Linear momentum, 59
Lines of force
  electric, 119
  magnetic, 140
Liquids
  heat energy in, 103
  viscosity of, 104
Lithium atom structure, 271
Litmus paper, 301
Longitudinal wave, 171
Los Angeles smog, 348
Low, weather, 340
Lyman series, 226, 269

Magma, 357
Magma chamber, 357
Magnet, 138
Magnetism, 138–155
  atomic interpretation, 143
Magnetic field, 139–141
  of earth, 140
  force on wire, 144
  sources, 140
Magnetic force on charge, 146
Magnetic materials, 138, 143
Magnetization on ocean floor, 370–371
Man, development of, 381–383
Mantle, of earth, 362
Marble, 356
Mars, planet, 31–32
Mass
  definition, 53–57
  of atoms and nuclei, 241–243
  relation to weight, 55
  relativistic, 214
  variation with speed, 155
Mass defect, 246
Mass-energy conversion, 214, 246
Mass spectrograph, 154, 241
Maxwell, James Clerk, 185
Maxwell speed distribution, 99
Mean free path, 112
Measurement of atomic mass, 153
Melting, 106
Mendeleev, Dimitri, 275
Mercury, planet, 31
Mesosphere, 331
Mesozoic era, 379
Metal bond, 296

Metal crystals, 296
Metamorphic rocks, 356, 360
Meteor, 37
Meteorite, 37
Meteorology, 332–348
Methane, 288, 309
Metric system, 56
Mid ocean ridge, 370
Milky Way galaxy, 4
Mineral crystals, 354–355
Mineral
  definition, 353
  in water, 301
mks system, 56
Moho, Mohorovičić discontinuity, 361
Molecular crystals, 296
Molecular, definition, 285
Molecule, definition, 285
Momentum angular, 5, 23, 228
  conservation of, 61
  and force, 59
  linear, 59
Monomer, 317
Monsoon, 342
Moon, 23–29
  free fall acceleration on, 50
  gravitation on, 24, 83
  phases of, 26
Motion equations, 45–49
Motion, relative, 209
Motor principle, 144
Mountains, 364–368
Muon and time dilation, 215
Musical instruments, 171–173
Mutations, 322

Neanderthal man, 382
Neon, 291
Neptune, planet, 31, 36
Neutralization reaction, 300
Neutrino, 259
Neutron, 243
Neutrons in nuclei, 244
Newton, Isaac, 51
Newton, unit of force, 57
Newton and orbiting ships, 85
Newton's first law, 51
Newton's gravitation law, 83
Newton's jet wagon, 62

Newton's second law, 53
Newton's third law, 52
Nitric acid, 299
Nitrogen
  molecule, 287
  liquid, 297
Noble gases, 291
Node, definition, 164–165
Noise pollution, 178
Nomenclature
  chemical, 301
  for organic molecules, 312
Nova, 10
Nova Herculis, 11
Nuclear atom, 221
Nuclear binding energy, 244–247
Nuclear bombardment, 254
Nuclear force, 244, 258
Nuclear masses, 241–243
Nuclear physics, 240–261
Nuclear reactor, 251
Nucleic acids, 318–323
Nucleon, 244
Nucleus size, 240
Nylon, 318

Obsidian, 356
Ocean basins, 368–370
Ocean currents, 343–345
Oceans, effect on climate, 343
Ohm, unit, 127
Ohm's law, 126
Orbital motion, 84–86
Organic acids, 299, 314
Organic chemistry, 309–326
Oven, radiation from, 216
Overtones, 168
Oxygen gas molecule, 287
Ozone in atmosphere, 331

Paleozoic era, 379
Pangaea, 370
Paraffin molecules, 309–312
Parallel circuit, 129
Parallax, method of, 16
Paricutin volcanoe, 364
Particle interference experiments, 230
Particle waves, 228–231

Particles, elementary, 258
Pauli, Wolfgang, 272
Pauli exclusion principle, 271–274
Peking man, 382
Pendulum, energy, 80
Perigee, moon, 23
Period of periodic table, 275
Period of vibration, 161
Periodic table, 274–277
Perseid meteor shower, 38
Phases of moon, 26
Phobos, moon of Mars, 32
Photoelectric effect, 219
Photons, 219–221
Photosynthesis, 323
Planck, Max, 217
Planck's constant, 218
Plane wave, 192
Planets, 30–36
   origin of term, 30
   data table, 30
   origin, 30
Plasma, 248
Plastic layer of earth, 362
Plastics, 317
Plates, crustal, 371
Pluto, planet, 31, 36
Polar molecules, 297
Polaris, pole star, 19–21
Pole star, 19–21
Poles, magnetic, 138
Pollution
   of atmosphere, 345–349
   by noise, 178
Polyethylene, 311, 317
Polymerization, 317
Polymers, 317–322
Polymethylene, 311, 317
Polypeptide, 319
Positron, 248
Potential difference, electric, 123
Potential, electric, 122
Potential energy
   electric, 122
   gravitational, 73–74
   of spring, 79
Power
   electric, 128
   mechanical, 78
Power of ten notation, 4, 83

Pre-Cambrian era, 379
Prefixes in metric system, 56
Pressure, 97
   of atmosphere, 101
   effect on boiling, 105
   and gas molecules, 96–100
Prevailing winds, 334
Primary coil, 152
Primeval fireball, 1
Principle of superposition, 163, 168
Prism and color, 194–196
Prism spectrometer, 195
Projectile motion, 85
Propane, 309–312
Proteins, 318–323
Proton, 243
   mass of, 241
Ptolemy, 38
Pulsar, 12
Pulse on a string, 160–163
Pumice, 357
P-wave, 360

Quantized vibrations, 218
Quantum
   of charge, 113
   light, 219–221
Quartz, 355
Quartzite, 356
Quasar, 12

Radar waves, 189
   and weather, 345
Radial force, 82
Radiation from hot object, 216
Radical, chemical, 317
Radio, 188
Radio waves, 185–189
   and ionosphere, 332
Radioactive dating of earth, 377
Radioactive series, 252
Radioactive tracers, 260
Radioactivity, 251–253
   uses of, 259
Radiocarbon dating, 261
Rare earths, 277
Rays of light, 191

Reactor, fission, 251
Recoil of gun, 61
Red giant star, 10
Red shift, doppler shift, 1, 175–177
Red spot, on Jupiter, 34
Reference frames, 210
Reflection of light, 191–193
Refraction of light, 193–196
Refractive index, 193
Relative humidity, 336
Relative motion, 209
Relativistic factor, 213
Relativity, 209–216
   postulates of, 211
Reproduction of cells, 321
Resistance, 122, 126
Resonance
   in molecules, 315
   sound waves, 171–173
   waves on string, 165–167
Resonance structure of atom, 272
Resonant circuit, 188
Rest mass, 155, 214
Rhyolite, 356
Right hand rule, 141, 144
Rings of Saturn, 36
RNA, 320
Roche's limit, 36
Rock types, 356
Rocket, 62
Rockies, 365
Rumford, Count, 92
Rumford and heat, 91
Rutherford, Ernest, 222

Safety, electrical, 130
Salt, see sodium chloride
Salts, inorganic, 300
San Andreas fault, 366
Sand dunes, 375
Sandstone, 356
Saturated air, 335
Saturn, 31, 34–36
Scalar, 47
Schist, 356
Schrödinger, Erwin, 232
Schrödinger's equation, 232
Science vs superstition, 44

Sea floor spreading, 371
Seasons, 20–23
   effects on weather, 342
Seat belts, use, 52, 62
Secondary coil, 152
Sedimentary rocks, 356, 360
Seismic waves, 360
Semiconductor, 116, 296
Sensitivity of the eye, 190
Series limit, 225
Shale, 356
Shell, atomic, 274
Shield region, 363
Shock, electrical, 130
Shock wave, 178
Sial, 363
Sierra Nevada range, 367
Silicon, 295
Sill, 358
Silver sulfide, 290
Sima, 362
Simple harmonic motion, 160
Slate, 356
Smog, 346–349
Smoke in atmosphere, 346–349
Sodium, 293
Sodium bicarbonate, 306
Sodium chloride, 289, 293, 298, 300
   crystal, 107
Sodium hydroxide, 300
sodium metal, 296
Solar system, 19–41
   composition of, 13
   data table, 30
   origin, 13
   position in galaxy, 8
Solenoid, 142
Solstice, 22
Sonar mapping of ocean floor, 368
Sonic boom, 178
Sound
   interference of, 174
   from moving source, 175–177
   of musical instruments, 173
   resonance of, 171–173
Sound waves, 169–177
Sources of magnetic fields, 140
Spaceships and weightlessness, 84–87
Space station, 89
Space, temperature of, 217

Spectra
   of complex atoms, 278
   continuous, 195
   picture of, 197
Spectral lines, 196
Spectrograph, mass, 154, 241
Spectrometer, prism, 195
Speed
   definition, 45
   of gas molecules, 99
   of light, 193
   and relativity, 209
   terminal, 77
Spin
   of electron, 270
   of particles, 259
Spiral galaxy, 5–8
Spring potential energy, 79
Squall line, 337
Stability of nuclei, 247
Standing waves, on string, 164
Star, nearest, 14
Star tracks, 20
Star types
   hydrogen, 9
   binary, 10
   neutron, 12
   red giant, 10
   white dwarf, 10
Stars
   energy source, 8–12
   formation of, 8–12
State, atomic, 273
Stearic acid, 314
Stellar energy source, 247
Step down transformer, 153
Step up transformer, 153
Stonehenge, 19
Stratosphere, 331
String
   resonance of, 165–167
   waves on, 160–169
Stripes, magnetic in ocean floor, 370–371
Sulfur dioxide, 290
Sulfur as a pollutant, 347
Sulfuric acid, 299
Summer solstice, 22
Superheating, 106
Supernova, 10–11
Superposition principle, 163, 168

Superstition cause, 44
S-wave, 361
Symbols, chemical, 301

Tacoma Narrows bridge, 182
Tarnish, 290
Temperature
   of atmosphere, 330
   of earth interior, 354
   and energy of molecules, 98
   measurement, 92–95
   molecular interpretation, 98
   of space, 217
   of universe, 16
Temperature scales, 93–95
Terminal speed, 77
Terminator, 20
Tesla, unit, 141
Tetons, 365, 376
Thermal expansion, 103
Thermionic emission, 125
Thermometers, 92–95
Thermosphere, 331
Thomson, J. J., 211
Tides, 29
Time dilation, 213–215
Time and relativity, 213
Tornado, 342
Total internal reflection, 207
Tracks of stars, 20
Transformer, 152
Transition elements, 277
Transverse wave, 169
Triple alpha process, 9
Troposphere, 331
Tsunami, 368
Tubes, resonance of, 172
Tunnel effect, 253
TV tube, 125
Twin paradox, 215
Typhoon, 341

Ultrasonic vibrations, 182
Ultraviolet catastrophe, 217
Ultraviolet light, 190
Uncertainty principle, 231
Universe, development, 1–14
Unsaturated compounds, 315

Uranium bomb, 249–251
Uranus, ecliptic plane, 42
Uranus, planet, 31, 36

Valence electron, 296
Valence of ions, table, 303
Van de Graaff generator, 255
van der Waals, J. D., 292
van der Waals force, 291
Vapor pressure, 105
Vector components, 63
Vector, definition, 47
Velocity
   definition, 47
   of escape, 2
Venus, planet, 31
Vernal equinox, 22
Vibratory systems, energy, 79
Vibration of complex systems, 168
Vibrator and quantized energy, 218
Victoria Falls, energy of, 90
Vinyls, 317
Viscosity, 104
Volcanoes, 363–365, 371
Voltage, 123

Voltmeter, 127, 146
Vulcanization, 317

Warm front, 336–339
Washing soda, 301
Water solutions, conductivity, 298
Waterspout, 342
Water wave interference, 198
Watt, unit, 78, 128
Wavelength
   definition, 162
   relation to frequency, 162
Waves, 160–178
   compressional, 169
   electromagnetic, 184–208
   longitudinal, 171
   resonance on string, 165–167
   sound, 169–177
   sound interference, 174
   on string, 160–169
   transverse, 169
Weather map, 338
Weather prediction, 345
Weight
   apparent, 86
   definition, 55
   in elevator, 86
Weightlessness, 84–87
White dwarf star, 10
Wind erosion, 374
Wind flow patterns, 332–334
Winter solstice, 22
Work
   definition, 69
   and kinetic energy, 72
   units of, 70
Work-energy theorem, 75

X-rays
   generation of, 270
   wavelength range, 190

Young's double slit experiment, 200
Yukawa, Hideki, 258

Zambezi river, energy of, 90
Zero, absolute, 94